José María Sánchez Montero, Andrés Rafael Alcántara León
**Biocatalysis and Biotransformations**

## Also of interest

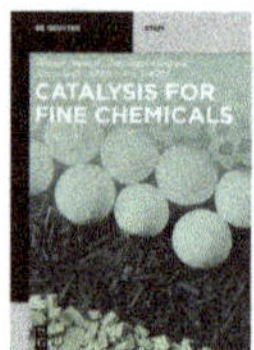


*Catalysis for Fine Chemicals*
Werner Bonrath, Jonathan Medlock, Marc-André Müller and Jan Schütz, 2024
ISBN 978-3-11-109609-4
e-ISBN 978-3-11-110267-2

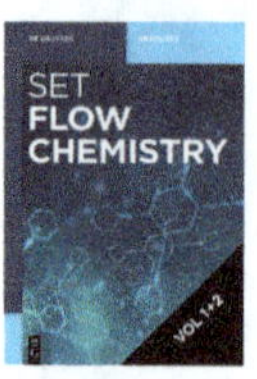


*Set Flow Chemistry, Vol 1+2.*
*Fundamentals and Applications*
Ferenc Darvas, György Dormán, Volker Hessel and Steven V. Ley (Eds.), 2021
ISBN 978-3-11-073679-3

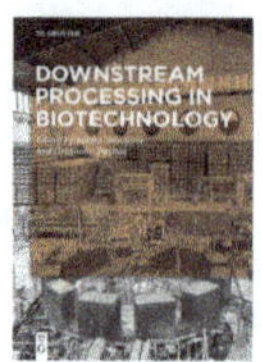


*Downstream Processing in Biotechnology*
Venko N. Beschkov and Dragomir Yankov (Eds.), 2021
ISBN 978-3-11-057395-4
e-ISBN 978-3-11-057411-1

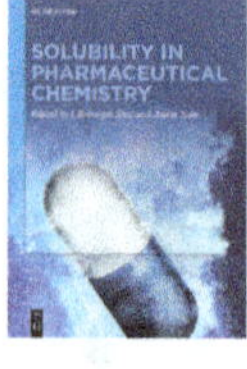


*Solubility in Pharmaceutical Chemistry*
Christoph Saal and Anita Nair (Eds.), 2020
ISBN 978-3-11-054513-5
e-ISBN 978-3-11-055983-5

José María Sánchez Montero,
Andrés Rafael Alcántara León

# Biocatalysis and Biotransformations

---

Enzymatic Synthesis for Biotechnological Applications

DE GRUYTER

**Authors**
Prof. José María Sánchez Montero
Faculty of Pharmacy
Complutense University
University City
Plaza De Ramon y Cajal S/N
28040 Madrid, Spain
jmsm@ucm.es

Prof. Andrés Rafael Alcántara León
Faculty of Pharmacy
Complutense University
University City
Plaza De Ramon y Cajal S/N
28040 Madrid, Spain
andalcan@ucm.es

ISBN 978-3-11-112284-7
ISBN 978-3-11-112299-1 (PDF)
ISBN 978-3-11-112400-1 (EPUB)
DOI https://doi.org/10.1515/9783111122991

Library of Congress Control Number: 2026932422

Bibliographic information published by the Deutsche Nationalbibliothek
The Deutsche Nationalbibliothek lists this publication in the Deutsche Nationalbibliografie; detailed bibliographic data are available on the Internet at http://dnb.dnb.de.



De Gruyter and Walter de Gruyter GmbH are part of De Gruyter Brill.
www.degruyterbrill.com

Questions about General Product Safety Regulation:
productsafety@degruyterbrill.com

Cover illustration: D3Damon/iStock/Getty Images Plus

# Contents

# Chapter 1
# Basic Aspects of Biocatalysis and Biotransformations

**Abstract:** This chapter introduces the fundamental aspects of biocatalysis and biotransformations, placing them within the broader context of biotechnology. It provides an overview of the global impact of this discipline – economic, social, and environmental – and describes its classification into various areas using a color code (e.g., red, green, and white biotechnology), illustrated with examples such as the development of NERICA variety to improve food security in Africa. The discussion then focuses on the crucial role of biocatalysts (enzymes or cells) as efficient and sustainable alternatives to traditional chemical methods, highlighting their key advantages: high selectivity (specifically enantioselectivity, regioselectivity, and chemoselectivity), mild reaction conditions, and lower environmental impact. Strategies to overcome the limitations of biocatalysts are explored, such as immobilization, directed evolution, and (meta)genomic approaches for novel enzyme discovery. Finally, the successful application of these systems in the industrial synthesis of chiral building blocks and active pharmaceutical ingredients is emphasized, underscoring the potential of biocatalysis to drive greener and more selective industrial processes in sectors like pharmaceuticals and fine chemicals.

**Keywords:** Biocatalysis, biotransformations, enzymatic selectivity, industrial biotechnology, environmental sustainability

## 1.1 Biocatalysis and Biotransformations

### 1.1.1 Definition and Biotechnological Context

Biotechnology means the use of living organisms or their parts and the techniques and processes derived from them to develop products, services, and applications that benefit society. This discipline combines the principles of biology, chemistry, genetics, and engineering to manipulate and use biological systems to solve problems and improve the quality of life.

Biotechnology is not a science in itself. It is a multidisciplinary approach that encompasses several disciplines and sciences, and represents a significant diversity in industrial activity.

Biotechnology, the science of harnessing the power of biological materials to create innovative products with a wide range of applications in agriculture, pharmaceuticals, medicine, and the environment, has recently received a lot of attention due to its unlimited potential to benefit humanity. The concept of industrial fermentation

 | https://doi.org/10.1515/9783111122991-001

originates from enzyme technology, which initially focused on brewing techniques, then gave birth to the concept of industrial fermentation and paved the way to the birth of biotechnology.

The beginning of biotechnology can be understood when simple fermentation techniques were used to make cheese, curd, and wine. First, the special feature of cheese is that it is one of the first products of biotechnology because it is made by adding an enzyme rent to sour milk. A pioneering event occurred in the 1940s when penicillin was discovered. Penicillin stood out as one of the most remarkable achievements of the last century. The advent of genetic engineering gave birth to biotechnology. The development of synthetic human insulin was an important milestone that sparked the rapid growth of the biotechnology industry. Genetic engineering continues to grow, especially in areas such as gene therapy, stem cell technology, and genetically modified organisms. Although many of these scientific advances are relatively recent, biotechnology's service to society goes back centuries. In short, biotechnology is a scientific and technical discipline that uses the knowledge and tools of biology and other fields to apply living organisms and their components for the benefit of society, which promotes scientific progress, economic development, and human well-being.

### 1.1.2 Global Impact of Biotechnology

The global impact of biotechnology can be analyzed on three main levels: economic, social, and environmental. Here's a look at how biotechnology affects each of these levels:

Financial impact:

a. Growth of industry: Biotechnology has created a multi-billion dollar global industry spanning medicine, agriculture, energy, and research, among others sectors. This contributed significantly to the economic growth of several countries, creating jobs in the respective fields.
b. Innovation and competitiveness: Biotechnology promotes innovation in various industries, enabling the development of more efficient and improved products and processes. Biotechnology companies often compete in global markets by researching and implementing technological advances.
c. Agriculture and food security: Genetically modified crops and genetic improvement techniques have increased food production, contributing to food security and reducing pressure on natural resources.
d. Pharmaceutical industry and healthcare: The production and marketing of biotechnology-derived medicines is an important part of the global health economy. These drugs often generate significant income and improve the health of the population.

Social impact:

a. Human health: Biotechnology has improved the diagnosis, treatment, and prevention of diseases and improved the quality and life expectancy of people worldwide. It also made it possible to treat rare and neglected diseases.
b. Education and employment: Biotechnology has increased the demand for highly qualified professionals in fields, such as molecular biology, genetic engineering, and bioinformatics, and creating jobs and opportunities for academic development.
c. Ethics and values: Biotechnology raises ethical and moral issues related to genetic modification, genetic privacy, and the manipulation of life. These debates influence how society views biotechnology research and applications.

Environmental impact:

a. Agricultural sustainability: Agricultural biotechnology can promote more sustainable agricultural practices by reducing the need for chemical pesticides and fertilizers and increasing resource efficiency.
b. Bioremediation: Biotechnology is used to mitigate environmental pollution by using genetically modified microorganisms to break down toxic pollutants in soil and water.
c. Biodiversity conservation: Biotechnology is used to protect endangered species, such as cloning and preserving genetic material to prevent endangered organisms from dying out.

In summary, biotechnology has a global impact that spans from economic growth and technological innovation to the improvement of human health and the mitigation of environmental impacts. However, it also raises ethical challenges and environmental risks that need to be addressed responsibly to maximize its global benefits.

### 1.1.3 Color Classification of Biotechnology

Currently, the different areas of biotechnology using a color code [1] in order to fractionate its complexity, starting with red [2], green [3], and white biotechnology [4]. These colors were chosen to establish an association with the corresponding field of study. With the advancement of biotechnology, different colors have been implemented. The colors of biotechnology serve as a way to categorize the various areas of biotechnology.

Biotechnology red: It is also known as medical biotechnology and it is responsible for improving human health from the perspective of disease prevention, diagnosis, and treatment.

Biotechnology green: It focuses on agriculture and food production.

Biotechnology blue: It focuses on marine products.

Biotechnology yellow: It focuses on increasing the production and processing of new and improved foods and beverages.

Biotechnology white: It focuses on the production of chemicals and pharmaceuticals.

Biotechnology brown: It focuses on waste disposal and energy production from waste.

Biotechnology gold: It focuses on functional genomics and nanotechnology.

Biotechnology gray: It focuses on environmental conservation and the detection, prevention, and elimination of pollutants.

Biotechnology black: It focuses on biological warfare and bioterrorism.

Biotechnology violet: It focuses on bioethical and intellectual property issues.
Biotechnology orange: It focuses on dissemination.

As we have discussed in the classification of agricultural biotechnology and as an extension of what has been said, it now deals with poverty, food insecurity, environmental conservation, and sustainable development.

An illustrative example would be the NERICA (an acronym for New Rice for Africa) program, which developed a variety of rice through hybridization between traditional African rice and high-yielding Asian rice. It was created with the aim of combining desirable characteristics of African rice, such as disease resistance and adaptability to adverse weather conditions, with the performance and quality advantages of Asian rice.

NERICA variety has demonstrated higher yields than traditional African rice varieties, making it an attractive option for African farmers. It has also shown resistance to common rice diseases and drought conditions, making it more adaptable to climate fluctuations and environmental challenges.

Furthermore, NERICA variety has been praised for its nutritional quality, as it contains high levels of vitamins and minerals important for health. This makes it a valuable food in combating malnutrition and improving food security in areas where rice is a staple crop.

The development and promotion of NERICA rice have been driven by international organizations such as the International Rice Research Institute and the World Food Program, with the goal of improving production and food security in Africa. Efforts have been made to disseminate the variety and provide training to farmers on its cultivation and proper management.

In summary, NERICA is a hybrid variety developed specifically for the agricultural conditions and challenges in Africa. It has shown higher yields, resistance to dis-

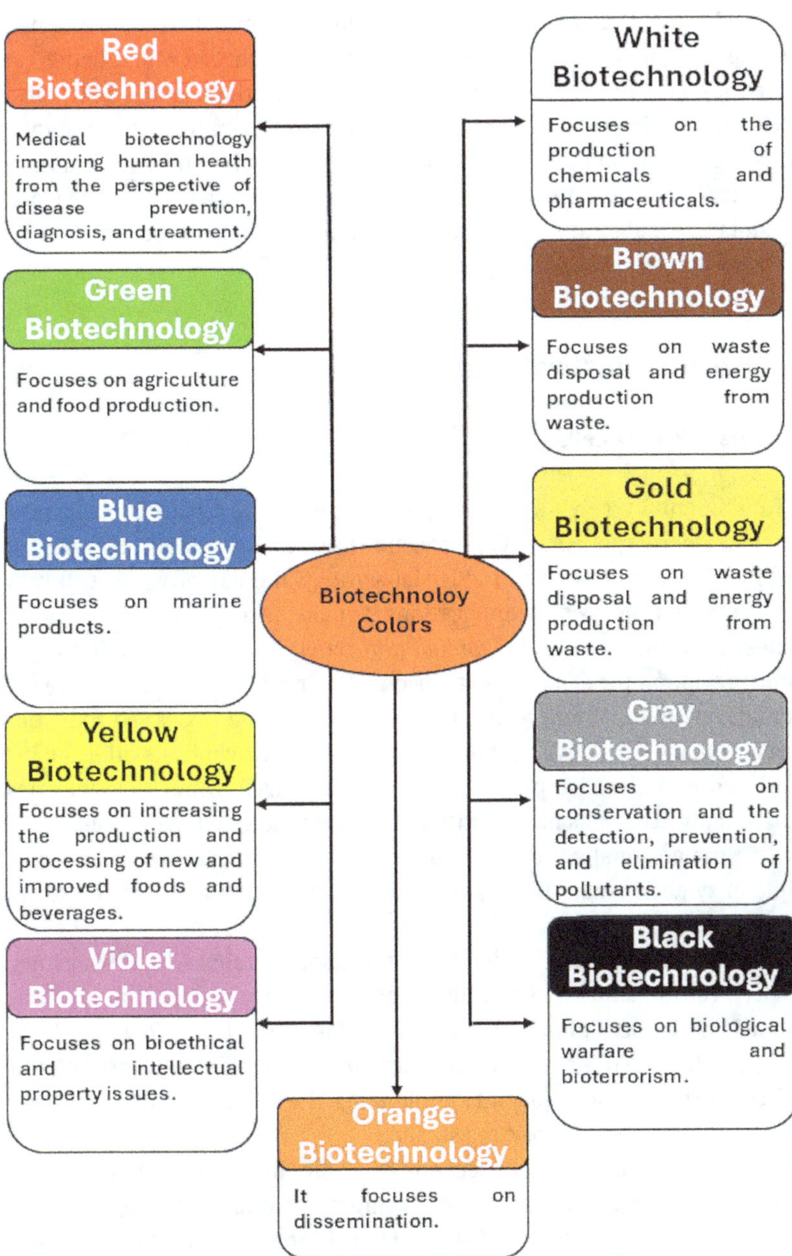


**Figure 1.1:** Colors of biotechnology and significance.

eases and adverse conditions, and nutritional benefits, making it a promising option to enhance rice production and food security in the African continent.

The need to reduce waste, due to current legislation, is demanding the implementation of more selective and efficient synthetic procedures, so that new and better catalysts are needed [5]. For this purpose, applied biocatalysis, "the use of a biocatalyst (enzymes or cells, free or immobilized) to achieve a biotransformation under suitable conditions, with the aim of obtaining a product of specific applied utility," and more specifically biotransformation, "processes in which biocatalysts are used for the transformation of nonnatural substrates for the biocatalyst in question,are becoming more prominent in the design of new strategies for drug synthesis [6].

## 1.2 Biocatalysts in Industrial Synthesis

### 1.2.1 Enzymes Versus Whole Cells

Bioprocesses fundamentally rely on biocatalysts, which can be isolated enzymes or whole cells, such as microorganisms. While reactions driven by isolated enzymes are widely applied in both laboratory and industrial settings, the use of whole-cell catalysts is typically reserved for specific applications. For instance, yeast whole-cell biocatalysts engineered to display CALB (*Candida antarctica* lipase B) on their surface present a promising option for synthesizing various flavor esters. This alternative can be more advantageous than commercial immobilized CALB due to lower production costs and superior enzymatic activity [7]. In general, whole-cell catalysts offer a more straightforward and cost-effective production route compared to purified enzymes. Historically, three main scenarios have justified choosing whole-cell over free-enzyme catalysis: (1) when the enzymatic reaction depends on a cofactor, (2) for processes requiring multiple enzymatic steps, and (3) when the enzyme is located inside the cell and is difficult to isolate.

Regarding fermentative process, primary metabolism contains all pathways necessary to allow cell reproduction and viability, while secondary metabolism produces cellular components not essential for cell life and not found in every growing cell. Sadeer et al. [8] report that the use of natural products in conjunction with conventional antibiotics offers a promising approach to prolonging the efficacy of existing treatments and curbing the rise of antibiotic resistance.

The biological activity of microbial secondary metabolites makes them very attractive for the pharmaceutical industry; these compounds often show high potency and selectivity when tested against multiple biological targets. Their chemical complexity generally does not allow chemical synthesis to replace fermentation as a production process.

## 1.3 Advanced Techniques in Biocatalysis

### 1.3.1 Improvement and Stabilization Strategies

An important impetus has been the stabilization of enzymes and cells [9, 10] to increase their resistance to organic solvents through immobilization [11–13] and modification techniques and genetic engineering. On the other hand, the knowledge of the reaction media and the control of the variables influencing the process took a real technological change [14]. In this sense, understanding the structure of proteins in an organic environment has broadened the applications of enzymology, synthetic organic chemistry and, more recently, structural biology [15].

Industrial biotechnology has become an important tool for obtaining medicines (and other bioproducts) that are difficult or impossible to obtain by traditional chemical methods.

Another interesting and very important point in the pharmaceutical industry is the ability of enzymes to separate racemic mixtures [16]. A large number of pharmaceutical active substances are molecules with one or more stereogenic centers, whose correct configuration depends on their bioactivity. Chemical synthesis methods are usually characterized by obtaining mixtures of isomers, which are also difficult to purify. In this sense, biocatalytic processes have a great advantage, because enzymes specifically and selectively produce only one of the possible isomers, and it can be concluded that producing good chemicals and medicines through biotechnological methods is an emerging field of research. Cost-effective biocatalytic production processes have already brought many benefits to sustainable chemistry and customer-centric value creation in the pharmaceutical, food, flavor, fragrance, vitamin, agrochemical, polymer, specialty, and fine chemical industries [17, 18].

The presence of the other enantiomer in racemic mixtures can be associated with various problems, since one of the enantiomers can sometimes cause side effects. In addition, they can interact differently with other drugs. Therefore, stereoselectivity can be considered a physical property that creates enzyme phenotypes [19]. The Food and Drug Administration encouraged the use of stereochemically pure drugs to improve the efficacy, tolerability, and safety of marketed drugs [20, 21].

This high degree of selectivity can manifest as regioselectivity (targeting a specific site) or chemoselectivity (targeting a particular functional group). In chemical synthesis, such precision is extremely valuable as it confers significant benefits. These include the minimization or complete elimination of protecting groups, a reduction in unwanted side reactions, simplified product purification, and a lower environmental impact. Furthermore, attributes like exceptional catalytic efficiency and mild reaction conditions make these processes highly appealing for industrial-scale applications [22].

The accuracy of enzymes in catalyzing chemical reactions is based on their ability to show different types of selectivity. These aspects of enzymatic selectivity are important in biotechnology and organic synthesis, where precision in the preparation of specific chemical compounds is of utmost importance.

### 1.3.2 Types of Enzymatic Selectivity

#### 1.3.2.1 Substrate Selectivity

Substrate selectivity with enzymes used in organic synthesis refers to the ability of the enzyme to preferentially catalyze the transformation of a given substrate in a chemical reaction, even if other substrates are present in the same mixture. This means that the enzyme clearly prefers a certain substrate and has a significantly higher catalytic activity for that substrate compared to other substrates. An example of substrate selectivity with organic synthetic enzymes is the enzyme glucose oxidase (GOx). GOx is an enzyme that catalyzes the oxidation of glucose to hydrogen peroxide and D-glucono-δ-lactone. In cells, it promotes the breakdown of sugars into metabolites. GOx is widely used for the determination and quantification of free glucose in biological fluids, such as blood and urine, in plant raw materials, and in the food industry. It also has many applications in biotechnology and is used in the production of glucose-sensitive biosensors. It is usually extracted from *Aspergillus niger*. This enzyme is very selective for glucose and does not effectively affect other sugars or similar compounds. For example, if you have a mixture of different sugars such as glucose, fructose and galactose, and you use the GOx enzyme, it will specifically select glucose and convert it to glucuronic acid and hydrogen peroxide, while the other sugars are not significantly affected. . This substrate selectivity with enzymes is valuable in organic synthesis because it enables the efficient and selective production of specific compounds that are essential in the production of ultrapure chemicals and in the pharmaceutical industry.

#### 1.3.2.2 Functional Group Selectivity

Functional group selectivity is related to the preference of the enzyme for a specific chemical group in a molecule, which allows for selective modification of these groups. An example of the selectivity of functional groups for enzymes used in chemical reactions is the enantioselective hydrolysis of esters. In this process, a specific enzyme like lipase can catalyze the hydrolysis of an ester, breaking the ester bond and releasing the corresponding alcohol and acid. What makes this interesting in terms of functional group selectivity is that an enzyme can be very selective about which ester functional group it attacks. For example, lipase can be very selective in hydrolyzing esters with a certain functional group, such as methyl ester, even if it does not significantly affect other functional groups in the molecule. This means that you can have a mixture of esters with different functional groups, and the enzyme will specifically target those with the functional group of interest. The selectivity of functional groups is very valuable in chemical synthesis and the production of ultrapure chemical compounds in the pharmaceutical industry and other fields where selective and controlled chemistry is required.

#### 1.3.2.3 Regioselectivity

Regioselectivity with enzymes used in organic synthesis refers to the ability of an enzyme to direct a chemical reaction towards the selective formation of a product in a specific region of the molecule rather than another region. In other words, the enzyme chooses a specific place in the substrate molecule to carry out the chemical transformation. An example of regioselectivity with enzymes is the hydroxylation reaction of a hydrocarbon molecule such as an alkane by an enzyme called monooxygenase. This enzyme can introduce a hydroxyl group (–OH) at a specific position in the alkane, resulting in the formation of an alcohol. The regioselectivity here lies in the enzyme's ability to selectively target a single carbon-hydrogen (C–H) bond instead of others available in the molecule.

For instance, if you have a molecule of alkane with multiple C–H bonds, the monooxygenase can catalyze the addition of a hydroxyl group to a specific carbon, such as the terminal carbon, rather than other carbons in the same compound. This leads to a regioselective alcohol product with the –OH group in a particular position.

Regioselectivity with enzymes is valuable in organic synthesis as it allows for the precise and selective production of specific chemical compounds, which is crucial in organic chemistry and in the manufacturing of chemicals and pharmaceutical products with specific molecular structures.

#### 1.3.2.4 Stereoselectivity

Enzymatic stereoselectivity applied to organic synthesis refers to the ability of an enzyme to guide a chemical reaction in such a way that a specific stereoisomer (a molecule with the same chemical formula but a different spatial arrangement of its atoms) is selectively formed over another. In other words, the enzyme promotes the formation of a particular stereoisomer during a chemical reaction, resulting in a higher proportion of that stereoisomer in the final product.

A classic example of stereoselectivity with enzymes is the reduction of a ketone or an aldehyde to form a chiral alcohol. In this process, an enzyme such as a reductase can catalyze the reduction of the ketone or aldehyde in a specific manner, favoring the formation of a particular stereoisomer of the alcohol. This means that the enzyme can produce one enantiomer (one of the two stereoisomers) in a greater amount than the other.

For example, if you have a ketone that can form two enantiomers (mirror images of each other) of the alcohol, the enzyme can be highly stereoselective and produce mainly one of the enantiomers. This is of great importance in the organic synthesis of pharmaceutical products and chemical compounds since stereoisomerism can have a significant impact on the biological and chemical properties of a molecule. Enzymatic stereoselectivity enables the controlled production of specific stereoisomers, which is essential for obtaining high-purity and effective chemical products.

Of course, there remain many enzymes and processes to explore, and thus the need for new enzymes that make possible these challenges. Genomics and (meta)genomics can be used as emerging tools to study gene coding enzymes at the level of single microorganisms and communities of uncultured and, so far, unexplored microorganisms

## 1.4 Metagenomics

(Meta)genomics is widely used in the characterization of new biocatalysts. It is the application of modern genomic techniques to study microbial communities directly in their natural environment, bypassing the need for isolation, and laboratory cultivation of individual species. Since then, (meta)genomics has revolutionized biotechnology, shifting the focus from single microbial isolates to the estimated 99% of microbial species that cannot currently be cultured.

## 1.5 Directed Evolution

Another strategy used in biotransformations is directed molecular evolution [23]. This technique is a powerful tool in protein engineering to improve known properties or create new enzyme functions not required in the natural environment [24]. In the laboratory, the key events of natural evolution (mutation, DNA recombination, and selection) can be recreated to design scientifically interesting and technically useful enzymes. Diversity mimicked by random mutagenesis and/or recombination in genes encoding a specific protein. Later, this genetic diversity is expressed and studied under conditions where one wants to improve the enzyme. Thus, the best performers of each generation were selected and used as parent types for a new round of evolution.

The process is repeated as often as necessary until a biocatalyst with the desired properties: stability at high temperatures or in organic solvents; improved catalytic activity; higher specificity, etc. Another important aspect to consider is process predictability since not all traditional chemical synthesis can be replaced by biocatalytic processes due to the complexity of process integration and the cost-benefit ratio.

## 1.6 Computational Modeling and Simulation

The use of modeling and simulation techniques is a powerful tool in this challenge, as alternative pathways and hypothetical changes to existing processes can be explored in silico before experimentation [25–28]. The models developed in biocatalysis can be classified into four different areas according to their application level and complexity: catalyst, reaction, reactor, and process [25].

This book presents the basic aspects and the fundamental principles of biocatalysis and biotransformations. It covers a range of advanced techniques that serve as effective tools for enhancing biotransformation processes and synthesizing novel products. These methods include structural enzyme improvement through protein engineering and directed evolution, engineering approaches like the use of organic media, water activity control, ionic liquids, and supercritical fluids, as well as physical stabilization via immobilization [29]. Additionally, the text highlights the growing significance of theoretical and computational chemistry in both deciphering the fundamental biological functions of enzymes and optimizing their application in biocatalysis [30].

The publication further provides numerous industrial application examples, showcasing the use of biocatalysts – primarily hydrolases and oxidoreductases – in the synthesis of pure chemical building blocks [31] pharmaceutically active compounds, and other valuable products. The substantial industrial relevance of these enzymes is underscored by bibliometric data; since the year 2000, searches for the terms "hydrolase" and "lipase" have yielded 35,687 and 60,094 entries, respectively, in the Web of Knowledge database. We analyze different strategies for the isolation of new enzymes and their direct application. (Meta)genomics of microbial communities and their enzymes enables a wide range of biomedical, industrial, and environmental applications.

## References

[1] De la Vega I, Requena J, Fernandez-Gomez R. The colors of biotechnology in Venezuela: A bibliometric analysis. Technol Soc 2015;42:123–134.

[2] Elsayed IG, Kanwugu ON, Ivantsova MN. Red biotechnology: A healthy world, In: Physics, technologies and innovation (PTI-2019) 2019.

[3] Barcelos MCS, Lupki FB, Campolina GA, Nelson DL, Molina G. The colors of biotechnology: General overview and developments of white, green and blue areas. Fems microbiology letters 2018;365.

[4] Heux S, Meynial-Salles I, O'Donohue MJ, Dumon C. White biotechnology: State of the art strategies for the development of biocatalysts for biorefining. Biotechnol Adv 2015;33:1653–1670.

[5] Alcalde M, Ferrer M, Plou FJ, Ballesteros A. Environmental biocatalysis: From remediation with enzymes to novel green processes. Trends Biotechnol 2006;24:281–287.

[6] Rosenthal K, Lutz S. Recent developments and challenges of biocatalytic processes in the pharmaceutical industry. Curr Opin Green Sustain Chem 2018;11:58–64.

[7] Shen JW, Cai X, Dou BJ, Qi FY, Zhang XJ, Liu ZQ, et al. Expression and characterization of a CALB-type lipase from Sporisorium reilianum SRZ2 and its potential in short-chain flavor ester synthesis. Front Chem Eng 2020;14:868–879.

[8] Sadeer NB, Mahomoodally MF. Antibiotic potentiation of natural products: A promising target to fight pathogenic bacteria. Curr Drug Targets 2021;22:555–572.

[9] Guisan JM, Lopez-Gallego F, Bolivar JM, Rocha-Martin J, Fernandez-Lorente G. The science of enzyme immobilization, In Guisan JM et al. editors. Immobilization of Enzymes and Cells: Methods and Protocols 2020, 4th Edition, p. 1–26.
[10] Guisan J, Bolivar J, López-Gallego F, Rocha Martín J. Immobilization of Enzymes and Cells Methods and Protocols: Methods and Protocols 2020.
[11] Fernandez-Lorente G, Rocha-Martin J, Guisan JM. Immobilization of lipases by adsorption on hydrophobic supports: Modulation of enzyme properties in biotransformations in anhydrous media, In Guisan JM et al. editors. Immobilization of Enzymes and Cells: Methods and Protocols 2020, 4th Edition, p. 143–158.
[12] Greifenstein R, Ballweg T, Hashem T, Gottwald E, Achauer D, Kirschhofer F, et al. MOF-Hosted enzymes for continuous flow catalysis in aqueous and organic solvents. Angew Chem Int Ed 2022;61.
[13] Hinzmann A, Adebar N, Betke T, Leppin M, Groger H. Biotransformations in pure organic medium: Organic- solvent-labile enzymes in the batch and flow synthesis of nitriles. Eur J Org Chem 2019;6911–6916. 2019.
[14] Sanchez Montero JM. White biotechnology and pharmaceutical industry, In Proceedings of Anales de la real academia nacional de farmacia. Real acad Nacional de Farmacia 9 y 11 2007, Madrid, Spain.
[15] AlQuraishi M. Protein-structure prediction revolutionized. Nature 2021;596:487–488.
[16] Fogassy E, Nógrádi M, Kozma D, Egri G, Pálovics E, Kiss V. Optical resolution methods. Org Biomol Chem 2006;4:3011–3030.
[17] Alcantara AR, De Maria PD, Littlechild JA, Schurmann M, Sheldon RA, Wohlgemuth R. Biocatalysis as key to sustainable industrial chemistry. Chemsuschem 2022;15.
[18] Choi JM, Han SS, Kim HS. Industrial applications of enzyme biocatalysis: Current status and future aspects. Biotechnol Adv 2015;33:1443–1454.
[19] Testa B. Types of stereoselectivity in drug metabolism: A heuristic approach. Drug Metab Rev 2015;47:239–251.
[20] Howland RH. Understanding chirality and stereochemistry three-dimensional psychopharmacology. J Psychosoc Nurs Ment Health Serv 2009;47:15–18.
[21] www.fda.gov [Internet].
[22] Tatta ER, Imchen M, Moopantakath J, Kumavath R. Bioprospecting of microbial enzymes: Current trends in industry and healthcare. Appl Microbiol Biotechnol 2022;106:1813–1835.
[23] Yadav D, Tanveer A, Yadav S. Metagenomics for novel enzymes: A current perspective, In Bharagava RN, editor. Environmental contaminants: Ecological implications and management 2019, p. 137–162.
[24] Thapa S, Li H, Ohair J, Bhatti S, Chen F-C, Nasr KA, et al. Biochemical characteristics of microbial enzymes and their significance from industrial perspectives. Mol Biotechnol 2019;61:579–601.
[25] Sin G, Woodley JA, Gernaey KV. Application of modeling and simulation tools for the evaluation of biocatalytic processes: A future perspective. Biotechnol Prog 2009;25:1529–1538.
[26] Sprenger KG, Pfaendtner J. Using molecular simulation to study biocatalysis in ionic liquids, In Voth GA, editor. Computational approaches for studying enzyme mechanism, pt a Vol. 2016, p. 419–441.
[27] Klem H, McCullagh M, Paton RS. Modeling catalysis in allosteric enzymes: Capturing conformational consequences. Top Catal 2022;65:165–186.
[28] Mendoza F, Masgrau L. Computational modeling of carbohydrate processing enzymes reactions. Curr Opin Chem Biol 2021;61:203–213.
[29] Caparco AA, Dautel DR, Champion JA. Protein mediated enzyme immobilization. Small 2022;18:2106425.
[30] Verma R, Mitchell-Koch K. In silico studies of small molecule interactions with enzymes reveal aspects of catalytic function. Catalysts 2017;7.
[31] De Gonzalo G, Alcántara AR, Domínguez de María P, Sánchez-Montero JM. Biocatalysis for the asymmetric synthesis of Active Pharmaceutical Ingredients (APIs): This time is for real. Expert Opin Drug Discov 2022;17:1159–1171.

# Chapter 2
# Types of Enzymes and Their Classification: Enzymes as Biocatalyst

**Abstract:** This chapter provides a comprehensive overview of enzyme types and their systematic classification, framing them as essential biocatalysts in both natural and industrial contexts. It begins by detailing the primary sources of enzymes – including bacteria, fungi, yeast, plants, and insects – and introduces the Enzyme Commission classification system, which categorizes enzymes into seven main classes based on the reactions they catalyze: oxidoreductases, transferases, hydrolases, lyases, isomerases, ligases, and translocases.

Each enzyme class is examined in depth, with sections dedicated to key representatives such as dehydrogenases, oxygenases, proteases, lipases, nucleases, glycosidases, esterases, phosphatases, and amidases. This chapter highlights the catalytic mechanisms, structural characteristics, substrate specificity, and biological roles of these enzymes. Special attention is given to their applications in biocatalysis and biotransformations, particularly in the synthesis of chiral compounds, pharmaceuticals, and fine chemicals.

The discussion also covers the importance of cofactors, enzyme selectivity (including stereoselectivity, regioselectivity, and chemoselectivity), and the industrial relevance of different enzyme classes. Examples of enzyme applications span medical diagnostics, therapeutic development, food processing, waste treatment, and sustainable chemical production.

In summary, this chapter serves as a foundational guide to enzyme diversity, structure, and function, emphasizing their growing role as versatile and sustainable biocatalysts in scientific and industrial advancements.

**Keywords:** Green chemistry, enzyme kinetics, bioprocess optimization, catalytic efficiency, process scalability

## 2.1 Enzyme Sources

1. Enzymes are typically globular proteins that can differ greatly in size. Their functionality is directly governed by their three-dimensional shape, which is itself dictated by the specific order of their constituent amino acids. However, although molecular architecture dictates activity, forecasting a new catalytic function using protein structure alone remains an unresolved challenge in the field of biochemistry. The vast majority of enzymes are significantly larger than the substrates they act upon, with only a small portion of the enzyme (approximately three to four amino acids) directly participating in catalysis. The area that contains these catalytic residues is known as the active site. Enzymes can also feature binding sites for

 | https://doi.org/10.1515/9783111122991-002

cofactors, which are sometimes crucial for the catalytic mechanism, or for small molecules like substrates or products (direct or indirect) of the reaction they catalyze. These interactions between the enzyme and its substrate or product can modulate enzymatic activity, leading to regulatory feedback mechanisms, either positive or negative. Like all proteins, enzymes are composed of a linear chain of amino acids that folds during translation to form a unique three-dimensional tertiary structure, which is sensitive to activity-affecting alterations. Each amino acid sequence is unique, resulting in a structure with specific properties. Sometimes, individual proteins can associate with other proteins to form complexes called protein quaternary structures. There are several organisms known for their ability to produce enzymes. Some of the main enzyme producers are [1]:

**Mushrooms**: These are widely used in the production of enzymes due to their secretory capacity. Fungal species such as *Aspergillus*, *Trichoderma*, and *Penicillium* can produce enzymes, including cellulases, xylanases, pectinases, and amylases [2].

**Yeast**: Some yeasts, such as *Saccharomyces cerevisiae*, are used to produce enzymes such as invertase, which are important in the food and fermentation industries [3].

**Plants**: These can also produce interesting enzymes. For example, papaya contains the enzyme papain, which is used as a meat tenderizer, and pineapple contains bromelain, an enzyme used in the food industry and in the production of dietary supplements [4].

**Insects:** Some insects, such as silkworms and bees, produce specific enzymes for their metabolism and biological functions. These enzymes can also have industrial applications such as silk production [5].

Enzymes are classified into six main groups based on the reaction they catalyze: oxidoreductases, transferases, hydrolases, lyases, isomerases, and ligases, and enzyme structures are classified by their Enzyme Classification (EC) by number (ENZYMA Data Bank) [6]. This database contains the known enzyme structures that have been deposited in the Protein Data Bank (PDB).

## 2.2 The Enzyme Classification (EC) System

The framework for the EC system was developed under the authority of the Nomenclature Committee of the International Union of Biochemistry and Molecular Biology. Regarding the structural representation of proteins, the PDB annotates protein chains with their corresponding Enzyme Commission numbers (Figure 2.1).

As of 2023, the PDB contains 205,101 enzyme records, which correspond to 38,378 distinct PDB files – some of these files are associated with more than one EC number. More information can be found at: https://www.rcsb.org/

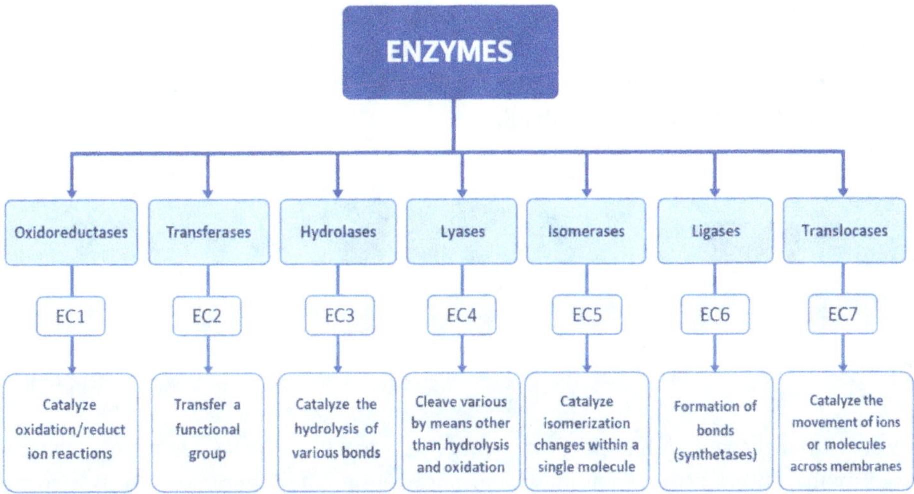


**Figure 2.1:** Classification of enzymes. These numerical codes describe the kind of chemical reaction the enzyme facilitates, along with the specific donors and acceptors of the chemical groups participating in the process. The assignment of an EC number relies on data sourced from multiple databases (such as UniProtKB, GenBank, and KEGG) as well as input from the original researchers. It is noteworthy that this browser specifically includes enzymes that have been assigned an EC number, offering a practical method for investigating and retrieving data on these proteins.

The scientific literature documents various methodologies for quantifying catalytic activity. Within the International System of Units, the derived unit for this purpose is the katal (symbol: kat). This unit is specifically used to express the catalytic activity of enzymes and other catalysts, providing a standardized measure of the reaction rate in enzymatic processes. Another commonly used measure is the "International Unit" (IU), where 1 IU is defined as the amount of enzyme that converts 1 micromole of substrate per minute. Unfortunately, other units, such as nmol/min or nmol/h, are frequently employed, often to make low catalytic activity numbers appear higher. Comparing the activity of different enzyme preparations is possible only if the analytical procedure is carried out in exactly the same way. Since most enzyme suppliers use their own experimental setups, it is rarely possible to estimate the cost/activity ratio of enzymes using published data from multiple commercial sources; therefore, activity data must be determined independently [8].

### 2.2.1 Oxidoreductases (EC 1)

These enzymes catalyze reactions where electrons are transferred from a donor molecule to an acceptor molecule (Figure 2.4). They are fundamental components of diverse metabolic pathways and cellular mechanisms, where they serve the essential function of facilitating and regulating electron transfer processes (Figure 2.2). A large proportion of oxidoreductases rely on nicotinamide cofactors, demonstrating a strong preference for either nicotinamide adenine dinucleotide **(NAD) or n**icotinamide adenine dinucleotide phosphate (NADP) [9].

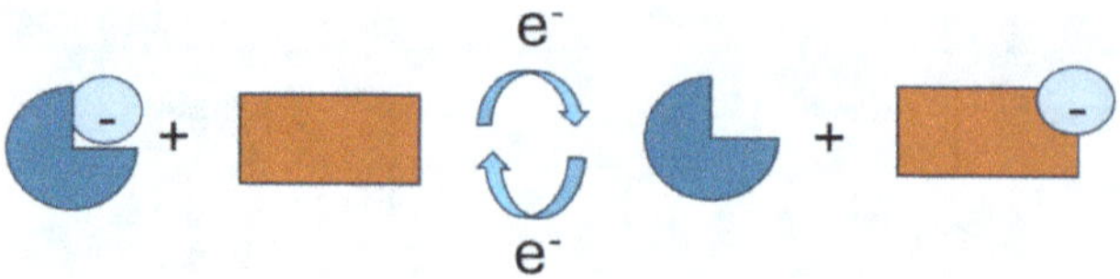


**Figure 2.2:** Scheme of action of oxidoreductases.

It is crucial to recognize that although a large number of enzymes are purely protein-based, many others require an additional nonprotein component, known as a cofactor, to achieve catalytic activity. This cofactor can be an organic molecule, referred to as a coenzyme, or an inorganic ion, often a metal like iron, manganese, cobalt, copper, or zinc. When a coenzyme is permanently and tightly bound to the enzyme's protein structure, it is typically designated as a prosthetic group.

In cases where enzymatic activity is contingent upon a cofactor, the inactive protein component by itself is known as an apoenzyme. The functional, complete enzyme is formed when the apoenzyme binds to its necessary cofactor, creating the active complex termed a holoenzyme (Figure 2.3).

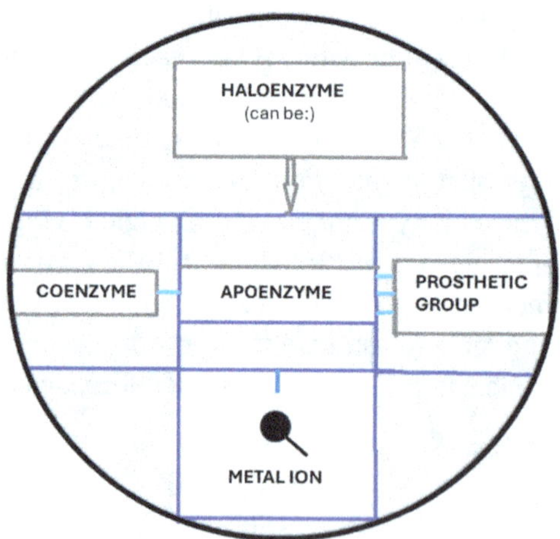


**Figure 2.3:** Components of holoenzyme.

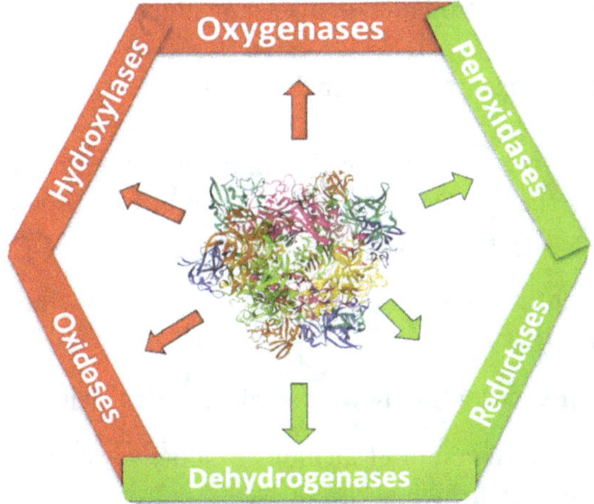


**Figure 2.4:** EC 1 Oxidoreductases. Several subclasses based on their specific reactions.

### 2.2.2 Dehydrogenases (EC 1.1.x)

Dehydrogenases represent a large and varied class of enzymes that are fundamental to cellular metabolism and biochemical processes. Their primary function is to catalyze redox reactions, specifically by extracting hydrogen atoms from a substrate molecule and donating them to an electron-accepting cofactor like NAD+ (nicotinamide adenine dinucleotide) or FAD (flavin adenine dinucleotide). This catalytic activity is vital for driving the electron transport chain and, consequently, for the production of cellular energy in the form of ATP (Figure 2.5).

**Figure 2.5:** General mechanism of a dehydrogenase enzyme. The enzyme catalyzes the transfer of hydrogen ($H_2$) from a substrate to a donor molecule, often a cofactor like $NAD^+$, enabling substrate oxidation and cofactor recycling for further catalytic cycles.

Dehydrogenases facilitate a wide range of chemical transformations, including ketone reduction, reductive amination, alkene reduction, and the oxidation of alcohols and aldehydes [10]. While these enzymes are capable of catalyzing both directions of a redox equilibrium, they are most commonly utilized for the reduction of carbonyl (C=O) or carbon-carbon double (C=C) bonds, as this process involves the creation of at least one stereocenter in the reaction product. When used in the oxidation direction, the goal is typically to achieve a regioselective oxidation of a specific functional group without affecting other similar groups.

Here are some key characteristics and examples of dehydrogenases:

**Function in energy metabolism**: Dehydrogenases participate in the oxidation of organic substrates, such as glucose, fatty acids, and amino acids, to produce energy in the form of ATP through processes like glycolysis and cellular respiration.

**Coenzymes**: Many dehydrogenases require specific coenzymes, such as $NAD^+$ and FAD, to carry out their electron transfer reactions.

Examples of dehydrogenases: Some of the most well-known dehydrogenases include:

*Alcohol dehydrogenase:* It catalyzes the conversion of alcohols to aldehydes or ketones, playing a crucial role in alcohol metabolism (Figure 2.6).

*Lactate dehydrogenase*: It converts lactic acid to pyruvate and vice versa, playing an important role in lactic acid fermentation and lactate homeostasis in the body [11].

*Glucose-6-phosphate dehydrogenase*: It participates in the pentose phosphate pathway, which is essential for nucleotide synthesis and protection against oxidative stress [12].

*Succinate dehydrogenase*: In the electron transport chain of cellular respiration, it catalyzes the conversion of succinate to fumarate [13].

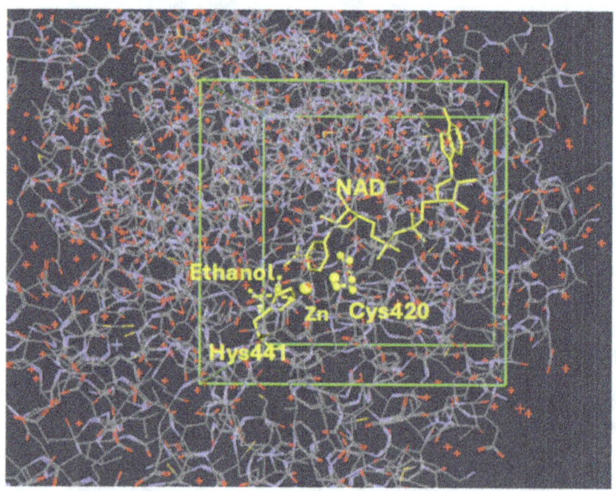


**Figure 2.6:** Section of structure of human alcohol dehydrogenase (PDB code: 1HSO) showing the interaction of $NAD^+$ with zinc (Zn), histidine 441 (His441), cysteine 420 (Cys420), and ethanol in the active site. The zinc ion is coordinated by Cys420 and His441, playing a critical role in stabilizing the enzyme's conformation for catalysis. $NAD^+$ participates in the hydride transfer from ethanol, which is polarized by its interaction with the zinc ion, facilitating its conversion to acetaldehyde.

**Regulation:** The activity of dehydrogenases can be regulated by various mechanisms, such as the availability of coenzymes, product feedback, or covalent modification.

In summary, dehydrogenases are key enzymes in cellular biochemistry and metabolism, playing a fundamental role in energy generation and the regulation of various metabolic pathways. Studying these enzymes is essential for understanding the functioning of biological systems and their relevance to health and disease.

From the perspective of their use in biotransformations, the main drawback to their industrial application, despite being highly attractive, is that they require cofactors (either free, such as adenine dinucleotides and nicotinamide ($NAD(P)^+/NAD(P)H$), or bound to the enzyme, such as adenine dinucleotide, FAD, flavin mononucleotide, or pyrroloquinoline quinone for their activity. These cofactors are consumed equimolarly with the substrate, and due to their high cost, systems need to be designed to facilitate their recycling [14]. For this reason, dehydrogenases are often used in the form of whole cells, as this enables the cellular metabolism to regenerate the cofactors. This approach is more economically advantageous, as it not only facilitates cofactor regeneration but also reduces the costs associated with enzyme isolation and purification.

### 2.2.3 Reductases (EC 1.6.x)

A reductase is an enzyme that catalyzes reduction reactions by decreasing the activation energy barrier required for the process to proceed. This class of enzymes falls under the broader oxidoreductase category. A reduction reaction is characterized by the gain of electrons and is always coupled with an oxidation event. Consequently, the species that is reduced experiences a decrease in its oxidation state.

Among other examples, the enzyme 5-alpha-reductase stands out, as it amplifies androgenic action by converting testosterone into dihydrotestosterone [15].

Other examples of reductases are dihydrofolate reductase [16], 5β-reductase [17], thioredoxin reductase [18], etc.

What sets reductase apart from oxidoreductase? Reductase and oxidoreductase are two enzyme categories. Reductase specifically facilitates reduction reactions, while oxidoreductase is involved in both oxidation and reduction reactions. This distinction is the primary difference between reductase and oxidoreductase. Reductases include enzymes such as dihydrofolate reductase, 5α-reductase, 5β-reductase, HMG-CoA reductase, methemoglobin reductase, ribonucleotide reductase, aldose reductase, and thioredoxin reductase. In contrast, oxidoreductases encompass a broader range of enzymes, including oxidases, dehydrogenases, peroxidases, hydroxylases, and oxygenases.

In summary, reductases are key enzymes in biology and biotechnology, as they catalyze essential reduction reactions in a wide range of biological processes and have important applications in the production of chemicals, bioplastics, bioremediation, and bioenergy, among other fields.

### 2.2.4 Peroxidases (EC 1.3.x)

Peroxidase is an enzyme in the group of oxidoreductases that catalyze the oxidation of various substrates using hydrogen peroxide ($H_2O_2$) as an electron acceptor [19] (Figure 2.7).

**PEROXIDASES**

$$2\ \text{Substrate-H} + H_2O_2 \longrightarrow 2\ \text{Substrate-O} + 2\ H_2O \longrightarrow \text{Substrate- Substrate}$$

$$\text{Substrate} + H_2O_2 \longrightarrow \text{Substrate-O} + H_2O$$

$$\text{ROOR}' + 2\ e^- + 2H^+ \xrightarrow{\text{Peroxidase}} \text{ROH} + \text{R}'\text{OH}$$

**Figure 2.7:** Catalytic mechanism of peroxidases. These enzymes utilize hydrogen peroxide ($H_2O_2$) to drive the oxidation of various substrates, leading to different outcomes: the formation of substrate dimers via free radicals, the direct oxygenation of a substrate molecule, or the reduction of organic peroxides into their corresponding alcohols.

They play a crucial role in detoxifying harmful reactive oxygen species (ROS) in cells and tissues. Peroxidases are involved in several biological processes, including the defense against oxidative stress, the breakdown of hydrogen peroxide, synthesis of various compounds, and according to recent studies, in mycotoxin degradation [20].

One well-known example of a peroxidase enzyme is catalase, which is found in many living organisms, including humans. Catalases are well-studied enzymes that play a critical role in protecting cells from the toxic effects of hydrogen peroxide. Catalase breaks down hydrogen peroxide into water and oxygen, thereby preventing the accumulation of toxic levels of hydrogen peroxide in cells [21].

Peroxidases also have important applications in biotechnology and research, where they are used in assays and experiments to detect specific molecules based on their catalytic activity with hydrogen peroxide. Catalase and peroxidases, as antioxidant nanozymes, hold therapeutic potential in diseases related to oxidative damage, such as cancer. Excess levels of ROS can contribute to cancer development by causing genetic mutations and abnormal protein functions. Antioxidant nanozymes, including platinum and cerium oxide nanoparticles, help regulate ROS levels, induce oxidative stress in cancer cells, and enhance antitumor therapies like photodynamic and photothermal therapy. They also protect healthy cells, demonstrating their versatility as both cytotoxic agents against cancer cells and protective agents for normal cells [22].

In summary, peroxidases are enzymes that play a significant role in managing oxidative stress and are involved in various biological processes and applications in science and biotechnology.

### 2.2.5 Oxygenases (EC 1.4.x)

Oxygenases are among the most intriguing oxidation biocatalysts (Figure 2.8). These enzymes catalyze the specific introduction of one or two oxygen atoms from $O_2$ into a substrate under very mild reaction conditions. They play a crucial role in the metabolism of a wide range of compounds and are involved in various biochemical pathways [23]. There are two main types of oxygenases:

**Monooxygenases:** It is also known as mixed-function oxidases. These introduce a single oxygen atom from $O_2$ into a substrate while reducing the other oxygen atom to water. They are involved in processes such as the hydroxylation of organic molecules, which can lead to the formation of various metabolites. Cytochrome P450 enzymes are a well-known example of monooxygenases and play a significant role in drug metabolism and detoxification.

**Dioxygenases:** These incorporate both oxygen atoms from $O_2$ into a substrate. They are often involved in the cleavage of aromatic or aliphatic rings in molecules. An example is catechol dioxygenase, which participates in the degradation of aromatic compounds like catechol.

Oxygenases proceed with high specificity and represent an intriguing synthetic strategy for oxyfunctionalization, primarily because there is generally no chemical alternative. The wide variety of oxygenases and their ability to specifically introduce oxygen atoms directly from $O_2$, without the need for destabilizing peroxides, has led to their widespread industrial use. Monooxygenases and dioxygenases introduce one or two oxygen atoms into the substrate, producing water as a co-product in the case of monooxygenases.

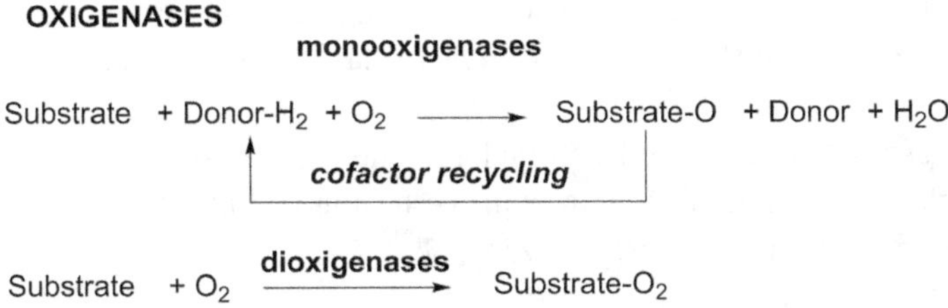


**Figure 2.8:** Mechanism of action of oxygenases. Monooxygenases (top) incorporate one atom from an oxygen molecule ($O_2$) into the substrate, requiring an electron donor (donor-$H_2$) and producing water. Dioxygenases (bottom) incorporate both atoms from the molecular oxygen directly into the substrate.

Oxygenases are essential in various biological processes, including the metabolism of xenobiotics (foreign compounds), the biosynthesis of important molecules like hormones and secondary metabolites, and the breakdown of environmental pollutants.

Their ability to introduce oxygen atoms into molecules makes them crucial for maintaining the balance of various biochemical pathways and responding to changes in cellular conditions. Given the diverse nature of the substrates they act upon, it is not surprising that numerous examples of their use in biotransformations have been documented in the literature

### 2.2.6 Oxidases (EC 1.2.x)

Oxidase enzymes primarily act upon three main types of substrates: amino acids, amines, and alcohols. These enzymes catalyze the oxidation of amino acids and amines into their corresponding imine derivatives, and convert alcohols into aldehydes or ketones. A key characteristic of many oxidases is their high enantioselectivity, which makes them valuable tools in synthetic chemistry for processes such as the desymmetrization of prochiral compounds, as well as the kinetic resolution and deracemization of racemic mixtures [24].

An important example of an oxidase is cytochrome c oxidase, a key enzyme in the utilization of oxygen for energy production and the final component of the electron transport chain. Other notable examples include glucose oxidase, monoamine oxidase, cytochrome P450 oxidase, nicotinamide adenine dinucleotide phosphate oxidase (NADPH), xanthine oxidase, L-gulonolactone oxidase, laccase, lysyl oxidase, polyphenol oxidase, and sulfhydryl oxidase.

### 2.2.7 Hydroxylases (EC 1.5.x)

Hydroxylases are a group of enzymes that catalyze oxidation reactions by incorporating one atom of molecular oxygen into the substrate and using the other atom to oxidize NADH or NADPH. There are different types of hydroxylases, such as steroid 21-hydroxylase.

Steroid 21-hydroxylase (P450c21) is a microsomal enzyme expressed in the adrenal gland. It catalyzes the conversion of 17-hydroxyprogesterone and progesterone to 11-deoxycortisol and deoxycorticosterone, respectively [25].

## 2.3 Transferases (EC 2)

Transferases are enzymes responsible for transferring functional groups (atoms or molecules) from a substrate acting as a donor to another substrate acting as a receptor. The nomenclature of transferase enzymes is typically based on the nature of the molecule that accepts the functional group in the reaction, such as DNA methyltransferase. Transferases are named according to the specific type of transfer they facilitate and are often categorized into various subclasses.

### 2.3.1 Kinases

These enzymes transfer a phosphate group (usually from ATP) to another molecule, often a protein, to activate or deactivate it. Protein kinases, for example, are involved in the regulation of many cellular processes by phosphorylating proteins [26]. These enzymes have several important biotechnological applications. Several types of cancer and diseases are known in humans in which mutations in kinase enzymes have been detected.

### 2.3.2 Methyltransferases

Methyltransferases (EC 2.1.x) transfer a methyl group ($CH_3$) from one molecule to another. They are important in processes like DNA methylation, which can influence gene expression [27]. Methyltransferases are pharmacological targets in the development of drugs to treat diseases related to DNA methylation, such as some types of cancer and for neuroscience research in the modification of molecules like histones and neurotransmitters.

### 2.3.3 Acyltransferases

Acetyltransferase or transacetylase (EC 2.3.x) is an example of an acyltransferase enzyme that transfers an acetyl group, typically using acetyl-CoA as a substrate. These enzymes transfer acyl groups (organic chemical groups derived from acids, often with a carbonyl group) from one molecule to another. Examples include fatty acid synthases, which are involved in lipid metabolism [28].

### 2.3.4 Glycosyltransferases

Glycosyltransferases (EC 2.4.x) are enzymes that facilitate the creation of glycosidic bonds by transferring a sugar unit, usually a monosaccharide, from one substrate (the donor) to another (the acceptor). While nucleotide sugars are the most common donors, sugar phosphates, sugar lipids, and saccharides can also serve as donors. Acceptors may include various molecules such as other saccharides, proteins, lipids, nucleic acids, as well as natural and synthetic compounds. These enzymes play a pivotal role in the synthesis of intricate carbohydrates and glycoproteins [29].

### 2.3.5 Aminotransferases

Aminotransferases (EC 2.6.x) catalyze the transfer of an amino group ($NH_2$) from one molecule to another in particular between an amino acid and an alpha-keto acid. They are essential in amino acid metabolism and the synthesis of various nitrogen-containing compounds. Transaminases need a coenzyme called pyridoxal phosphate (derived from pyridoxine or vitamin B6) to perform their function; acts as a transporter of the amino group between substrates, alternating its structure between the aldehyde form (pyridoxal phosphate, PLP) and the aminated form (pyridoxamine-5-phosphate, PMP), The applications of aminotransferases in biocatalysis are: synthesis of amino acids, production of chemical intermediates, synthesis of asymmetric chemical substances, production of aromas and fragrances, synthesis of alkaloids, and production of biotransformations [30] (Figure 2.9).

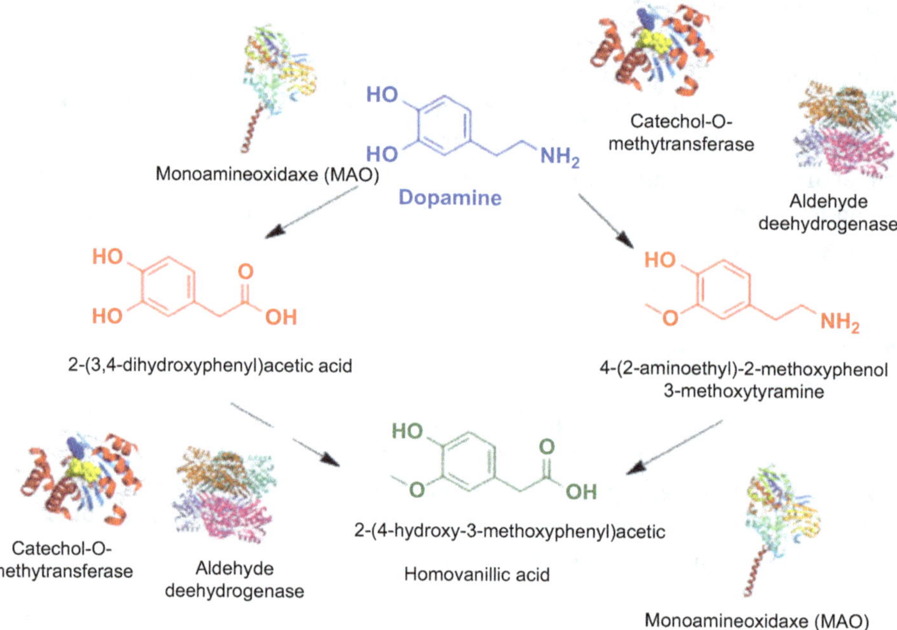


**Figure 2.9:** Biodegradation of dopamine via catechol-*O*-methyltransferase (along with other enzymes.

## 2.4 Hydrolases (EC 3)

Hydrolases are the largest group of chemical catalysts used in industry. These enzymes catalyze hydrolysis reactions, which involve the cleavage of chemical bonds through the addition of water molecules. Hydrolases are classified according to the type of bond they hydrolyze.

These enzymes play a crucial role in various biochemical processes by breaking down complex molecules into simpler components. They are involved in processes such as digestion, cellular waste disposal, and the breakdown of various biomolecules.

Several subclasses of hydrolases are specialized in hydrolyzing specific types of substrates. Some common examples of hydrolase subclasses include:

### 2.4.1 Proteases

These enzymes catalyze the hydrolysis of peptide bonds in proteins, breaking them down into individual amino acids. Proteases are involved in processes like protein digestion and regulation [30, 31]. They not only perform several physiological functions in living organisms but are also of great importance in various industries, providing significant economic benefits.

The classification of proteases is quite extensive, but here are some of the main categories:

#### 2.4.1.1 Classification by Catalytic Mechanism

a. Serine proteases: These enzymes use a serine residue as a nucleophile in their catalytic mechanism. They hydrolyze peptide bonds in proteins and peptides, relying on a catalytic triad – typically consisting of serine, histidine, and aspartate – for their activity. Serine proteases play crucial roles in various biological processes, including digestion, blood coagulation, immune response, and cellular regulation. Examples include trypsin, chymotrypsin, and subtilisin [32].
b. Cysteine proteases: These enzymes, also known as thiol proteases, are named after the nucleophilic cysteine residue that forms a covalent bond with the carbonyl group of the scissile peptide bond in their substrates, a key interaction that drives their catalytic activity. The most abundant group within this class is the C1 family, which predominantly consists of papain-like proteases. These enzymes play a crucial role in biological processes by cleaving peptide bonds, utilizing the thiol (–SH) group in the cysteine side chain. Examples include caspases [33].
c. Aspartic proteases: These enzymes, also known as aspartyl proteases, are a class of proteolytic enzymes that use aspartic acid residues in their active sites to catalyze the hydrolysis of peptide bonds in proteins and peptides. These enzymes are named after the amino acid aspartic acid, which has an acidic side chain and plays a crucial role in their catalytic mechanism. Typically, they contain two highly conserved aspartate residues in the active site and exhibit optimal activity under acidic pH conditions.

   Notable examples of aspartic proteases include pepsin, found in the stomach and essential for the digestion of dietary proteins, and renin, which regulates blood pressure and electrolyte balance.

In industrial applications, aspartic proteases face limitations such as low catalytic efficiency, conversion rates, thermostability, and enzyme performance. However, modifying their structure and expression systems can result in highly selective and efficient enzymes. Microbial aspartic proteases are particularly valued in industry for their robustness and eco-friendliness, regulating specific processes while offering distinct advantages [34].

d. Metalloproteases: These enzymes, also known as metal-dependent proteases or metalloenzymes, are a diverse group of proteolytic enzymes that rely on metal ions for their catalytic function. These enzymes play a crucial role in various biological processes by cleaving peptide bonds in proteins and peptides, with the metal ions serving as essential catalysts for hydrolysis.

The MEROPS database (http://www.merops.co.uk/merops/merops.htm) is a comprehensive resource that categorizes proteases into families and clans based on sequence similarities. It classifies metalloproteases into 46 distinct families, offering a detailed representation of their diversity. Additionally, the elucidation of three-dimensional structures for numerous metalloproteases has unveiled insights into the mechanisms governing peptide bond hydrolysis by these enzymes. Notably, thermolysin and carboxypeptidase A stand out as extensively studied members of this enzyme class [35].

#### 2.4.1.2 Classification by Structural Characteristics

a. Endopeptidases: These proteases cleave peptide bonds within the interior of a protein or peptide.
b. Exopeptidases: These proteases cleave peptide bonds at the terminal ends (either N-terminus or C-terminus) of a protein or peptide.

#### 2.4.1.3 Classification by Specificity

a. Trypsin-like proteases (TLPs): These proteolytic enzymes share similar characteristics and mechanisms of action with trypsin, a well-known serine protease. These are primarily characterized by their substrate specificity for basic residues, such as lysine and arginine, at the P1 position. They are involved in the cleavage of peptide bonds in proteins and peptides and play essential roles in various biological processes.
b. Chymotrypsin-like proteases: These enzymes cleave peptide bonds following hydrophobic amino acids, such as phenylalanine, tyrosine, and tryptophan.
c. Elastase-like proteases: These enzymes are specialized for cleaving peptide bonds adjacent to small, uncharged amino acids like alanine and glycine.
d. Aminopeptidases: These enzymes cleave amino acids from the N-terminus of peptides or proteins.

#### 2.4.1.4 Classification by Evolutionary Relationships

Proteases can be classified based on their evolutionary relationships and sequence homology.

#### 2.4.1.5 Classification Function

Proteases can also be categorized by their functional roles, such as:

Digestive enzymes (e.g., pepsin and trypsin)
Proteases involved in blood clotting (e.g., thrombin)
Proteases involved in programmed cell death (e.g., caspases)

The classification of proteases is complex due to the diversity of these enzymes and the multiple criteria used for categorization. Ongoing research continues to discover and characterize new proteases, adding to this complexity. However, understanding these categories is critical for studying their functions and designing specific inhibitors or activators for various applications in biotechnology, medicine, and research.

### 2.4.2 Lipases

Lipases (EC 3.1.1.3) are a class of enzymes that play a crucial role in the digestion and metabolism of lipids (fats). Lipases are widely distributed in nature and are found in various organisms, including animals, plants, and microorganisms (Figure 2.10).

Lipases belong to the hydrolase group, catalyzing the hydrolysis of esters formed from glycerol and long-chain fatty acids in a regio- and enantioselective manner. Due to their stability in organic media, their lack of requirement for cofactors, high regio-, chemo-, and enantioselectivity, and their broad substrate spectrum, lipases are often the biocatalyst of choice for the preparation of chiral organic molecules [13]. . They are the enzymes most commonly used in the kinetic resolution of racemic mixtures of alcohols, acids, esters, and amines. They are preferably used in the resolution of secondary alcohols [14], through hydrolysis of the corresponding esters or transesterification processes, due to their high enantioselectivity toward these types of substrates. In summary, the enantioselectivity of the process depends on the origin of the lipase, the structure of the substrate, and the experimental conditions of the reaction.

The versatility of these enzymes enables them to carry out the hydrolysis of other unnatural substrates, as well as the synthesis of esters and other compounds in organic media.

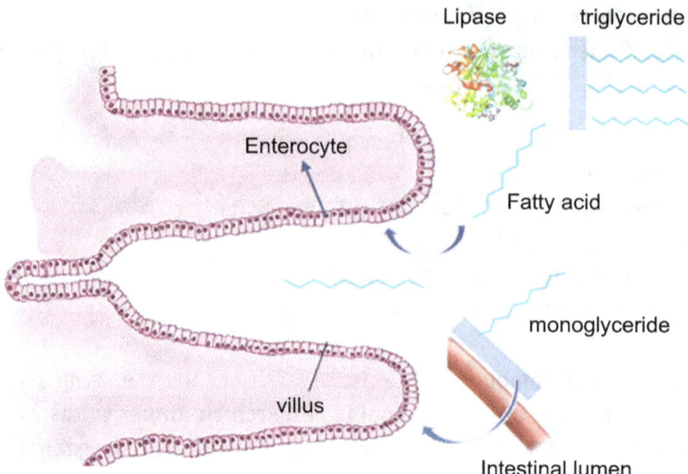


**Figure 2.10:** Absorption of triglycerides through the digestive tract wall. Pancreatic lipase breaks down triglycerides into monoglycerides and free fatty acids.

### 2.4.3 Nucleases

Nucleases are responsible for the hydrolysis of phosphodiester bonds in nucleic acids (DNA and RNA). They are essential for DNA replication, repair, and RNA degradation. Due to their importance in various sectors of industry, we will devote special attention to these enzymes in this chapter dedicated to the industrial applications of enzymes.

### 2.4.4 Glycosidases (EC 3.2.1)

These are enzymes that facilitate the hydrolysis of glycosidic bonds within carbohydrate molecules, resulting in the breakdown of complex sugars into simpler monosaccharides. They play essential roles in biological systems, including the digestion of dietary carbohydrates and the modification of glycans on the cell surface. Specifically, these enzymes are responsible for cleaving oligosaccharides in the small intestine and for processing cell surface oligosaccharides, which are critical for mediating cell-to-cell recognition in processes like infection, cancer metastasis, and immune system activation.

Historically, glycosidases – enzymes specialized in carbohydrate hydrolysis – have seen limited use in industrial synthesis and were primarily employed to remove undesired carbohydrates. However, current research has revealed promising new applications. These include the synthesis of precisely defined oligosaccharides, the production of valuable galacto-oligosaccharides, and the creation of specialized compounds like glycosylated flavonoids. Additional synthetic pathways, such as those for producing human milk oligosaccharides or engineered antibodies, are also emerging [36].

### 2.4.5 Esterases (EC 3.1.x)

These are a class of enzymes that catalyze the hydrolysis of ester bonds. Ester linkages are found in many molecules, including lipids (such as triglycerides and phospholipids), carbohydrates (such as acetylcholine), and various organic compounds (such as ester-based drugs and pesticides). They can be classified into different categories based on their substrate specificity and mechanism of action. Some of the more common types of esterases include carboxylesterases, cholinesterases, and phospholipases.

Phenomenon of interfacial activation: Lipases possess a hydrophobic domain around their catalytic site, which confers a preference for long-chain triglyceride substrates. In contrast, esterases feature a distinct acyl-binding pocket. These enzymes are integral to numerous biological functions and have significant utility in both industrial and research settings. Due to their unique biocatalytic properties, esterases contribute to environmentally sustainable design solutions, including biomass degradation, food and feed, dairy, clothing, agrochemicals (herbicides and pesticides), bioremediation, biosensor development, anticancer and antitumor therapies, gene therapy, and diagnostics [37].

### 2.4.6 Phosphatases

These are a group of enzymes that catalyze the removal of phosphate groups from molecules through a process called dephosphorylation. Because a phosphatase enzyme catalyzes the hydrolysis of its substrate, it is classified as a subcategory of hydrolases. Phosphatases play crucial roles in various biological processes by regulating the levels of phosphorylation in proteins, nucleic acids, and other cellular components. They are essential for maintaining cellular homeostasis and signaling pathways.

Phosphatases have important applications in the field of biocatalysis, which involves using enzymes to catalyze chemical reactions for various industrial and research purposes. In biocatalysis, phosphatases are particularly valuable due to their ability to catalyze the dephosphorylation of molecules, leading to a wide range of applications.

Phosphatases can also catalyze the regio- and stereoselective phosphorylation of alcohols using high-energy inorganic phosphate donors, such as di-, tri-, and polyphosphates [38].

### 2.4.7 Amidases (EC 3.5.1.x)

These are a group of hydrolases belonging to a family of enzymes that act on carbon-nitrogen bonds other than peptide bonds, specifically targeting linear amides. These versatile enzymes catalyze the hydrolysis of amide bonds, breaking them into their constituents, usually amino and carboxylic acid groups. Amide bonds are present in a variety of organic substances, including proteins, peptides, and many synthetic and naturally occurring molecules.

Amidases play an important role in various biological processes, including protein turnover and digestion. Additionally, they are of considerable importance in industrial applications, especially in the production of pharmaceuticals and fine chemicals. Amidases are essential tools in both the synthesis and hydrolysis of amide bonds in many chemical reactions, contributing to the production of antibiotics, plastics, and many other chemical compounds.

Their inherent hydrolytic and acyl transfer activities directed at CN bonds make amidases highly adaptable biocatalysts for the synthesis of chiral carboxylic acids, α-amino acids, and amides. This versatility makes them valuable assets in enzyme synthesis, playing a key role in the precise and efficient production of diverse and complex chemical compounds [39].

In summary, hydrolases are a type of enzyme, which play an important role in biocatalysis and biotransformations. These enzymes are known for their high selectivity, mild reaction conditions, and sustainable nature.

Hydrolases are capable of catalyzing the hydrolysis of various substrates and altering reaction conditions for synthesis, which can be useful in various biotechnological processes, such as the production of chemical intermediates and pharmaceuticals [40].

Biocatalysis and biotransformations offer advantages such as high specificity, efficiency, pollution reduction, and energy savings. These characteristics make hydrolases suitable for use in the development of new sustainable production processes. Furthermore, biocatalysis has established itself as a practical, effective, and environmentally compatible alternative in various synthetic routes.

## 2.5 Lyases (EC 4)

Lyases are enzymes that catalyze the formation and cleavage of bonds using mechanisms distinct from hydrolysis or redox reactions. Lyases catalyze the scission of C–C, C–N, C–O, and other bonds through elimination to form double bonds or, conversely, catalyze the addition of groups to double bonds. These enzymes do not require cofactor recycling, exhibit absolute stereospecificity, and have the potential to achieve 100% conversion, whereas enantiomeric resolution is only 50%. As a result, lyases have garnered significant interest as biocatalysts for producing optically active compounds and have already found applications in several large-scale commercial processes.

Lyases can be categorized into seven subclasses:

**EC 4.1** includes lyases that cleave carbon-carbon bonds, such as decarboxylases (EC 4.1.1), aldehyde lyases (EC 4.1.2), and oxo acid lyases (EC 4.1.3).

**EC 4.2** contains lyases that cleave carbon-oxygen bonds, such as dehydratases.

**EC 4.3** includes lyases that break carbon-nitrogen bonds.

**EC 4.4** includes lyases that break carbon-sulfur bonds.

**EC 4.5** contains lyases that break carbon-halide bonds.

**EC 4.6** includes lyases that cleave phosphorus-oxygen bonds, such as adenylyl cyclase and guanylyl cyclase.

**EC 4.99** includes other lyases, such as ferrochelatase.

Various enzymes can catalyze the formation of C–C bonds, and many of these biocatalysts are lyases. Notably, aldolases and hydroxynitrile lyases (oxynitrilases) are prominent for their use in organic synthesis. Several applications include enzymatic aldol synthesis, acyloin condensation, and cyanohydrin reactions [41].

Aldol synthesis is a valuable tool in organic chemistry, enabling chemists to create complex carbon frameworks and construct intricate molecules. It is widely employed in the production of pharmaceuticals, agricultural chemicals, and other fine chemicals. The stereoselectivity and regioselectivity of aldolases make them particularly useful for controlling the outcome of these reactions, allowing for the synthesis of chiral compounds with high specificity (Figure 2.11).

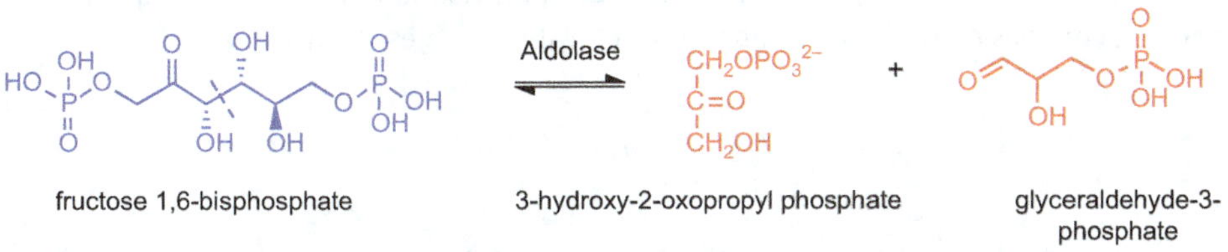

**Figure 2.11:** The enzyme aldolase, an example of a lyase, catalyzes the cleavage of fructose 1,6-bisphosphate into dihydroxyacetone phosphate and glyceraldehyde-3-phosphate. This reaction is a key step in the glycolytic pathway.

### 2.5.1 Benzaldehyde Lyase (BAL, EC 4.1.2.38)

It is an enzyme demonstrated to facilitate a wide variety of carbon-carbon (C–C) bond-forming reactions. Its synthetic utility is well-established, particularly in the production of acyloin compounds, including aromatic-aromatic, aromatic-aliphatic, and nonfunctionalized aliphatic-aliphatic derivatives.

## 2.6 Isomerases (EC 5)

Isomerases are a class of enzymes that catalyze isomerization reactions, involving the rearrangement of atoms within a molecule to form isomers. They play a crucial role in various biological processes and industrial applications.

Isomerases that interconvert optical, geometric, or positional isomers are interesting biocatalysts for the synthesis of pharmaceuticals and pharmaceutical intermediates. These enzymes can transform inexpensive biomolecules into valuable isomers with suitable stereochemistry, which is useful in synthetic medicinal chemistry. Notable examples include the production of L-ribose, D-psicose, lactulose, and D-phenylalanine [42].

Here are some key points about isomerases:

***Isomerases*** are classified into several subclasses based on the type of isomerization reaction they catalyze.

### 2.6.1 Racemases

Racemases catalyze the conversion of one enantiomer (chiral isomer) into its mirror-image counterpart. This process is known as racemization and is often important in the biosynthesis of amino acids and other chiral molecules (Figure 2.12).

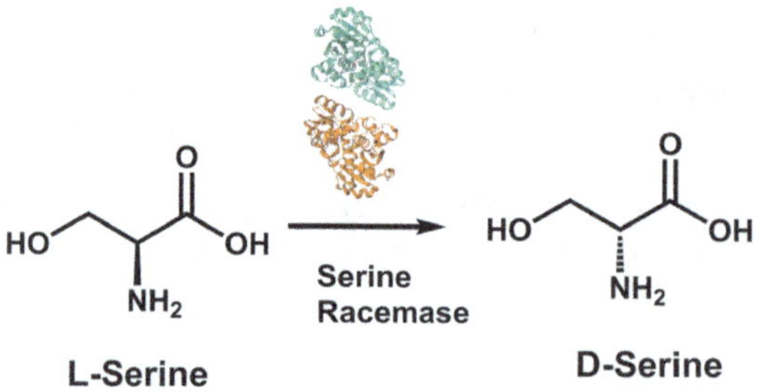

**Figure 2.12:** Transformation of L-serine into D-serine by the enzyme serine racemase (PDB code 5X2L).

### 2.6.2 Epimerases

Epimerases catalyze the conversion of one epimer into another. Epimerization reactions are important in the interconversion of sugars and other biomolecules.

### 2.6.3 *Cis-Trans* Isomerases

These enzymes catalyze the conversion of *cis* isomers into *trans* isomers or vice versa. They are involved in processes such as the conversion of *cis*-unsaturated fatty acids to *trans*-unsaturated fatty acids in lipid metabolism.

### 2.6.4 Intramolecular Transferases

Intramolecular transferases catalyze the transfer of a functional group from one part of a molecule to another within the same molecule [43]. These enzymes are involved in reactions like intramolecular phosphoryl transfer in nucleic acid metabolism.

*Industrial applications*: Isomerases have industrial applications in the production of various chemicals, including high-fructose corn syrup, which involves the isomerization of glucose to fructose. They are also used in the food and beverage industry to convert starches into syrups with improved sweetness and stability.

A distinct subclass of isomerases has been established to categorize enzymes that induce conformational changes in proteins and nucleic acids. The official enzyme list is continuously updated with newly validated enzymes. Furthermore, if new research impacts the classification of an existing enzyme, a fresh EC number is assigned, while the previous one is permanently retired and not reassigned.

## 2.7 Ligases (EC 6)

Ligases are a class of enzymes that catalyze the binding or joining of two molecules. By definition in biology, ligases catalyze the binding of two macromolecules by forming new bonds such as C–O, C–N, and C–S. An example is DNA ligase, which links two fragments of DNA by forming a phosphodiester bond. Ligase enzymes are synonymous with synthetase enzymes.

Ligases are classified as EC 6 in the EC number classification of enzymes. They can be further classified into six subclasses:

**EC 6.1**: Ligases that form carbon-oxygen bonds

**EC 6.2**: Ligases that form carbon-sulfur bonds

**EC 6.3:** Ligases that form carbon-nitrogen bonds (including argininosuccinate synthetase).

**EC 6.4:** Ligases that form carbon-carbon bonds, such as acetyl-CoA carboxylase

**EC 6.5:** Ligases that form phosphoric ester bonds, such as DNA ligase

**EC 6.6**: Ligases that form nitrogen-metal bonds, as in chelatases

Some examples of ligases include ubiquitin ligases (C–N bond), succinyl coenzyme A synthetase (C–S bond), acetyl-CoA carboxylase (C–C bond), DNA ligase (phosphoric ester bond), magnesium chelatase (nitrogen-metal bond), and DNA synthetase [44].

## 2.8 Translocases (EC 7)

Translocases are a class of enzymes that catalyze the transport, movement, or translocation of ions or molecules across cell membranes or their separation within membranes. They are fundamental to the functioning of cells and play a crucial role in regulating ion balance and cellular homeostasis.

Within the EC 7 class, there are several subclasses that correspond to different types of transport mechanisms. These transporters may be involved in the transport of ions such as sodium, potassium, calcium, and protons, or in the transport of organic molecules such as amino acids, sugars, and lipids.

Some examples of EC 7 subclasses include:

**EC 7.1**: Hydron translocation

**EC 7.2**: Inorganic cation translocation

**EC 7.3**: Inorganic anion translocation

**EC 7.4**: Translocases that transport amino acids and peptides across membranes

**EC 7.5**: Carbohydrates and their derivatives translocation

**EC 7.6**: Translocation of other compounds

Translocases are essential for the functioning of biological systems, as they facilitate the movement of substances necessary for growth, cell regulation, and energy generation in cells. Their study is fundamental to understanding cellular physiology and biochemistry [45] (Figure 2.13).

Enzymes play a central role as biocatalysts in research, diagnostics, medicine, and industrial biotechnology. Recent advances in microbial processes have enabled the sustainable production of enzymes with tailored catalytic properties, such as esterases, which are particularly relevant for applications in the pharmaceutical and

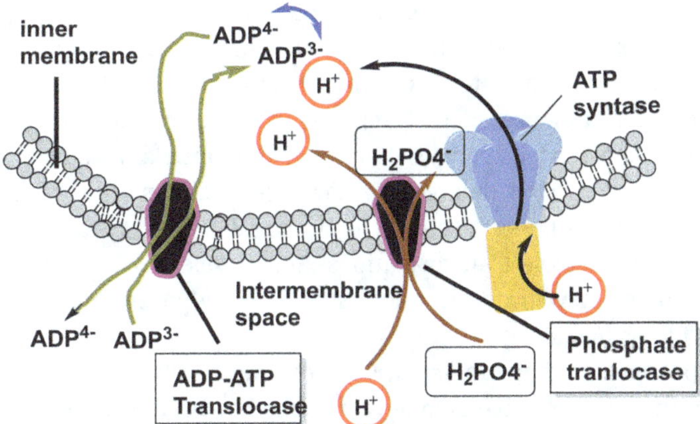


**Figure 2.13:** (A) ATP-ADP translocase: protein responsible for the 1:1 exchange of intramitochondrial ATP for ADP produced in the cytoplasm. (B) Phosphate translocase: the transport of $H_2PO_4^-$ together with a proton are produced by symport $H_2PO_4^-/H^+$.

food industries [46]. In parallel, the integration of biocatalysis into synthetic chemistry has allowed the development of more efficient and environmentally responsible routes for the synthesis of active pharmaceutical ingredients [47].

The diagnostic value of enzymes has been recognized for decades. Early clinical studies established the relevance of serum enzymes as indicators of liver function, providing a foundation for enzyme-based diagnostics [48]. More recently, enzymatic approaches have been applied to investigate atypical metabolic pathways involved in drug and drug-candidate biotransformations, offering insight into their pharmacokinetic behavior [49]. Enzymes have also been incorporated into healthcare and consumer products, such as denture-care formulations, where their catalytic activity improves cleaning efficiency and material compatibility [50]. In oncology-related research, antioxidant enzymes have been explored as potential biomarkers for cancer diagnosis and patient monitoring [51].

Modern medical diagnostics rely heavily on enzyme-based biosensing technologies. Continuous glucose monitoring systems represent a prominent example of enzyme-driven biosensors used in clinical practice [52]. Similarly, enzyme-based platforms have been developed for the determination of urea concentrations in serum samples [53] and for the selective detection of cholesterol in biological matrices [54]. Immunoenzymatic techniques, particularly ELISA-based methods, further illustrate the analytical versatility of enzymes in sensitive protein quantification [55, 56]. Advances in nanotechnology have extended these concepts to electrochemical nanokits capable of measuring enzyme activity at the single-cell level [57].

Beyond diagnostics, enzymes have gained importance as therapeutic targets. In particular, modulation of lipid-metabolizing enzymes has been proposed as a promising approach for the treatment of neurodegenerative diseases. In this context, inhibition of monoacylglycerol lipase has been identified as a potential strategy for slowing disease progression in disorders such as Alzheimer's disease [58].

Enzymes are also widely applied as catalysts in industrial processes, where they offer high selectivity and reduced environmental impact. Lipases, especially those of fungal origin, are among the most extensively used enzymes in industrial biotechnology, with applications in food processing, detergents, pharmaceuticals, and fine chemical synthesis [59]. A major contribution of enzymes to industrial organic synthesis is their ability to perform stereoselective transformations. Enantioselective enzymatic desymmetrization has become a key strategy for accessing optically pure compounds under mild reaction conditions [60, 61].

Oxidoreductases represent another important class of industrially relevant enzymes. Their application in stereoselective ketone reductions has been examined in detail, with studies highlighting the influence of substituent effects on catalytic activity and enantioselectivity [62]. Broader process-oriented analyses have demonstrated the feasibility of enzyme-catalyzed regio- and enantioselective reductions at preparative and industrial scales [63]. Advances in protein engineering have further expanded the scope of industrial biocatalysis, enabling the creation of highly enantioselective nitrilases through targeted mutagenesis strategies [64].

Oxidative biocatalysis has emerged as a rapidly evolving area within enzyme technology. Early efforts focused on the development of biomimetic oxidation systems inspired by enzymatic catalysis [65], while subsequent studies employing directed evolution significantly improved the performance of cytochrome P450 monooxygenases for selective oxidation of nonactivated substrates [66]. These enzymes have been successfully applied to asymmetric dihydroxylation reactions of alkenes [67] and to regioselective oxidation of aromatic compounds using recombinant dioxygenase systems [68]. Continued advances in cytochrome P450 engineering have further expanded their applicability in synthetic chemistry and industrial biocatalysis [69].

Baeyer-Villiger monooxygenases have attracted increasing attention due to their capacity to catalyze oxygen insertion reactions with high regio- and enantioselectivity. Comprehensive reviews have outlined recent advances and remaining challenges in this enzyme family [70], while structural studies have provided detailed insights into substrate binding and reaction mechanisms through the characterization of substrate-bound enzyme complexes [71].

Additional enzyme classes are gaining industrial relevance due to their distinctive catalytic properties. Epoxide hydrolases display considerable diversity and have demonstrated strong potential for biocatalytic ring-opening reactions [72]. Enzyme-mediated routes to pharmaceutically relevant diols have also been reported, highlighting greener alternatives to traditional chemical synthesis [73]. Halohydrin formation catalyzed by enzymes has been explored as an efficient route to epoxides and related intermediates [74], and comprehensive treatments of enzymatic biotransformations continue to underscore their importance in organic synthesis [75]. Carbon-carbon lyases, although less widely used, have found specialized industrial applications due to their ability to catalyze C–C bond formation and cleavage under mild conditions [76].

Finally, polysaccharide-modifying enzymes such as alginate lyases have enabled significant advances in industrial and biomedical applications. Extensive studies have examined their sources, structural features, and functional properties [77]. The controlled enzymatic modification of alginate has expanded its use in biomedical materials, where clear relationships have been established between polymer composition and resulting physicochemical properties [78].

## References


[1] Ali S, Khan SA, Hamayun M, Lee IJ. The recent advances in the utility of microbial lipases: A review. Microorganisms 2023;11(2).

[2] Kumar A, Verma V, Dubey VK, Srivastava A, Garg SK, Singh VP, et al. Industrial applications of fungal lipases: A review. Front Microbiol 2023;14.

[3] De Brabander P, Uitterhaegen E, Delmulle T, De Winter K, Soetaert W. Challenges and progress towards industrial recombinant protein production in yeasts: A review. Biotechnol Adv 2023;64.

[4] Bibi F, Ilyas N, Saeed M, Shabir S, Shati AA, Alfaifi MY, et al. Innovative production of value-added products using agro-industrial wastes via solid-state fermentation. Environ Sci Pollut Res Int 2023.

[5] Paloi S, Kumla J, Paloi BP, Srinuanpan S, Hoijang S, Karunarathna SC, et al. Termite Mushrooms (Termitomyces), a potential source of nutrients and bioactive compounds exhibiting human health benefits: A review. Journal of fungi 2023;9(1).

[6] Bairoch A. The Enzyme data bank. Nucl Acids Res 1994;22(17):3626–3627.

[7] McDonald AG, Tipton KF. Enzyme nomenclature and classification: The state of the art. Febs journal 2023;290(9):2214–2231.

[8] Faber K. Biocatalytic applications. In Faber K, editor. Biotransformations in Organic Chemistry: A Textbook 2018, Cham: Springer International Publishing; p. 31–313.

[9] Younus H. Oxidoreductases: Overview and practical applications. In Husain Q, Ullah MF, editors. Biocatalysis: Enzymatic Basics and Applications 2019, Cham: Springer International Publishing; p. 39–55.

[10] Hollmann F, Opperman DJ, Paul CE. Biocatalytic reduction reactions from a chemist's perspective. Angew Chem Int Ed Engl 2021;60(11):5644–5665.

[11] Adeva M, González-Lucán M, Seco M, Donapetry C. Enzymes involved in L-lactate metabolism in humans. Mitochondrion 2013;13(6):615–629.

[12] Ramírez-Nava EJ, et al. Molecular cloning and exploration of the biochemical and functional analysis of recombinant Glucose-6-Phosphate Dehydrogenase from Gluconoacetobacter diazotrophicus PAL5. Int J Mol Sci 2019;20(21).

[13] Huang S, Millar AH. Succinate dehydrogenase: The complex roles of a simple enzyme. Curr Opin Plant Biol 2013;16(3):344–349.

[14] Velasco-Lozano S, Benítez-Mateos AI, López-Gallego F. Co-immobilized Phosphorylated Cofactors and Enzymes as self-sufficient Heterogeneous Biocatalysts for chemical processes. Angew Chem Int Ed Engl 2017;56(3):771–775.

[15] Srivilai J, Minale G, Scholfield CN, Ingkaninan K. Discovery of natural Steroid 5 Alpha-reductase inhibitors. Assay Drug Dev Technol 2019;17(2):44–57.

[16] Herrmann L, Bockau U, Tiedtke A, Hartmann MW, Weide T. The bifunctional dihydrofolate reductase thymidylate synthase of Tetrahymena thermophila provides a tool for molecular and biotechnology applications. BMC Biotechnol 2006;6:21.

[17] Li Y, et al. Identification of key sites determining the cofactor specificity and improvement of catalytic activity of a steroid 5β-reductase from Capsella rubella. Enzyme Microb Technol 2020;134:109483.

[18] Hägglund P, et al. Seed thioredoxin h. Biochim Biophys Acta 2016;1864(8):974–982.

[19] De Oliveira FK, Santos LO, Buffon JG. Mechanism of action, sources, and application of peroxidases. Food Res Int 2021;143:110266.

[20] Wang X, Li S, Li L, Ding Y, Xu T, Liu J, et al. Mycotoxins degradation by reactive oxygen species in plasma: Reactive molecular dynamics insights. LWT 2023;184:115058.

[21] Goyal MM, Basak A. Human catalase: Looking for complete identity. Protein & Cell 2010;1 (10):888–897.

[22] Thao NTM, et al. Antioxidant Nanozymes: Mechanisms, activity manipulation, and applications. Micromachines (Basel) 2023;14(5).

[23] Frey PA, Hegeman AD, Frey PA, Hegeman AD. Oxidases and Oxygenases. In Enzymatic Reaction Mechanisms 2007, Oxford University Press; 0.

[24] Turner NJ. 7.12 Oxidation: Oxidases. In Carreira EM, Yamamoto H, editors. Comprehensive Chirality 2012, Amsterdam: Elsevier; p. 256–274.

[25] White PC, Tusie-Luna M-T, New MI, Speiser PW. Mutations in Steroid 21-Hydroxylase (CYP21). Hum Mutat 1994;3(4):373–378.

[26] Newton AC. Protein kinases in tune. IUBMB Life 2019;71(6):670–671.

[27] Jurkowska RZ, Jeltsch A. Mechanisms and biological roles of DNA Methyltransferases and DNA Methylation: From past achievements to future challenges. Adv Exp Med Biol 2022;1389:1–19.

[28] Günenc AN, Graf B, Stark H, Chari A. Fatty acid synthase: Structure, function, and regulation. Subcell Biochem 2022;99:1–33.

[29] Palcic MM. Glycosyltransferases as biocatalysts. Curr Opin Chem Biol 2011;15(2):226–233.

[30] Cutlan R, De Rose S, Isupov MN, Littlechild JA, Harmer NJ. Using enzyme cascades in biocatalysis: Highlight on transaminases and carboxylic acid reductases. BBA Proteins Proteomics 2020;1868 (2):140322.

[31] Barrett AJ. Proteases. Curr Protoc Protein Sci 2000;21(1):21.1.1–1.12.

[32] Goettig P. Reversed Proteolysis-Proteases as Peptide Ligases. Catalysts 2021;11(1).

[33] Brömme D. Papain-like Cysteine Proteases. Curr Protoc Protein Sci 2000;21(1):21.2.1–2.14.

[34] Herman RA, Ayepa E, Zhang WX, Li ZN, Zhu X, Ackah M, et al. Molecular modification and biotechnological applications of microbial aspartic proteases. Crit Rev Biotechnol 2023.

[35] Nagase H. Metalloproteases. Curr Protoc Protein Sci 2001;Chapter 21: 21.4.1–4.13.

[36] Mészáros Z, Nekvasilová P, Bojarová P, Křen V, Slámová K. Reprint of: Advanced glycosidases as ingenious biosynthetic instruments. Biotechnol Adv 2021;51:107820.

[37] Rafeeq H, Hussain A, Shabbir S, Ali S, Bilal M, Sher F, et al. Esterases as emerging biocatalysts: Mechanistic insights, genomic and metagenomic, immobilization, and biotechnological applications. Appl Biochem Biotechnol 2022;69(5):2176–2194.

[38] Tasnádi G, Lukesch M, Zechner M, Jud W, Hall M, Ditrich K, et al. Exploiting acid Phosphatases in the synthesis of Phosphorylated Monoalcohols and Diols. Eur J Org Chem 2016;2016(1):45–50.

[39] Wu Z, et al. Amidase as a versatile tool in amide-bond cleavage: From molecular features to biotechnological applications. Biotechnol Adv 2020;43:107574.

[40] De Gonzalo G, Alcántara AR, Domínguez de María P, Sánchez-Montero JM. Biocatalysis for the asymmetric synthesis of Active Pharmaceutical Ingredients (APIs): This time is for real. xpert Opin Drug Discov 2022;17(10):1159–1171.

[41] Brovetto M, Gamenara D, Mendez PS, Seoane GA. C-C bond-forming lyases in organic synthesis. Chem Rev 2011;111(7):4346–4403.
[42] Trincone A. Application-oriented marine Isomerases in Biocatalysis. Marine Drugs 2020;18(11).
[43] Cuesta SM, Rahman SA, Thornton JM. Exploring the chemistry and evolution of the isomerases. Proc Natl Acad Sci USA 2016;113(7):1796–1801.
[44] Holliday GL, Rahman SA, Furnham N, Thornton JM. Exploring the biological and chemical complexity of the ligases. J Mol Biol 2014;426(10):2098–2111.
[45] Balaji S. The transferred translocases: An old wine in a new bottle. Appl Biochem Biotechnol 2022;69 (4):1587–1610.
[46] Dahiya D, Nigam PS. Sustainable biosynthesis of esterase enzymes of desired characteristics of catalysis for pharmaceutical and food industry employing specific strains of microorganisms. Sustainability 2022;14(14).
[47] Simic S, Zukic E, Schmermund L, Faber K, Winkler CK, Kroutil W. Shortening synthetic routes to small molecule active pharmaceutical ingredients employing biocatalytic methods. Chem Rev 2022;122 (1):1052–1126.
[48] Zimmerman HJ. Serum enzymes in diagnosis of hepatic disease. Gastroenterology 1964.
[49] Isin EM. Unusual biotransformation reactions of drugs and drug candidates. Drug Metab Dispos 2023;51(4):413–426.
[50] Kim J-H, Lee H-N, Bae S-K, Shin D-H, Ku B-H, Park H-Y, et al. Development of a novel denture care agent with highly active enzyme, arazyme. BMC Oral Health 2021;21(1):365.
[51] Cecerska-Heryć E, Surowska O, Heryć R, Serwin N, Napiontek-Balińska S, Dołęgowska B. Are antioxidant enzymes essential markers in the diagnosis and monitoring of cancer patients – A review. Clinical Biochemistry 2021;93:1–8.
[52] Lee I, Probst D, Klonoff D, Sode K. Continuous glucose monitoring systems – Current status and future perspectives of the flagship technologies in biosensor research. Biosens Bioelectron 2021;181:113054.
[53] Vardar G, Attar A, Yapaoz MA. Development of urea biosensor using non-covalent complexes of urease with aldehyde derivative of PEG and analysis on serum samples. Prep Biochem Biotechnol 2019;49(9):868–875.
[54] Narwal V, Deswal R, Batra B, Kalra V, Hooda R, Sharma M, et al. Cholesterol biosensors: A review. Steroids 2019;143:6–17.
[55] Tan X, Chen Q, Zhu H, Zhu S, Gong Y, Wu X, et al. Fast and reproducible ELISA laser platform for ultrasensitive protein quantification. ACS Sensors 2020;5(1):110–117.
[56] Tan XT, Chen QS, Zhu HB, Zhu S, Gong Y, Wu XQ, et al. Fast and reproducible ELISA laser platform for ultrasensitive protein quantification. ACS sensors 2020;5(1):110–117.
[57] Pan RR, Jiang DC. Nanokits for the electrochemical quantification of enzyme activity in single living cells. In Allbritton NL, Kovarik ML, editors. Enzyme activity in single cells Vol. 6282019, p. 173–189.
[58] Chen C. Inhibiting degradation of 2-arachidonoylglycerol as a therapeutic strategy for neurodegenerative diseases. Pharmacol Ther 2023;244:108394.
[59] Kumar A, Verma V, Dubey VK, Srivastava A, Garg SK, Singh VP, et al. Industrial applications of fungal lipases: A review. Front Microbiol 2023;14:1142536.
[60] García-Urdiales E, Alfonso I, Gotor V. Enantioselective enzymatic desymmetrizations in organic synthesis. Chem Rev 2005;105(1):313–354.
[61] Garcia-Urdiales E, Alfonso I, Gotor V. Enantioselective enzymatic desymmetrizations in organic synthesis. Chem Rev 2005;2005(105):313–354.
[62] Zhu D, Rios BE, Rozzell JD, Hua L. Evaluation of substituent effects on activity and enantioselectivity in the enzymatic reduction of aryl ketones. Tetrahedron: Asymmetry 2005;16(8):1541–1546.

[63] Müller M, Wolberg M, Schubert T, Hummel W. Enzyme-Catalyzed Regio- and Enantioselective Ketone reductions. In Kragl U, editor. Technology Transfer in Biotechnology: From lab to Industry to Production 2005, Berlin, Heidelberg: Springer Berlin Heidelberg; p. 261–287.
[64] DeSantis G, Wong K, Farwell B, Chatman K, Zhu Z, Tomlinson G, et al. Creation of a productive, highly enantioselective nitrilase through Gene Site Saturation Mutagenesis (GSSM). J Am Chem Soc 2003;125(38):11476–11477.
[65] Ricoux R, Sauriat-Dorizon H, Girgenti E, Blanchard D, Mahy JP. Hemoabzymes: Towards new biocatalysts for selective oxidations. J Immunol Methods 2002;269,:39–57.
[66] Farinas ET, Schwaneberg U, Glieder A, Arnold FH. Directed evolution of a cytochrome P450 monooxygenase for alkane oxidation. Adv Synth Catal 2001;343:601–606.
[67] Boyd DR, Sharma ND, Brannigan IN, Groocock MR, Malone JF, McConville G, et al. Biocatalytic asymmetric dihydroxylation of conjugated mono- and poly-alkenes to yield enantiopure cyclic cis-diols. Adv Synth Catal 2005;347:1081–1089.
[68] Yildirim S, Franco TT, Wohlgemuth R, Kohler H-PE, Witholt B, Schmid A. Recombinant chlorobenzene dioxygenase from Pseudomonas sp. P51: A biocatalyst for regioselective oxidation of aromatic nitriles. Adv Synth Catal 2005;347:1060–1072.
[69] Urlacher VB, Girhard M. Cytochrome P450 monooxygenases: An update on perspectives for synthetic application. Trends Biotechnol 2012;30:26–36.
[70] Torres Pazmiño DE, Dudek HM, Fraaije MW. Baeyer–Villiger monooxygenases: Recent advances and future challenges. Curr Opin Chem Biol 2010;14(2):138–144.
[71] Yachnin BJ, Sprules T, McEvoy MB, Lau PCK, Berghuis AM. The Substrate-Bound Crystal Structure of a Baeyer–Villiger Monooxygenase Exhibits a Criegee-like Conformation. J Am Chem Soc 2012;134:7788–7795.
[72] van Loo B, Kingma J, Arand M, Wubbolts MG, Janssen DB. Diversity and biocatalytic potential of epoxide hydrolases identified by genome analysis. Appl Environ Microbiol 2006;72(4):2905–2917.
[73] Meena VS, Banerjee UC. Biocatalytic route for the synthesis of active pharmaceutical diol: A greener approach. Indian J Biotechnol 2011;10:452–457.
[74] Geigert J, Neidleman SL. Enzymes and Immobilized Cells in Biotechnology: Enzymic Synthesis of Halohydrins and their Conversion to Epoxides 1985, Butterworth-Heinemann.
[75] Faber K. Biotransformations in Organic Chemistry 2011, 6 th ed, Springer.
[76] Hilterhaus L, Liese A. Industrial application and processes using carbon carbon lyases. In Sons JW. editor, Enzyme Catalysis in Organic Synthesis 2012, third ed, Wiley-VCH Verlag GmbH & Co. KGaA; p. 991–1000.0
[77] Wong TY, Preston LA, Schiller NL. Alginate Lyase: review of major sources and enzyme characteristics, structure-function analysis, biological roles, and applications. Annu Rev Microbiol 2000;54:289–340.
[78] Lee KY, Mooneya DJ. Alginate: Properties and biomedical applications. Prog Polym Sci 2012;37:106–26.

# Chapter 3
# Immobilization of Enzymes

**Abstract:** This chapter provides a comprehensive overview of enzyme immobilization, a fundamental technique in biocatalysis that enhances the practicality and economic viability of enzymes for industrial applications. Immobilization involves confining enzymes to solid supports, creating heterogeneous catalysts that can be easily separated and reused. This chapter begins by discussing the advantages of this approach, including improved enzyme stability, reusability, and compatibility with nonaqueous media, while also acknowledging potential challenges such as activity loss or diffusion limitations.

A significant focus is placed on the critical choice of support materials, which are systematically classified into inorganic (e.g., mesoporous silica, magnetic nanoparticles, and carbon nanotubes), organic (e.g., biopolymers like alginate and chitosan, and synthetic polymers), and advanced hybrid materials (e.g., metal-organic frameworks, covalent organic frameworks, and cross-linked enzyme aggregates). This chapter details various immobilization methodologies, encompassing covalent binding, adsorption, affinity interactions, cross-linking, and physical entrapment within gels, polymers, membranes, or ionic liquids.

Furthermore, innovative techniques such as 3D printing, electrospraying, and Pickering emulsion encapsulation are highlighted as emerging strategies for creating tailored immobilization systems. This chapter also covers magnetic immobilization for facile biocatalyst recovery and chemical modification of enzymes to tailor their properties. By integrating traditional methods with cutting-edge approaches and materials, this chapter underscores how enzyme immobilization serves as a powerful tool to develop robust, efficient, and sustainable biocatalysts for diverse sectors including pharmaceuticals, fine chemicals, food technology, and biosensing.

**Keywords:** Whole-cell catalysis, reaction optimization, biocatalytic cascades, chemoenzymatic synthesis, enzyme stability

## 3.1 Introduction

The utilization of biocatalytic applications is steadily increasing across a wide range of industrial sectors, from fine and pharmaceutical chemistry to the food and chemical industries. However, a notable challenge arises from the costs associated with enzymes, which often necessitate their repeated use to ensure economic viability. Moreover, due to the water-soluble nature of these enzymes, their recovery typically presents a complex challenge.

 | https://doi.org/10.1515/9783111122991-003

To address this issue, the strategy of immobilizing enzymes onto solid substrates has been implemented, leading to the creation of heterogeneous catalysts. This can be achieved by using existing supports or even generating novel support structures during the immobilization process. Given that enzymes do not dissolve in organic media, immobilization emerges as a viable alternative for their application in such environments [1].

Many biotechnological processes rely on enzymes or cells, making this field brimming with promising prospects. Enzyme-driven processes can experience significant enhancements when enzymes are effectively reused through multiple reaction cycles. In this context, enzyme immobilization encompasses any methodology that facilitates the recurrent use or continuous deployment of enzymes. On the one hand, the development of cost-efficient and straightforward immobilization techniques is essential. On the other hand, the potential improvement of enzyme characteristics, such as activity, selectivity, and stability, adds layer of interest. In this chapter, we will explore various approaches to enzyme immobilization and methods for enhancing the functional properties of enzymes.

"Enzyme immobilization" involves physically confining soluble protein-based enzyme molecules through different types of interactions onto a supporting material within a restricted area. This support is usually an insoluble matrix that can be separated from the reaction mixture through simple techniques such as filtration, centrifugation, self-aggregation, or sieving. The performance of immobilized enzymes is largely determined by three key elements: the enzyme's nature and type, the properties of the support material, and the specific conditions applied during the immobilization process.

Immobilized biocatalysts refer to enzymes, whole cells, or cellular organelles (or combinations thereof) in a reusable state [2].

## 3.2 Advantages and Disadvantages of Enzyme Immobilization

*Reusability*: One of the key advantages of immobilization is the ability to reuse enzymes or biocatalysts multiple times. This reduces costs associated with acquiring new enzymes, making processes more economically sustainable.

*Increased stability:* Immobilization can enhance the stability of enzymes by protecting them from adverse conditions such as changes in pH, temperature, or the presence of inhibitors. This helps preserve enzyme activity in challenging environments.

*Facilitates separation and purification*: Immobilization simplifies the separation of enzymes or biocatalysts from the final product, as they are attached to a solid support. This streamlines product purification.

*Improved selectivity:* In some cases, immobilization can enhance the stereoselectivity, specificity, and regiospecificity of enzymes.

*Compatibility with nonaqueous media:* Immobilized enzymes are more compatible with nonaqueous media, expanding their potential applications in chemical and pharmaceutical processes [3].

### 3.2.1 Disadvantages of Immobilization

*Loss of activity*: In some instances, enzyme immobilization may reduce enzymatic activity because the proteins can become deactivated or denatured due to limited mobility on the solid support.

*Initial costs*: Immobilization may involve initial investment in support preparation and immobilization processes, which can be costly.

*Variable reproducibility:* The formation of enzyme aggregates or immobilization on solid surfaces can lead to variable and less reproducible results compared to enzymes in solution.

*Diffusion limitations:* Immobilization can sometimes restrict substrate diffusion to the enzyme, reducing process efficiency [4].

*Potential loss of selectivity:* In specific cases, immobilization may decrease enzyme selectivity, limiting their applicability in certain reactions [5].

Despite these drawbacks, enzyme immobilization has managed to overcome many of these disadvantages, enabling numerous biocatalyzed industrial processes to become viable and economically profitable today.

Recent progress in biotechnology, nanobiotechnology, and electronics is driving a swift increase in the need for biosensors that utilize enzymes. For example, creating biosensors with lipase enables the analysis of their traits and features, highlighting the present scope, prospects, and hurdles in biosensor technology, which holds significant potential for profitability [6].

## 3.3 Choice and Characteristics of Support

The enzyme immobilization process involves three main components: the enzyme, the support, or matrix, and the mode of interaction between the enzyme and its matrix. The choice of support or matrix is critical, as it can enhance the functional stability of the immobilization system [7]. While there is no universal matrix, there are several characteristics that any material considered for immobilization should possess.

A wide range of materials, most commonly inert polymers and inorganic compounds, can serve as matrices or carriers for enzyme immobilization. An ideal support matrix should possess the following characteristics:

- Economical: The acquisition and operating costs of the matrix should be reasonable and sustainable, ensuring that it remains economically viable.
- Inertness: The matrix must not chemically interact with the substances involved in the process. This property is essential to avoid unwanted reactions that could reduce the efficiency of the matrix as a support for enzymes or other biological agents.
- Stability is another important requirement of a carrier matrix. It must maintain its structural and functional integrity during the process or application to prevent premature degradation or degradation. This ensures robustness and consistency of matrix performance.
- Physical strength: The matrix must withstand mechanical and environmental stress without damage or deformation. This is particularly crucial in industrial applications where significant physical stress may be present.
- Enhancement of enzyme activity: The matrix should ideally increase the specificity and activity of the enzymes or other biological substances immobilized on it. This can be achieved by modifying the matrix surface to selectively interact with target molecules, improving reaction efficiency.
- Renewability: The matrix should be reusable, making the process more cost-effective and environmentally friendly. The ability to regenerate the matrix and restore its original functionality after use helps reduce costs and waste production.
- Reduction of product inhibition: The matrix should not inhibit the formation or release of products in the enzymatic process, allowing for optimal efficiency.
- Prevention of nonspecific adsorption and contamination: The matrix should minimize nonspecific adsorption and prevent bacterial contamination. This is crucial to maintaining product purity, especially in medical, pharmaceutical, food, and biotechnology applications.

In summary, an ideal support matrix should combine economic, chemical, physical, and biological properties, making it suitable for various applications that require the controlled interaction of biological agents, such as enzymes, with their environment. These characteristics ensure optimal efficiency and effectiveness in industrial and scientific processes.

In enzyme immobilization, the functional groups and surface charges of the carrier dictate their interaction with the enzyme. Furthermore, the matrix's surface area and porosity are essential parameters that govern the loading capacity of the enzyme.

## 3.4 Classification of Support for Enzyme Immobilization

The supports used for enzyme immobilization can be broadly classified into inorganic and organic categories, with further subcategories based on their properties and composition. Common classifications include:

### 3.4.1 Inorganic Supports

These include materials such as glass, silica gel, zeolite (molecular sieves), ceramics (including metal oxides like $Al_2O_3$, $TiO_2$, $ZrO_2$, and $SnO_2$) [8], activated carbon, and charcoal. These materials are valued for their thermal and mechanical resistance and their resistance to microorganisms, as they do not serve as substrates for bacterial or fungal growth. They also offer rigidity and controlled porosity, ensuring consistent pore diameter and volume. For example, silica supports feature nanoscale structures, high surface area, ordered arrangement, and excellent stability against chemical and mechanical forces. Overall, these supports are highly attractive for industrial applications due to their mechanical robustness, thermal stability, resistance to organic solvents, bacterial contamination, and chemical degradation [9]. They are also easy to handle, have a long lifespan, and can be regenerated through pyrolysis.

Organic supports: These can be divided into two groups: (i) synthetic materials, mainly polymers and (ii) renewable materials derived from natural sources (biopolymers).

Magnetic supports: Magnetic particles are commonly used as a support matrix for enzyme immobilization. This approach involves attaching enzyme molecules to the surface of magnetic iron oxide nanoparticles (MNPs), allowing for easy separation of the biocatalytic system using an external magnetic field. Magnetic supports simplify the recovery of biocatalysts from the reaction medium after the catalytic process [10].

A representative case is glucose oxidase adsorbed onto unmodified MNPs. This biocatalytic system was applied to decolorize Acid Yellow 12, maintaining more than 90% of its original activity after 15 consecutive reaction cycles [11].

#### 3.4.1.1 Mesoporous Materials

Two primary strategies for immobilizing enzymes onto mesoporous substances are covalent attachment and pore encapsulation. Both approaches confine the enzyme inside the porous network of the carrier, which serves to stabilize the protein's three-dimensional (3D) conformation and typically maintains its catalytic functionality. A key consideration with this pore confinement, however, is the emergence of diffusional constraints, since the movement of substrate molecules into, and product molecules out of, the pores can be hindered.

A common application involves mesoporous silica carriers like SBA-15 and MCM-41. These supports have been utilized to anchor lipases sourced from *Candida antarctica* and *Candida rugosa* via direct, non-covalent adsorption, eliminating the requirement for a linking agent. Once immobilized, these enzymes act as efficient catalysts in organic synthesis reactions and demonstrate a markedly improved capacity for reuse (Figure 3.1).

The category of mesoporous supports includes materials such as zeolites, carbon-based substances, sol-gel matrices, along with various precipitated and ordered mesoporous oxides [12].

Notably, mesoporous silica nanoparticles have been utilized as carriers for lipase immobilization in biofuel synthesis, an application leveraged from their high chemical and thermal stability [13].

Furthermore, SBA-15 and MCM-41 silica types have been applied for the adsorption and immobilization of alkaline protease. Mesoporous silica MCM-41 has also served as a support for tyrosinase in phenol detection applications. Similarly, mesoporous silica spheres have proven effective for immobilizing enzymes including catalase, protease, and peroxidase [14] (Figure 3.1).

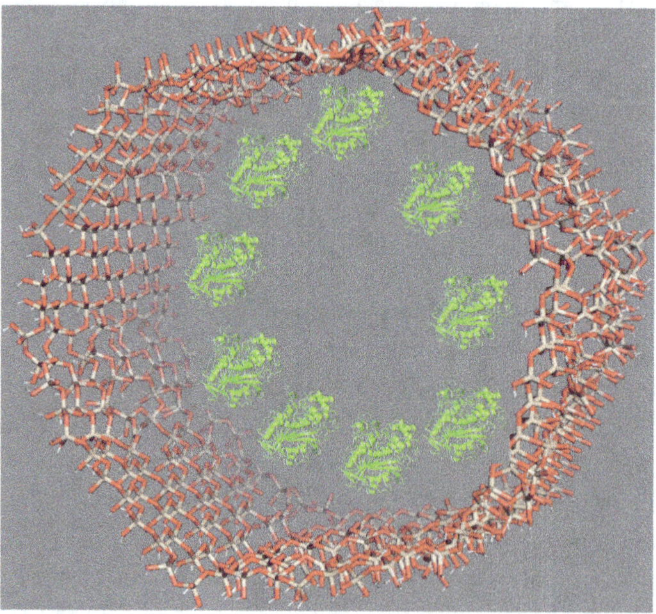

**Figure 3.1:** Illustration of SBA 15 (approximate pore diameter 30 nm) with *Candida rugosa* lipase (approximate pore diameter 5 nm) attached through nonspecific interactions.

#### 3.4.1.2 Nanoparticles

Nanoparticles provide a substantial surface area for the attachment of enzymes, which enables a greater quantity of enzyme to be loaded onto the carrier's surface and improves the overall yield of the immobilization process. A key benefit of using nanoparticles, compared to other inorganic supports, is their capacity to significantly reduce diffusion-related constraints.

Materials such as gold oxide nanoparticles and graphene can serve as effective carriers for enzyme immobilization. Titanium nanoparticles have also been used to immobilize carbonic anhydrase, with glutaraldehyde (GA) acting as a cross-linking agent. In a further illustration, *Rhizomucor miehei* lipase was successfully attached via covalent bonds to silica nanoparticles that had been functionalized with octyltriethoxysilane and glycidoxypropyltrimethoxysilane [15].

#### 3.4.1.3 Ceramic Materials

Inorganic ceramic supports, including alumina, zirconia, titania, silica, iron oxide, and calcium phosphate, are employed for immobilizing enzymes. These substances can also be applied as ceramic foams or integrated into composite ceramic membranes (e.g., $TiO_2/Al_2O_3$).

Hydroxyl groups are predominantly present on the surface of these materials, as is the case with mesoporous materials, promoting enzyme adsorption and immobilization primarily through nonspecific interactions.

#### 3.4.1.4 Carbon Nanotubes

Carbon nanotubes are cylindrical nanostructures composed of carbon atoms organized in a hexagonal lattice configuration. They are primarily categorized into two varieties: single-walled carbon nanotubes and multi-walled carbon nanotubes (Figure 3.2).

The immobilization of enzymes on carbon nanotubes can be achieved by either binding the enzymes to the surface of the nanotubes or encapsulating them within the nanotube structure. This process allows enzymes to retain their catalytic activity while benefiting from a high surface area, enhanced stability and durability, ease of functionalization, improved sensitivity and response, and biocompatibility.

They are frequently used in the fabrication of biosensors designed to detect a range of chemical substances, which include phenol along with its mono- and multi-substituted derivatives, bisphenols, and certain pharmaceutical drugs such as diclofenac and tetracycline. Furthermore, these nanostructures find significant application in medical diagnostics and clinical testing [16].

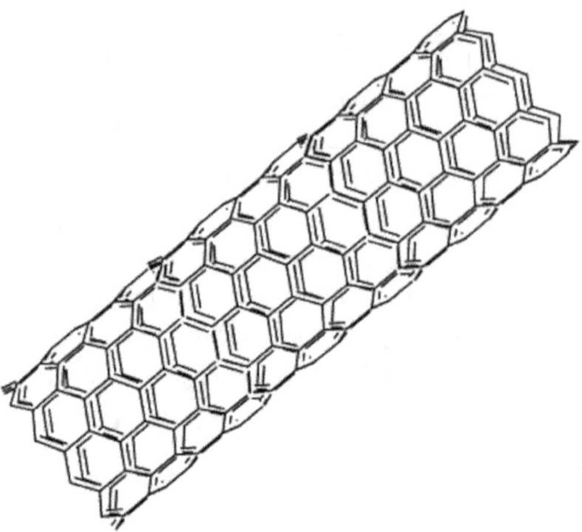

**Figure 3.2:** Illustration of carbon nanotube.

### 3.4.2 Organic Supports

The most widely utilized organic supports are categorized into natural substances – including polysaccharides, proteins, and carbon-based materials and synthetic polymers.

Natural supports: A wide variety of natural polymers, mainly water-insoluble polysaccharides, such as collagen, chitosan [1], chitin[2], alginate[3], cellulose[4], agarose[5], carrageenan[6], and starch[7] (Figure 3.3), have been used as support matrices for enzyme immobilization [17]. In general, they are referred to as biopolymers.

**Chitosan**[1] is a biopolymer derived from chitin[2], and its chemical structure is characterized by the deacetylation of chitin. Chitin is a long-chain polymer composed of *N*-acetylglucosamine units, whereas chitosan results from the partial deacetylation of chitin, where some of the acetyl groups are removed. The specific composition of chitosan can vary based on the degree of deacetylation, which refers to the extent to which acetyl groups have been removed from the chitin molecule. The key components of chitosan include glucosamine units, acetyl groups, and amine groups.

**Alginate**[3] is a negatively charged polysaccharide that, in the presence of divalent metal ions, particularly calcium ($Ca^{2+}$), cross-links to form a biopolymer. It is composed of long chains of sugar molecules and is a linear copolymer consisting of two types of monosaccharides: β-D-mannuronic acid (M) and α-L-guluronic acid (G). The ratio of M and G residues in the alginate molecule can vary depending on its source (typically obtained from brown seaweed) and the processing methods used. The specific composition and structure of alginate influence its properties and applications [18].

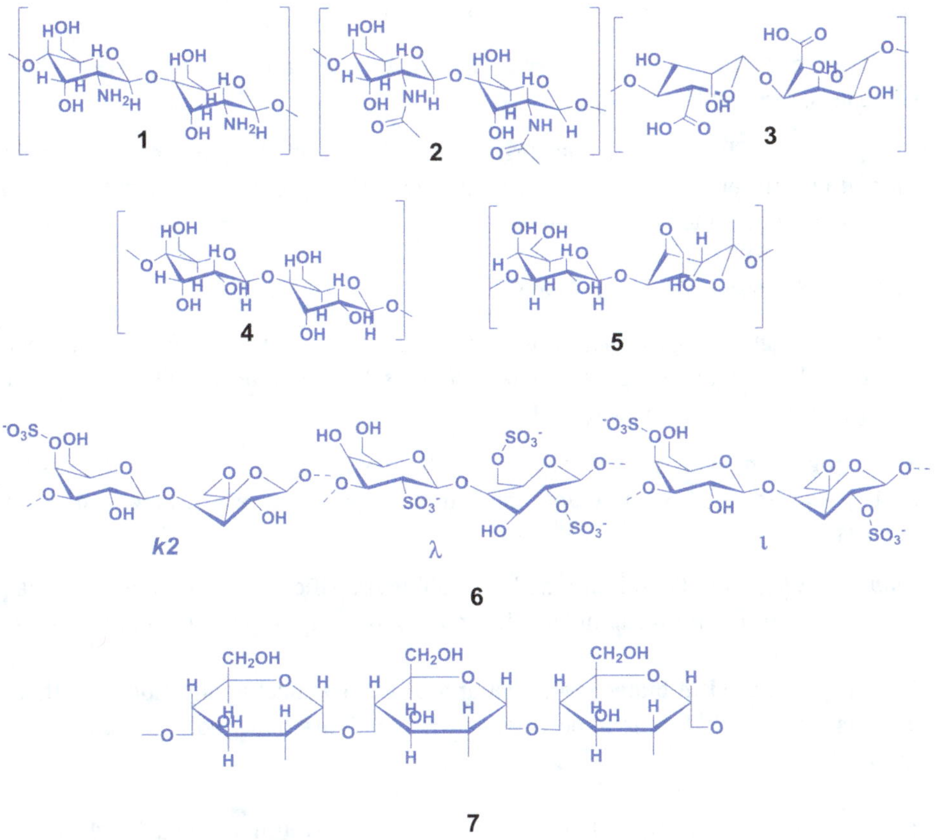

Figure 3.3: Some examples of biopolymers.

**Cellulose**[4] is composed of repeating units of the simple sugar glucose. Its chemical structure consists of linear chains of glucose molecules linked by β(1→4) glycosidic bonds.

**Agarose**[5] is a polysaccharide extracted from certain red seaweeds. It is a linear polymer made up of repeating agarobiose units, a disaccharide consisting of D-galactose and 3,6-anhydro-L-galactopyranose. Agarose is one of the two main components of agar and is purified from agar by removing its other component, agaropectin. Agarose can be modified physically, biologically, or chemically, allowing it to perform optimally in various environments [19].

**Carrageenan**[6] is a natural hydrocolloid extracted from various species of red seaweed (algae). It is composed of a mixture of sulfated linear polysaccharides that can be categorized into three main types based on their chemical structure and solubility: kappa, iota, and lambda carrageenan.

The primary components of carrageenan are:

Galactose units: These units are linked through alternating α-(1→3) and β-(1→4) glycosidic bonds, forming linear chains.

Sulfate groups: The position and degree of sulfation on the galactose units vary among the different types of carrageenan. The sulfate groups give carrageenan its ability to form gels and interact with proteins.

The types of carrageenan are as follows:

*Kappa carrageenan*: Characterized by a high degree of sulfate esterification, it forms strong, rigid gels in the presence of potassium ions. It is commonly used in dairy products like ice cream and chocolate milk.

*Iota carrageenan*: With a lower degree of sulfate esterification, it forms soft, elastic gels in the presence of calcium ions, and is used in gel-based desserts and some dairy products.

*Lambda carrageenan*: Containing the least sulfate esterification, it does not form gels but acts as a thickening agent and stabilizer, often used in sauces and dairy items [19].

They are often used in biotechnological and pharmaceutical applications. Synthetic supports can include synthetic polymers such as polyurethane, polyacrylamide, and agarose.

**Starch**[7] is a carbohydrate polymer made up of glucose units linked by glycosidic bonds. It typically consists of two main components: amylose and amylopectin. Amylose is a linear polymer of glucose units linked by α-1,4-glycosidic bonds, while amylopectin is a branched polymer, containing both α-1,4-glycosidic bonds (as in amylose) and α-1,6-glycosidic bonds.

Synthetic polymers such as ion-exchange resins are characterized by their porous surfaces and insolubility [20]. These polymers, due to their porous structure, can immobilize enzymes strongly and are resistant to microbial attack. Examples include polystyrene, polyvinyl chloride (PVC), polyacrylate, polyamide, polypropylene, diethylaminoethyl cellulose (DEAE cellulose), UV-activated polyethylene glycerol, and GA-activated nylon.

For instance, PVC has been used for the covalent immobilization of oxalate oxidase, demonstrating its effectiveness as an inert support in the determination of urinary oxalate [21].

Polymer structures often contain a variety of chemical functional groups, such as carbonyl, carboxyl, hydroxyl, epoxy, amine, and diol groups, along with hydrophobic alkyl groups and trimethylamine fragments. Organic supports are frequently chosen for their versatility and capacity for chemical modification to meet specific immobilization requirements [22].

Membranes and gels are also used as supports for enzyme immobilization. These materials have pores that retain enzymes while allowing the diffusion of substrates and products.

### 3.4.3 Mixed Supports

Some supports combine inorganic and organic materials, merging the properties of both. This combination can offer greater stability and adaptability to different working conditions. For example, the immobilization of laccase on chitosan-coated magnetite ($Fe_3O_4$) nanoparticles is particularly advantageous due to its high mechanical strength, good biocompatibility, and high loading capacity [23].

The choice of support depends on the specific characteristics of the enzyme, the working environment, and the intended applications. Each type of support has its own advantages and disadvantages, making the proper selection crucial for the success of immobilization in a given application.

### 3.4.4 Covalent Organic Frameworks (COFs)

*Covalent organic frameworks* ***(COFs)*** are porous, crystalline organic polymers known for their precise pore structures, chemical stability in organic solvents, and resistance to harsh conditions [24]. Their lack of metals enhances their stability, making them suitable for applications in gas storage, separation, catalysis, and other tasks, while also allowing them to withstand high temperatures.

COFs are synthesized through a controlled solvothermal process that involves high temperature and pressure, and their properties can be tailored to meet specific needs (Figure 3.4).

COFs serve as ideal supports for enzyme immobilization due to their stability, porous structure, and large surface area. Enzymes can be incorporated via physical adsorption or covalent bonding, offering structural control, efficient mass transfer, and stable catalysis. Current research shows promising results, such as the efficient immobilization of lipase in biodiesel production using magnetic COF composites.

In summary, COFs are versatile materials with a range of applications. Their unique properties – such as a porous structure, stability, and functional groups – enable various interactions, making them highly valuable in biocatalysis and related fields [25, 26].

### 3.4.5 Metal-Organic Frameworks (MOFs)

**Metal-organic frameworks (MOFs)** are porous materials composed of metal nodes and organic linkers [27]. MOFs allow precise control over their composition, structure,

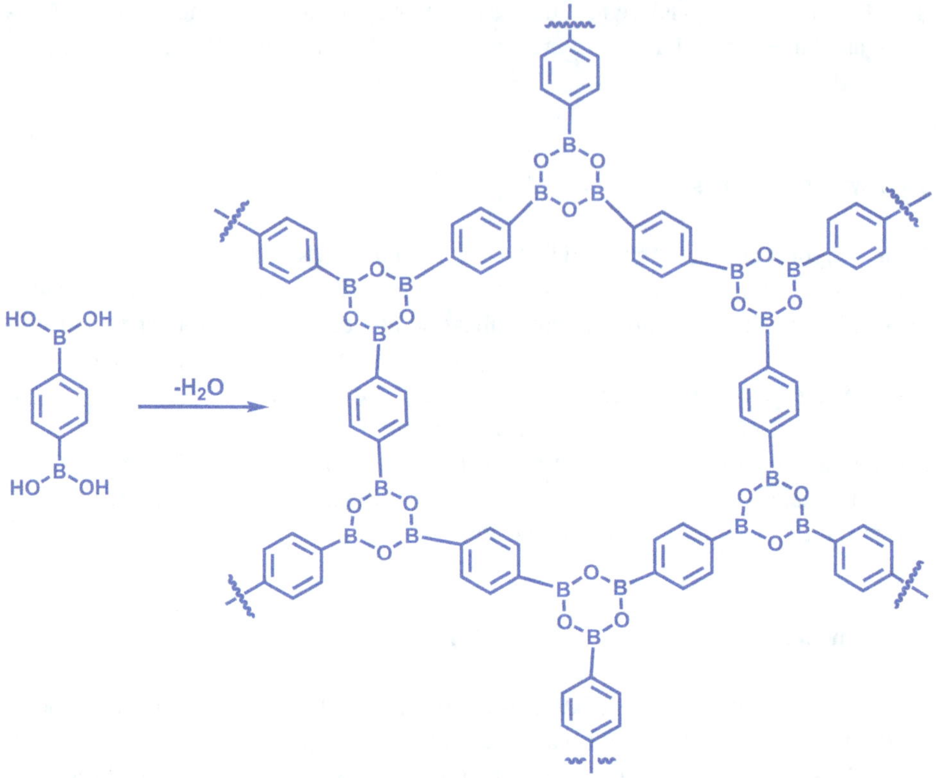

Figure 3.4: Scheme of COF-1 synthesis via boron condensation reaction, where three boronic acid molecules form a six-membered $B_3O_3$ (boroxine) ring through molecular dehydration, resulting in the elimination of three water molecules [26].

and functionality, enhancing their performance in various applications. They can be synthesized using different metals and organic compounds. Catalysts such as nanoparticles, metal complexes, or biomolecules can be incorporated into MOFs, providing stability and serving as size-selective catalytic supports (Figure 3.5).

MOFs, with their high surface area and tunable porosity, enable greater enzyme loading compared to traditional supports. Their structure protects and stabilizes enzyme conformations, enhancing overall stability. Enzyme immobilization in MOFs occurs through encapsulation, surface immobilization, or incorporation into pre-synthesized MOFs. Enzymes encapsulated in these porous materials show improved stability under harsh conditions.

MOFs are particularly well-suited for enzyme immobilization due to their high specific surface area, customizable properties, thermal and chemical stability, and mechanical strength. By allowing variations in functionalization, size, morphology, and electrostatic potential, MOFs improve enzyme stabilization and activity, even under extreme conditions, making them promising for industrial applications [28].

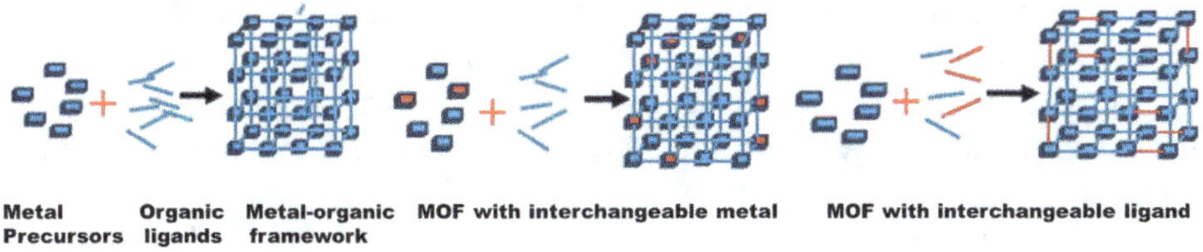

Metal Precursors
Organic ligands
Metal-organic framework
MOF with interchangeable metal
MOF with interchangeable ligand

**Figure 3.5:** Examples of MOFs and modified MOFs.

### 3.4.6 Cross-Linked Enzyme Aggregates (CLEAs)

Cross-linked enzyme aggregates (CLEAs) have gained significant interest in recent years as an effective immobilization technique, valued for their straightforward preparation and operational stability. This method removes the requirement for a solid carrier and facilitates high catalytic density, rendering it an attractive option for scientific investigations. The enzyme aggregates are stabilized using a cross-linking reagent, most commonly GA (Figure 3.6).

GA is a homobifunctional cross-linker that reacts with amine groups and forms the fundamental basis for creating CLEAs. In the CLEA methodology, enzyme molecules are first precipitated and then interconnected covalently, leading to the production of a solid, insoluble biocatalyst. This outcome can be attained via chemical cross-

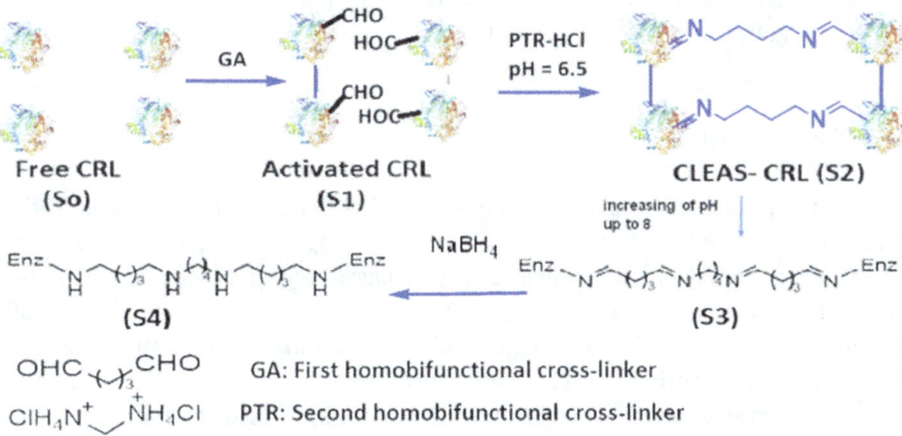


**Figure 3.6:** Illustration of different steps of CLEA-like copolymer preparation. S0–S4 are different states (or forms) of catalyst: S0, CRL-free enzyme suspension; S1, GA-activated enzyme; S2, copolymer as liquid copolymer before pH jump; S3, formation of solid copolymer induced by pH rising (amine activation); S4, $NaBH_4$-treated CLEA-LC.

linking, as demonstrated by the application of GA, or through the physical aggregation of enzymes employing techniques such as spray drying.

CLEAs offer numerous advantages over traditional enzyme preparations, including enhanced stability, extended shelf life, and suitability for continuous flow reactions [19].

### 3.4.7 3D Printing

Three-dimensional printing, or additive manufacturing, has emerged as a significant technology for enzyme immobilization and bioengineering. This technology enables the creation of objects and 3D structures layer by layer (LbL) from various materials, including polymers, metals, and ceramics, based on computer-generated designs.

In the field of enzyme immobilization, 3D printing technology enables the creation of tailored supports or matrices designed to efficiently entrap or immobilize enzymes. To improve biocompatibility, numerous techniques and materials have been devised for 3D printing hydrogels. These hydrogel structures are capable of encapsulating enzymes without requiring specific modifications, while also offering protection from organic solvents and modulating reaction kinetics due to mass transfer restrictions inherent in their typically dense printed forms.

In summary, 3D printing offers significant advantages in terms of custom design, process control, and adaptability to a wide range of applications in fields such as biomedicine, catalysis, and biotechnology [29].

### 3.4.8 Electrospraying

The electrospray technique, often referred to as electrohydrodynamic atomization, is an emerging method comparable to electrospinning that enables the fabrication of polymeric nanoparticles or bioactive fiber-based materials for diverse applications. Traditionally, the electrospray technique is defined as the atomization of a liquid through electrical forces (Figure 3.7). The key distinction between electrospraying and conventional electrification processes is the need to adjust both the solution's characteristics – such as solvent content and viscosity – and the operational parameters, including flow rate, needle-to-collector distance, and, most critically, the applied voltage.

This process generates micro- and nanoparticles with high loading capacity and a uniform particle size distribution. One of the key advantages of this technique for enzyme immobilization is the ability to design and adjust the shape and size of nanoparticles simply by modifying experimental conditions. For instance, the immobilization efficiency of lipases on electrosprayed fibers can be improved by incorporating rein-

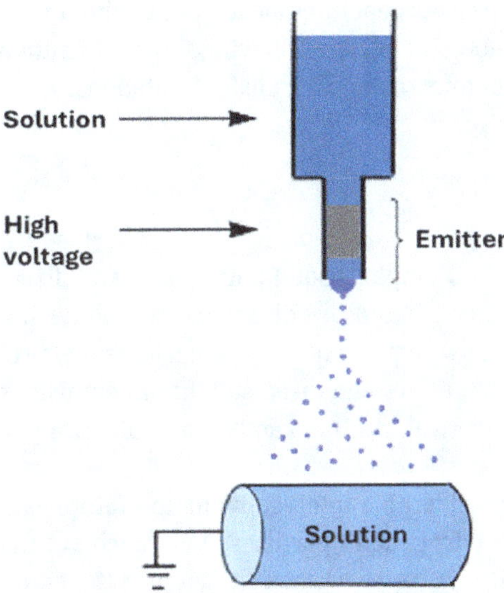


**Figure 3.7:** The spraying process involves pumping a solution while applying voltage to create an electric field. This leads to the formation of a Taylor cone when electrostatic repulsion overcomes surface tension, ejecting a charged liquid jet toward various collectors, such as flat plates or rotating drums.

forcement materials such as *P*(GMA-co-MA)-*g*-PEO (poly(glycidyl methacrylate-co-methyl acrylate)-*g*-polyethylene oxide) [30].

### 3.4.9 Hybrid Nanoflowers

Hybrid nanoflowers represent a unique and innovative category of materials used as a supporting framework for enzyme immobilization. These nanoflowers are formed from a combination of organic and inorganic constituents, and display a hierarchical 3D structure that imitates the petal-like arrangement of a natural flower [31]. They have garnered significant attention across various industries due to their exceptional properties and versatility in enzyme immobilization applications.

The principal benefits of hybrid nanoflowers encompass exceptional thermal stability, robust mechanical strength, and a substantially enlarged surface area. These combined characteristics render them particularly appropriate for applications in enzyme immobilization. Using these supports enhances enzyme tolerance to various reaction media, including organic solvents, and facilitates the reusability of the biocatalyst across multiple reaction cycles.

The synthesis of hybrid nanoflowers is typically straightforward, involving a reaction between active organic enzymes or molecules and metal ions in an aqueous phosphate buffer, usually at pH 7 and 25 °C. This process generates a compound with a hierarchical structure and a substantial surface-to-volume ratio, allowing for the retention of a significant portion of biomolecules without notable mass transfer limitations.

### 3.4.10 Pickering Emulsion Encapsulation

Pickering emulsion encapsulation: This method employs solid particles as stabilizers at the oil-water interface, where droplets of one liquid are dispersed in another immiscible liquid. The solid particles at the interface prevent droplet coalescence, providing stability to the emulsion. This approach offers enhanced stability to emulsions and is particularly useful for biphasic systems (water-oil) in catalytic and biocatalytic processes [32] (Figure 3.8).

Heterogeneous catalysis has attracted considerable interest owing to multiple advantages, such as simplified product recovery from solid particles, low toxicity of the particulates, and high catalytic efficiency. Recent research has shown that the emulsion system exhibits exceptional stability in various organic reaction media, including hexane, methanol, acetonitrile, and tetrahydrofuran.

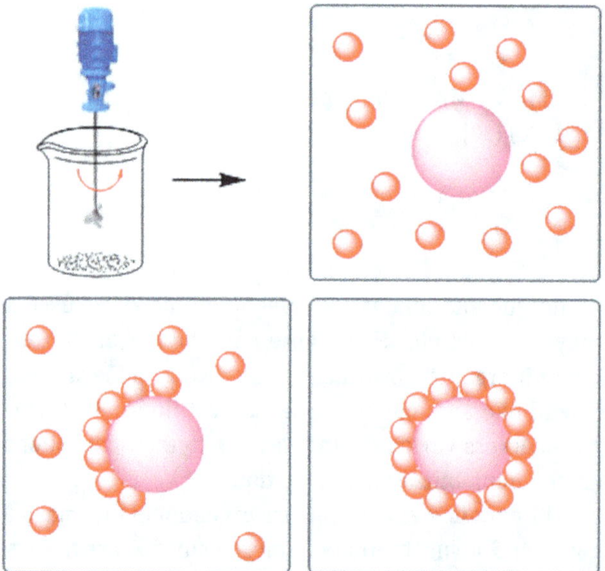

**Figure 3.8:** Apparatus used to obtain Pickering emulsions: schematic of the homogenization process with a rotor-stator, resulting in stable spherical particles at the oil interface.

This method allows for the encapsulation of both lipophilic and hydrophilic substances within the dispersed droplets in the emulsion. However, the production of enzymatic lipase biocatalysts via Pickering emulsions faces certain challenges. Enhancements in stability, activity, and durability are necessary to make large-scale, reproducible, and economically viable industrial applications feasible.

The past decade has seen a focus on hybrid and composite materials that integrate properties from both types of precursor materials, maximizing their advantages. These materials, derived from both organic and inorganic sources, are known for their exceptional thermal and chemical stability, as well as their excellent mechanical properties. Consequently, enzymes immobilized on these supports exhibit improved catalytic performance, resulting in higher purity and superior quality of the resulting products when compared to those generated through processes using conventionally immobilized enzymes.

## 3.5 Synthetic Polymers

The selection of monomers for constructing synthetic polymers is a critical step in enzyme immobilization, as it allows for a tailored approach to meet specific requirements. This careful choice aligns with both the demands of the enzyme itself and the particulars of the intended application [41, 42]. The type and quantity of monomers play a pivotal role in shaping the chemical structure and properties of the resulting polymer. The composition of these monomers significantly influences key attributes such as solubility, porosity, stability, and mechanical strength within the polymer matrix.

Polymeric materials in various physical forms, such as films, beads, coatings, and fibers, are employed as support matrices for enzyme immobilization. As a result, a diverse range of polymeric materials can serve as effective carriers, enhancing the properties of the immobilized enzyme [33]. This improvement extends to areas such as thermal stability and reusability, making these synthetic polymers invaluable in biotechnological and industrial applications.

Several synthetic polymers have been utilized as support materials for enzyme immobilization:

Polyacrylates: Synthetic polymers with carboxyl (–COOH) functional groups. These groups provide anchor points for enzyme attachment and have been used for immobilizing enzymes such as lipases, amylases, and proteases.

Polyvinyl alcohol (PVA): A water-soluble synthetic polymer that contains hydroxyl (–OH) groups, facilitating enzyme binding. PVA-based supports have been used for immobilizing enzymes in biosensors and other applications.

Polyurethane: Polyurethane foams are employed for covalent immobilization of enzymes, offering good mechanical stability and flexibility, making them suitable for various applications, including bioconversion processes.

Polyamide (nylon): Polyamide polymers, such as Nylon 66, can be used for immobilizing enzymes. The presence of amide (NH) and carbonyl (C=O) groups in their structure allows for covalent bonding with enzymes.

Polystyrene: Polystyrene is a hydrophobic synthetic polymer that has been utilized for enzyme immobilization. It is often activated with epoxy groups to enable covalent enzyme attachment. In a hydrophobic environment, enzymes like lipases can exhibit enhanced catalytic properties.

Polypropylene: Polypropylene can be functionalized to incorporate amino and carboxyl groups, enhancing its affinity for enzymes. It has been applied in various enzyme immobilization applications, including biosensors.

PVC: This can be modified to include functional groups that facilitate enzyme attachment. Modified PVC supports have found use in biosensors and biocatalysis.

Polyethylene glycol (PEG): This is a synthetic polymer recognized for its biocompatibility. It can be functionalized for enzyme immobilization and enzyme modification in applications where biocompatibility is essential.

Polyaniline: Polyaniline, a conductive polymer, has been employed for enzyme immobilization. It provides both physical and electrochemical support for enzymes, particularly in biosensor applications.

Polycarbonate: Polycarbonate polymers can be modified with suitable functional groups to support enzyme attachment. These modified polymers are utilized in various enzyme immobilization techniques.

Polyethyleneimine (PEI): This is a cationic polymer characterized by amino groups. It has been used for enzyme immobilization and as a support material in gene delivery systems.

Poly(ethylene terephthalate) (PET): This can be modified to include functional groups such as hydroxyl and carboxyl groups. These functionalized PET supports are employed for enzyme immobilization in diverse applications.

Polycaprolactone (PCL): This is a biodegradable synthetic polymer that has been utilized for enzyme immobilization. It is particularly suitable for applications where biodegradability is important.

Poly(*N*-isopropylacrylamide) (PNIPAM): This is a thermoresponsive polymer that undergoes conformational changes with temperature fluctuations. It has been employed for temperature-sensitive enzyme immobilization and release.

Poly(ethylene terephthalate) glycol (PETG): This is a modified version of PET that incorporates glycol groups. It is utilized for enzyme immobilization in biocatalytic processes.

Polyacrylonitrile (PAN): This can be functionalized to introduce various functional groups for enzyme attachment. Modified PAN supports have been successfully used to immobilize enzymes such as lipases and proteases.

Poly(sulfone): Poly(sulfone) membranes and materials have found application in enzyme immobilization within bioprocesses and filtration systems.

Poly(hydroxyethyl methacrylate) (PHEMA): This is a hydrophilic polymer widely used for enzyme immobilization in biosensors and biocatalysis applications.

Poly(glycidyl methacrylate) (PGMA): This serves as a support for covalent enzyme immobilization due to the presence of epoxy groups in its structure.

Poly(4-vinylpyridine) (P4VP): This is a cationic polymer utilized for enzyme immobilization, particularly in biocatalytic applications.

Poly(ethylene imine) (PEI): PEI, recognized for its cationic nature and amino groups, is employed in various biotechnological applications for enzyme immobilization.

Polypropylene carbonate (PPC): This is a biodegradable polymer suitable for enzyme immobilization in contexts where biodegradability and enzyme stability are critical.

These examples demonstrate the wide range of synthetic polymers available for enzyme immobilization. The choice of polymer depends on the specific needs of the enzyme, the desired application, and the desired properties, such as mechanical strength, biocompatibility, and chemical reactivity.

**Graphene and carbon-based graphene oxide (GO)** have garnered considerable attention as matrices for enzyme immobilization owing to their distinctive characteristics. These include biocompatibility, a planar two-dimensional geometry, an extensive surface area, and significant porosity, combined with remarkable thermal and chemical resilience. Furthermore, their surfaces contain diverse oxygen-containing functional groups – including carboxylic (–COOH), hydroxyl (–OH), and epoxide moieties – which promote effective enzyme-matrix binding without the need for additional coupling agents or surface pretreatment (Figure 3.9).

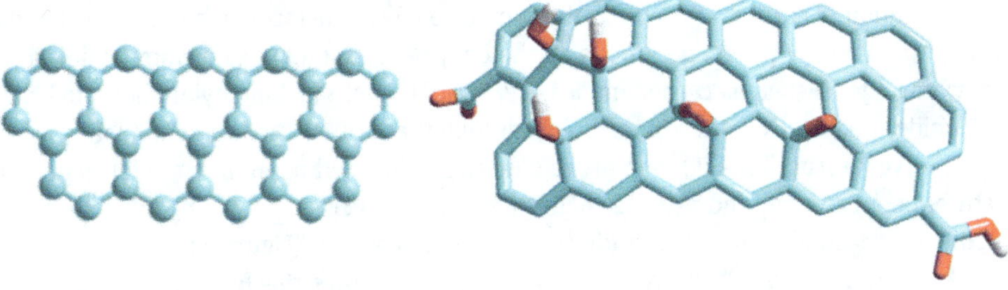

**Figure 3.9:** Schematic representation of graphene and graphene oxide.

These graphene-based materials can even enhance the biocatalytic activity of enzymes. Consequently, enzymes such as lipases and peroxidases can be immobilized on graphene surfaces primarily through adsorption, covalent bonding, or encapsulation [34].

## 3.6 Methods of Enzyme Immobilization

### 3.6.1 Chemical Retention

Chemical retention in enzyme immobilization refers to forming chemical bonds or specific interactions between the enzyme and the support material, ensuring a stable and long-term attachment.

The bibliography contains various instances of this methodology; however, given the introductory nature of this chapter, we will only highlight those we deem most pertinent.

#### 3.6.1.1 Covalent Bonds

Covalent enzyme immobilization is a methodology that ensures the stable attachment of enzymes to solid supports by forming irreversible covalent bonds. These bonds are established between reactive groups on the enzyme – such as amino ($-NH_2$), carboxyl ($-COOH$), hydroxyl ($-OH$), or thiol ($-SH$) groups – and functional groups previously introduced onto the support's surface. This approach stands out for its excellent chemical and mechanical stability, minimal enzyme leaching, and the ability to reuse the biocatalyst through multiple reaction cycles.

Below are some examples of methods for activating supports for enzyme immobilization. Diazonium salts enable azo coupling reactions with aromatic rings present in tyrosine or histidine residues of enzymes, forming stable covalent bonds through a direct arylation process on the activated support. The mechanism involves electrophilic coupling between the diazonium group on the support and an aromatic nucleophile on the enzyme. This method offers high specificity toward aromatic residues and operates under mild reaction conditions, making it particularly suitable for immobilizing enzymes with a high content of exposed aromatic residues on their surface (Figure 3.10).

Activation of carboxyl groups on the support enables the formation of peptide bonds through reactions with amino groups on the enzyme, resulting in stable amide linkages. Activating agents such as carbodiimides (e.g., EDC [1-ethyl-3-(3-dimethylaminopropyl)carbodiimide] or DCC [*N*,*N*′-dicyclohexylcarbodiimide]) are often used in combination with NHS (*N*-hydroxysuccinimide) to improve efficiency. This method mimics the formation of natural peptide bonds and provides high stability. However, care must be taken to avoid modifying essential amino acid residues near the catalytic site (Figure 3.11).

Activation of carboxyl groups on the support allows the formation of peptide bonds through reactions with amino groups on the enzyme, resulting in stable amide

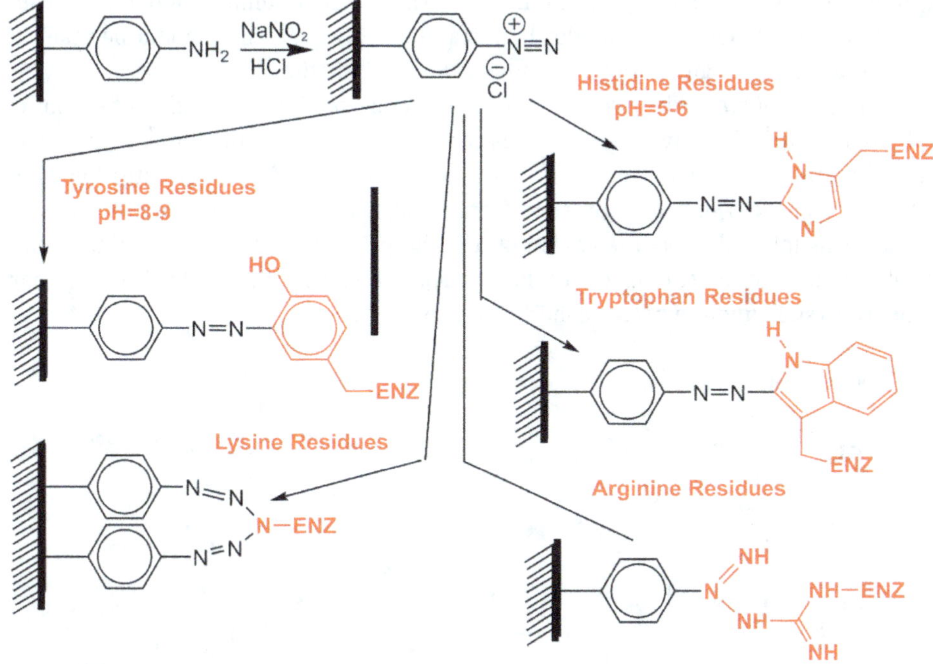


**Figure 3.10:** Immobilization via diazonium salts.

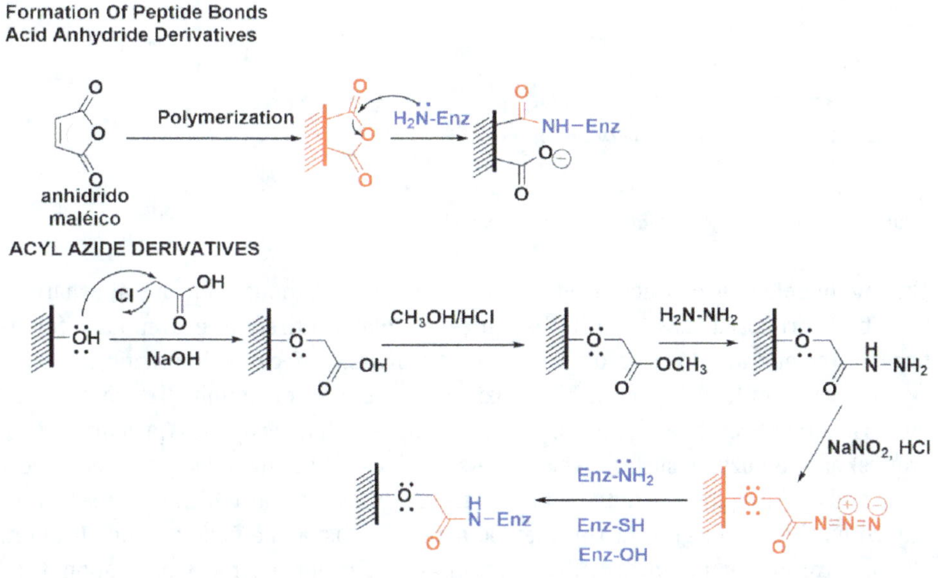

Formation Of Peptide Bonds
Acid Anhydride Derivatives

O
O
O
anhidrido
maléico
Polymerization
O
O
O
$H_2N$-Enz
O
NH−Enz
O
O

ACYL AZIDE DERIVATIVES

Cl
OH
O
OH
NaOH
O
OH
O
$CH_3OH/HCl$
O
$OCH_3$
O
$H_2N$-$NH_2$
O
H
N−$NH_2$
O
$NaNO_2$, HCl
O
N=N=N
O
Enz-$NH_2$
Enz-SH
Enz-OH
O
H
N−Enz
O

**Figure 3.11:** Immobilization via formation of peptide bonds.

linkages. Activating agents such as carbodiimides (e.g., EDC or DCC) are often used in combination with NHS to enhance efficiency. This method mimics the natural formation of peptide bonds and provides high stability. However, care should be taken to avoid modifying essential amino acids near the catalytic site.

Imidocarbonate derivatives, such as those generated by the reaction of carbonyldiimidazole, form reactive intermediates capable of reacting with amino, hydroxyl, or thiol groups. Bromocyanogen (BrCN) is a classical reagent that activates hydroxyl groups on surfaces such as agarose to form covalent bonds with amino groups of enzymes. This method offers the advantage of mild pH and temperature conditions and applies to a wide range of enzymes; however, the bonds formed by BrCN can be partially reversible under extreme conditions (Figure 3.12).

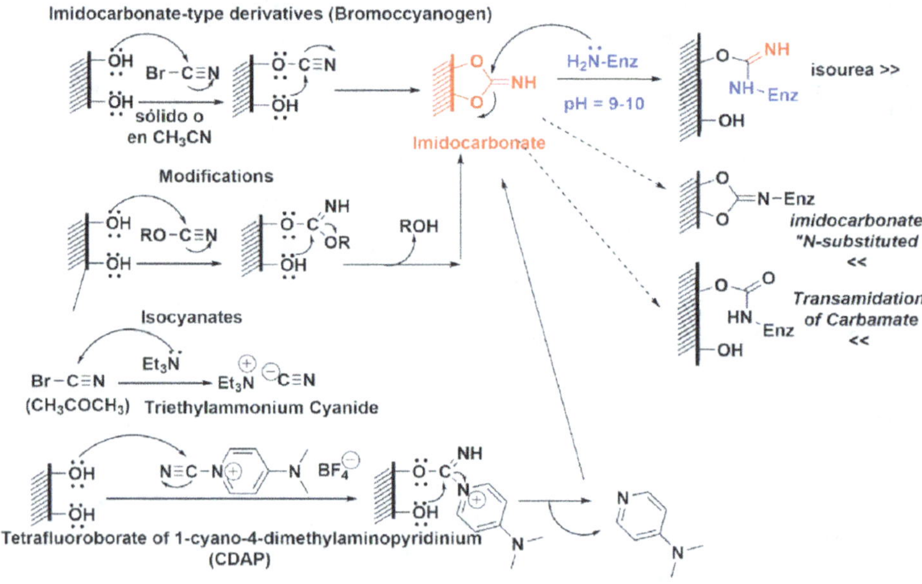


**Figure 3.12:** Immobilization via imidocarbonate derivatives.

This methodology is characterized by its high toxicity. Although various alternatives have been proposed, as shown in the figure, none are entirely free from harmful effects for the operator. As a result, its use has significantly declined in practice.

Activation of the support with acyl azides enables enzyme immobilization through the formation of reactive acylnitrenes, which couple with nucleophiles (primarily $-NH_2$ groups) on the enzyme surface. The process involves activating the support as an acyl azide and applying heat or radiation to generate the acylnitrene, which then reacts rapidly with nucleophilic groups. This method allows for fast and efficient immobilization, although careful temperature control is required to prevent enzyme denaturation.

Support activation with isocyanates (–N=C=O) or isothiocyanates (–N=C=S) allows for the immobilization of biomolecules through the formation of stable covalent bonds with nucleophiles, mainly free amino groups (–$NH_2$) present in proteins or enzymes.

The support is functionalized with isocyanate or isothiocyanate groups, which react directly with nucleophiles such as amines to form urea-type bonds (in the case of isocyanates) or thiourea-type bonds (in the case of isothiocyanates), without the need for external activation.

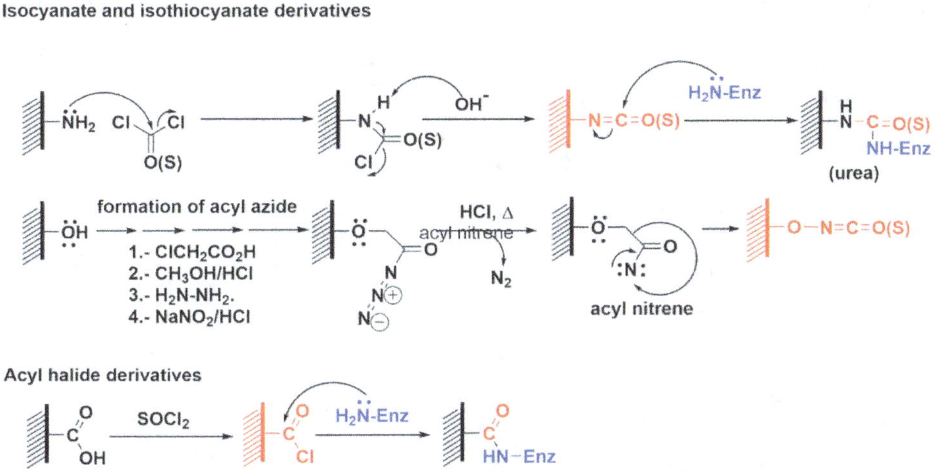

Isocyanate and isothiocyanate derivatives

formation of acyl azide
1.- $ClCH_2CO_2H$
2.- $CH_3OH/HCl$
3.- $H_2N\text{-}NH_2$.
4.- $NaNO_2/HCl$
HCl, Δ
acyl nitrene
acyl nitrene
(urea)

Acyl halide derivatives

$SOCl_2$

**Figure 3.13:** Immobilization via isocyanates or isothiocyanates.

The process provides and highly stable immobilization under mild conditions, suitable for sensitive biological systems; commonly used in biosensors, chromatographic matrices, and controlled release systems (Figure 3.13).

Cyclic carbonates can be used to activate supports by reacting with nucleophilic groups on the enzyme (especially –$NH_2$ and –OH), resulting in stable urethane or carbonate linkages. They have the advantage of being less toxic than other activators and provide efficient reactions under aqueous conditions. This strategy has proven effective in systems with biocompatibility requirements (Figure 3.14).

Direct modification of enzymes by alkylation or arylation is possible using activated alkyl or aryl halides on the support. These reactions preferentially occur with nucleophilic residues such as cysteine (–SH) or lysine (–$NH_2$). Examples include benzyl halides, methanesulfonyl chloride, and arylsulfonyl fluoride derivatives. Controlled conditions are required to avoid excessive modification or loss of enzymatic activity (Figure 3.15).

The use of coupling agents with multiple reactive functions, such as GA, toluene diisocyanate (TDI), divinyl sulfone, or polycarbodiimides, allows for more robust immobilization and, in some cases, partial cross-linking of the enzyme, thereby increas-

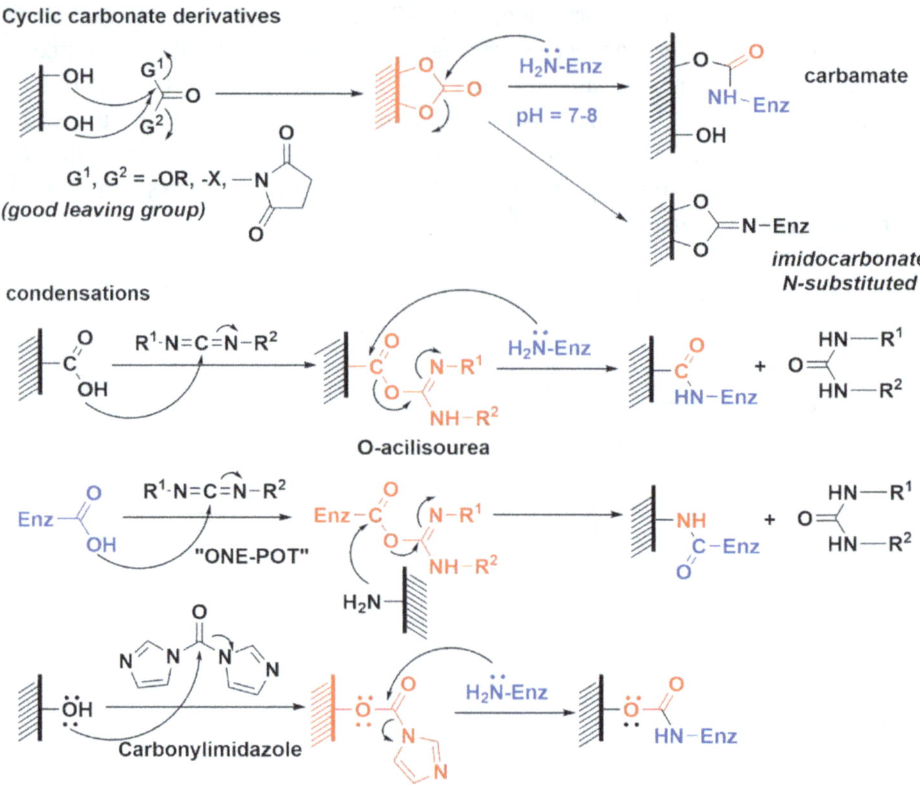


Figure 3.14: Immobilization via cyclic carbonates.

ing its stability. These agents can react simultaneously with multiple enzyme sites, creating 3D networks. The use of flexible spacers can improve substrate access to the active site of the immobilized enzyme.

The use of coupling agents with multiple reactive functions, such as GA, TDI, divinylsulfone, or polycarbodiimides, enables more robust immobilization and, in some cases, partial cross-linking of the enzyme, increasing its stability. These agents can react simultaneously with several enzymatic sites, creating 3D networks. The use of flexible spacers can improve substrate access to the active site of the immobilized enzyme (Figure 3.16).

### 3.6.1.2 Non-covalent Bonds: Adsorption

Enzyme immobilization through adsorption is a straightforward, direct, and cost-effective methodology. This methodology operates fundamentally under the influence of van der Waals forces, hydrophobic effects, and electrostatic interactions. Notably, no chemical bonds are formed between polymers and enzymes, and polymer functionali-

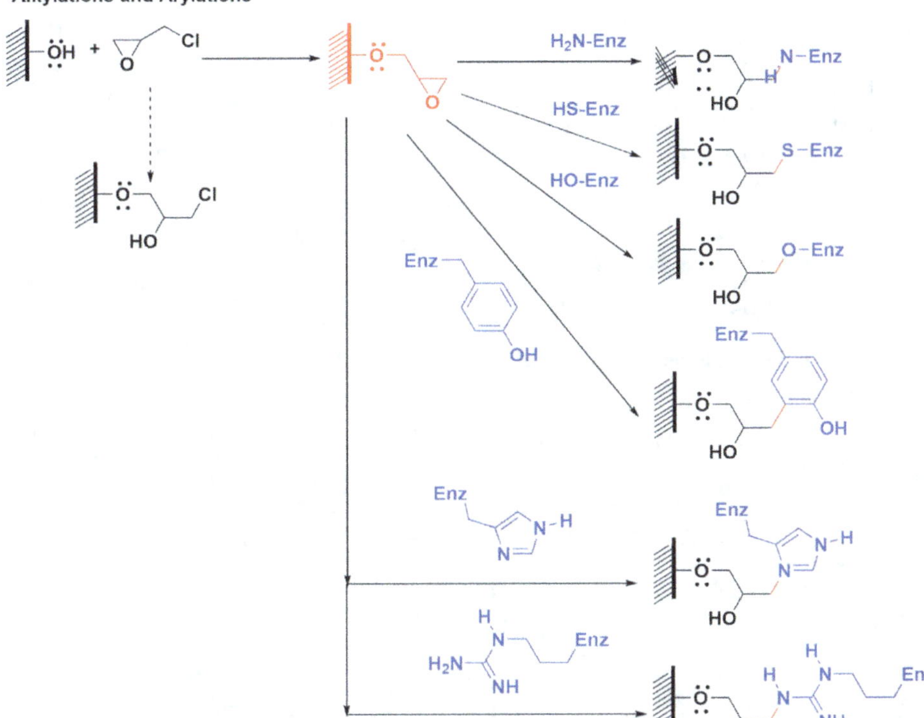


**Figure 3.15:** Immobilization via alkylation or arylation.

zation is not required. This implies minimal changes, if any, in the configuration of the enzymes, which can help optimize the preservation of their enzymatic activity.

To effectively carry out enzyme immobilization through van der Waals interactions, both the supports and the polymers must possess lipophilic surface areas. This enhances the attraction between the enzyme and the support, allowing for more effective immobilization.

Despite its advantages, this immobilization technique has notable drawbacks. The strength of van der Waals interactions is relatively weak, which can lead to inadequate retention of enzymes on their supports, particularly in fluctuating environmental conditions such as changes in pH and temperature. This limitation can result in low stability, hindering its application in industrial settings.

In cases of electrostatic interactions, enzymes with an electrical charge are absorbed onto supports that have opposite charges. Two techniques developed under

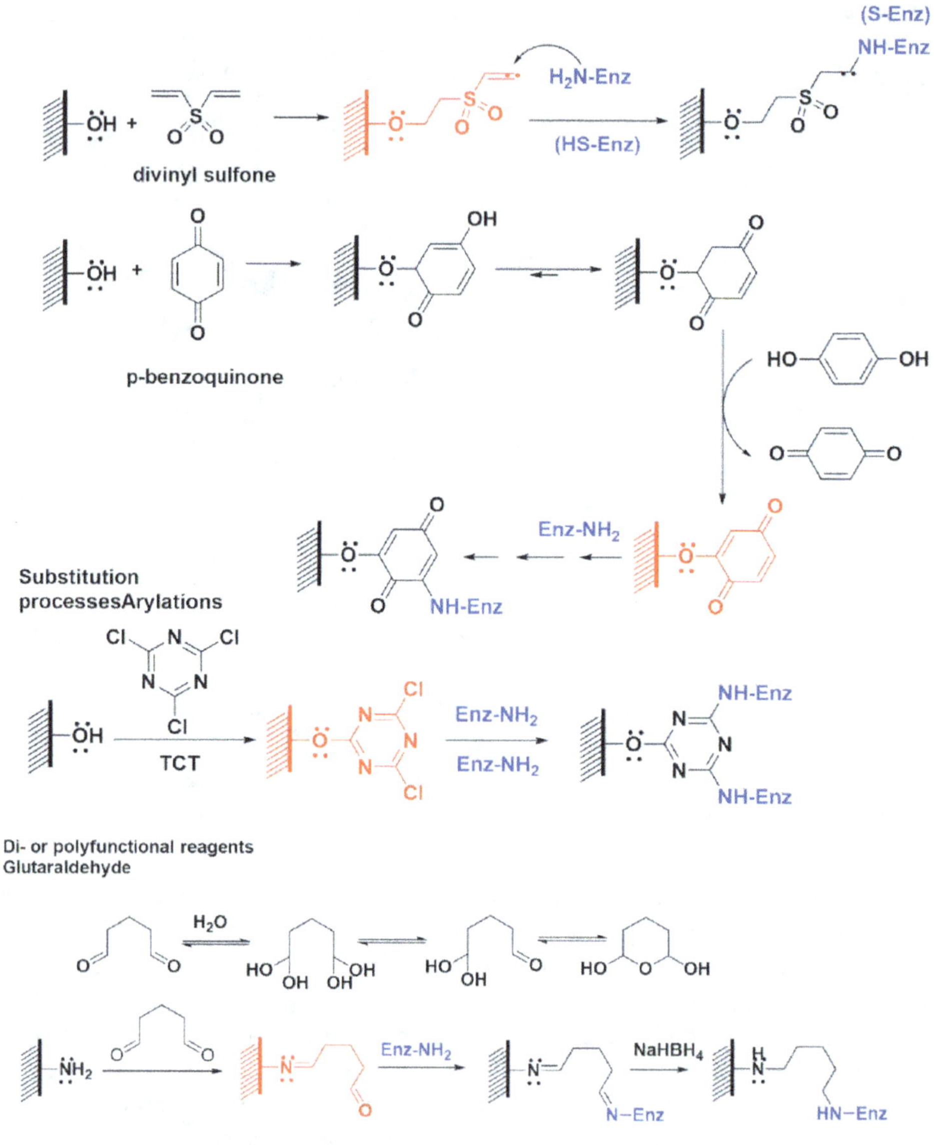

**Figure 3.16:** Immobilization via bifunctional reagents.

this method include LbL deposition, introduced in 1991, and electrochemical doping. The latter involves immersing the polymer film in the enzyme solution, followed by electrochemical oxidation, which facilitates the attachment of the enzymes to the support [35].

#### 3.6.1.2.1 Layer-by-Layer (LbL) Deposition for Enzyme Immobilization

The LbL deposition technique is employed to create thin films through the sequential adsorption of layers composed of different materials onto a specific substrate. In this process, the substrate is immersed in a polyelectrolyte solution whose charge is opposite to that of the enzyme or the material designed for immobilization. The enzyme electrostatically adheres to the substrate's surface due to the interaction between the charges of the enzyme and the polyelectrolyte.

Once the first layer is applied, the modified substrate is immersed in a solution of another polyelectrolyte with a charge opposite to that of the previously deposited layer. This procedure is repeated alternately with polyelectrolyte solutions until the desired number of layers is achieved. In certain layers, enzymes can be incorporated directly or in the form of complexes before applying the next layer, thereby facilitating the immobilization of enzymes within the stratified structure.

#### 3.6.1.2.2 Electrochemical Doping for Enzyme Immobilization

The electrochemical doping technique for enzyme immobilization involves coating an electrode with a polymer, immersing it in an enzyme solution, and subsequently applying electrochemical oxidation. This process modifies the polymer, enhancing the interaction and binding of enzymes to the surface. Both adsorptive and covalent bonds can be formed during this process, which significantly improves the stability of enzyme immobilization. Following oxidation, a washing step is performed to remove any unbound enzymes.

Electrochemical characterization is conducted to assess the effectiveness of immobilization and the performance of the modified electrode. This technique leverages the electrochemical properties of the polymer to create a favorable environment for enzyme immobilization.

It is essential to consider factors such as pH conditions, ionic strength, and compatibility of the materials used in each layer to ensure effective and stable construction of the thin film.

### 3.6.1.3 Affinity Binding

Affinity immobilization involves the attachment of enzymes to a support matrix through specific interactions that capitalize on the enzyme's inherent specificity for binding under various physiological conditions. This approach can be executed through two primary strategies:

- Precoupling the support matrix with an affinity ligand specific to the target enzyme
- Modifying the enzyme to possess an affinity for the support matrix

Affinity adsorbents are also commonly employed in the purification of enzymes. When enzymes are immobilized on complex support matrices, such as multilayered concanavalin A bound to agarose or alkali-stable chitosan-coated porous silica beads, they can exhibit increased quantities, thereby enhancing both efficiency and stability.

The bioaffinity stratification method further enhances enzyme binding and reusability through non-covalent forces, including van der Waals interactions, Coulombic attractions, and hydrogen bonding. Common affinity pairs used in this context include avidin-biotin and protein A-protein G, which facilitate effective enzyme immobilization [17] (Figure 3.17).

#### 3.6.1.4 Cross-Linking

The stabilization of enzymes via cross-linking involves the formation of covalent bonds either within a single enzyme molecule (intramolecular cross-linking) or between different enzyme molecules (intermolecular cross-linking). This strategy enhances enzyme stability, allows for their reuse, and facilitates separation from substrates or reaction media (Figure 3.18).

Cross-linking reactions can be implemented using both in vivo and in vitro strategies. Although these reactions can selectively target specific natural amino acids, accessibility is often constrained by the residue's spatial positioning and abundance on the protein surface. Residues such as serine, glutamine, glutamic acid, glycine, lysine, and aspartic acid are predominantly surface-exposed, displaying high average surface accessibility. This accessibility enables these amino acids to be recognized and react with the cross-linking catalyst [36].

Enzyme cross-linking techniques are as follows:

1. One-point modification: This technique involves attaching a cross-linking agent to a specific site on the enzyme molecule, usually targeting a particular amino acid. This selective modification can enhance the enzyme's stability and functionality.
2. Intermolecular cross-linking: In this method, covalent bonds are formed between different enzyme molecules, creating larger complexes or aggregates. This process increases the stability of the enzymes and their resistance to denaturation. Cross-linking agents, such as GA, react with functional groups on multiple enzyme molecules, linking them together to form stable aggregates.
3. Intramolecular cross-linking: This technique focuses on forming covalent bonds within the same enzyme molecule. By reinforcing the enzyme's tertiary or quaternary structure, it prevents unfolding or conformational changes under varying environmental conditions.

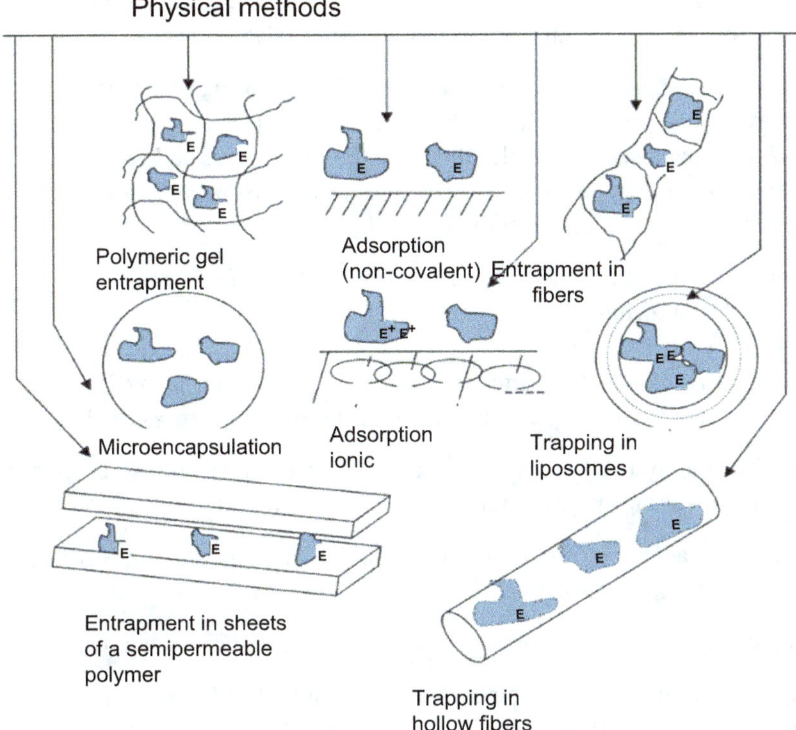


**Figure 3.17:** Schematic of enzyme immobilization methods by adsorption, illustrating the process of enzyme attachment to support surfaces for enhanced stability and reusability.

Key derivatives in biocatalysis are as follows:

Cross-linked enzyme crystals (CLECs): This method involves the formation of CLECs. The entwining of crystals is achieved through cross-linking, which enhances their stability and allows for repeated use. High purity of the enzyme is essential to prevent interference in the crystallization process.

Cross-linked enzyme aggregates (CLEAs): This method focuses on creating CLEAs. In this approach, enzymes are aggregated and cross-linked, often using chemical agents like GA, to enhance their stability. While purity is important, CLEA can tolerate certain impurities, making it versatile for various applications.

Cross-linked spray-dried enzymes (CSDEs): This method combines the advantages of spray drying with cross-linking to stabilize enzymes in a powdered form, facilitating easy handling and application in various processes.

Cross-linked enzymes (CLEs): This general category refers to enzymes that have been cross-linked to improve their stability and reusability in catalytic processes [37].

Cross-linked spray-dried enzymes (CSDEs): In the CSDE approach, enzymes are subjected to a spray-drying process followed by cross-linking. This sequence improves the stability of the enzymes while facilitating their handling in various applications. Ensuring high enzyme purity during this process is crucial, as it helps preserve the quality and activity of the enzymes post-drying.

Cross-linked enzymes (CLEs): These involve the cross-linking of enzymes using various methods, which can include:

Chemical cross-linking: This method employs chemical agents to form covalent bonds between enzymes and a solid matrix. Common agents used in this process include:

Glutaraldehyde: A widely used cross-linking agent that enhances the stability and functionality of the resulting enzyme complexes.

Carbodiimides: These agents facilitate the formation of stable amide bonds between carboxyl and amine groups in proteins.

Disulfide bridges: Formed between cysteine residues to stabilize the enzyme structure.

Maleimide: Specifically used for the selective chemical modification of cysteine residues in proteins, promoting stability and functionality [38]. Enzymatic cross-linking: In this approach, specific enzymes catalyze the formation of cross-links between other enzymes or substrates, enhancing stability and reusability.

Physical cross-linking: Techniques such as radiation or exposure to light can induce cross-linking, creating stable bonds between enzymes and their matrices without altering their inherent properties.

The resulting bonds formed through these methods significantly enhance enzyme stability, allowing for their reuse in various catalytic processes.

Enzymatic cross-linking: This method involves the use of specific enzymes, such as transglutaminase, which catalyze the formation of covalent bonds between proteins in the presence of particular substrates. This process can effectively stabilize enzyme structures by linking them to each other or to a support matrix, enhancing their stability and functionality. Transglutaminase, for instance, catalyzes the formation of isopeptide bonds between the amino acids lysine and glutamine, leading to cross-linked protein networks that can improve enzyme reusability in biocatalytic applications.

Non-covalent cross-linking: This utilizes ionic or affinity interactions to bind enzymes to a support matrix with complementary functional groups. This method leverages various non-covalent forces, such as:

Ionic interactions: These occur between charged groups on the enzyme and oppositely charged groups on the support matrix, facilitating stable enzyme attachment without covalent bond formation.

Affinity interactions: These interactions involve specific binding pairs, such as antigen-antibody or receptor-ligand pairs, allowing for precise enzyme immobilization. For example, biotin can be used to bind to streptavidin-coated surfaces, creating a strong and specific interaction.

While non-covalent cross-linking generally offers simpler and milder conditions for enzyme immobilization, it may result in lower stability compared to covalent methods, especially under varying environmental conditions (e.g., changes in pH or temperature).

### 3.6.2 Physical Retention = Entrapment

When we talk about enzyme entrapment, particularly non-covalent entrapment, we are typically referring to a method of immobilizing enzymes within a specific structure or matrix without forming covalent bonds between the enzyme and the supporting material. This method of immobilization offers various advantages, including enhanced stability, reusability, and controlled release of the enzyme.

#### 3.6.2.1 Entrapment in Matrices

An example supporting the aforementioned concept is the "entrapment of enzymes through the polymerization of acrylamide." This method involves confining enzymes within a polymeric matrix, with the polymerization process utilizing acrylamide.

In the process of entrapping enzymes in polyacrylamide gels, the concentrations of acrylamide and the amount of cross-linking agent can vary across different methodologies. A significant drawback of this approach is the tendency of the enzyme to escape from the gel. This issue can be addressed by increasing the degree of cross-linking in the gel matrix, which reduces the pore size. However, this increase in cross-linking can lead to potential mass transfer or diffusion limitations. The impact of these limitations can be improved by reducing the size of the trapping polymer particles. This process may encompass various techniques, including free radical polymerization or other methods that create a 3D network.

Free radical polymerization is widely employed to synthesize various polymers, such as polyethylene, polypropylene, and polystyrene. The polymerization of acrylamide is a commonly utilized method for forming hydrogels, which can be applied to immobilize enzymes within the resulting polymer matrix. During this process, acrylamide monomers polymerize to generate a 3D network, and enzymes may become trapped within this developing polymer structure.

This method is frequently used to enhance the stability and activity of enzymes in applications such as biocatalysis, biosensors, and controlled drug release. The acrylamide polymer matrix provides a supportive environment that protects the enzymes and enables controlled release based on the specific requirements of the application

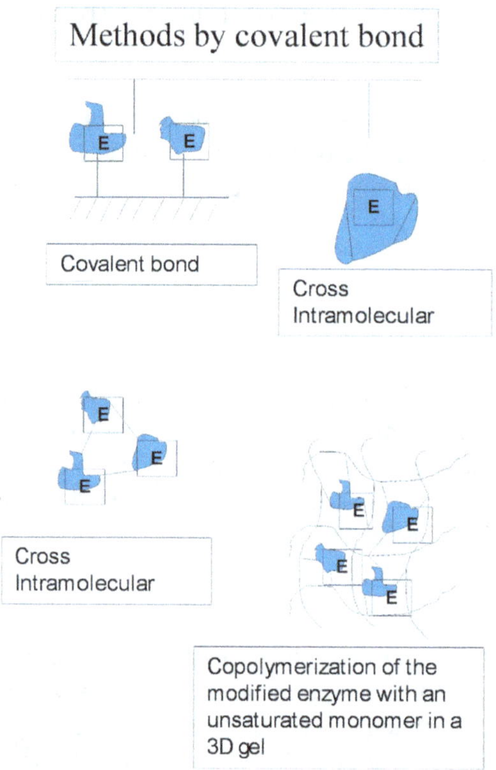


**Figure 3.18:** Schematic of enzyme immobilization methods by covalent bond crosslinking and copolymerization.

For example, in the entrapment of α-chymotrypsin into a cross-linked copolymer of acrylamide and 2-hydroxyethyl methacrylate, the entrapped enzyme demonstrated enhanced stability during storage and at elevated temperatures, experiencing only a 50% loss in activity after nine cycles. Notably, no conformational changes were detected in the enzyme during the entrapment process. The hydrolysis rate of the substrate was found to depend on the molecular weights of the substrate, which can be attributed to variations in the extent of internal mass transfer [39].

#### 3.6.2.2 In Gels or Polymers

Among the various gels and matrices used as supports for enzyme entrapment, agar stands out as one of the most attractive options. Agar forms a gel matrix that provides a stable and porous structure, allowing enzymes to be entrapped while maintaining their activity.

Enzymes can be effectively immobilized within an agarose gel due to its biocompatibility and nontoxic nature, distinguishing it from synthetic polymers like polyacrylamide. Additionally, agarose exhibits robust solidification properties. Similar to other gel matrices, the concentration of agar is crucial for protein immobilization, as the porosity of the agar matrix depends on the specific agar concentration used. Furthermore, the optimal pH and temperature conditions for the enzyme may change following immobilization in agar.

Agarose powder is liquefied by heating, and gelation occurs as the temperature decreases, during which the enzyme becomes trapped in the gel. The structure of agarose and other polymers is detailed in Section 3.4.2.

A wide variety of natural polymers, primarily water-insoluble polysaccharides such as collagen, chitosan, chitin, alginate, cellulose, agarose, carrageenan, and starch, can also be used for enzyme immobilization.

Alginate, an anionic polymer extracted from brown algae cell walls, consists of alginic acid salts – primarily calcium, magnesium, and sodium. It has found extensive application in enzyme immobilization, commonly being shaped into alginate beads, alginate-polyacrylamide composites, or calcium-alginate capsules.

Alginate is a linear copolymer consisting of β-D-mannuronic acid (M) and α-L-guluronic acid (G) residues. These M and G units can be organized in varying sequences and ratios, generating homopolymeric regions of G blocks, homopolymeric M blocks, and mixed heteropolymeric MG segments.

As a biocompatible and biodegradable polysaccharide derived from seaweed, alginate can be utilized to create hydrogels. The entrapment of enzymes or cells within these hydrogel networks helps protect them from degradation and enhances their stability.

Alginate can be structured into diverse configurations such as spheres, films, fibers, and chemically modified variants. Of these, spherical gel beads are the predominant form for enzyme immobilization and controlled release, owing to their simple fabrication process and mild operating conditions. The standard technique involves extruding alginate droplets into a solution of divalent cations, which induces bead formation through ionic cross-linking. This approach remains one of the most prevalent methods for entrapping enzymes and bioactive compounds [23]. Critically, key characteristics of the alginate beads, including their dimensions, pore structure, and release kinetics, can be precisely modulated by optimizing parameters such as pH [24] and alginate precursor concentrations [25].

However, this methodology faces challenges, including low encapsulation efficiency, enzymatic leaching, and unsatisfactory enzyme recycling.

To address these issues, an additional coating layer can be applied to the outside of the alginate beads. For instance, the activity and encapsulation efficiency of pig liver esterase can be improved through a chitosan coating, while the structural and mechanical stability of encapsulated recombinant enzymes can be enhanced using silica-coated alginate beads [40] (Figure 3.19).

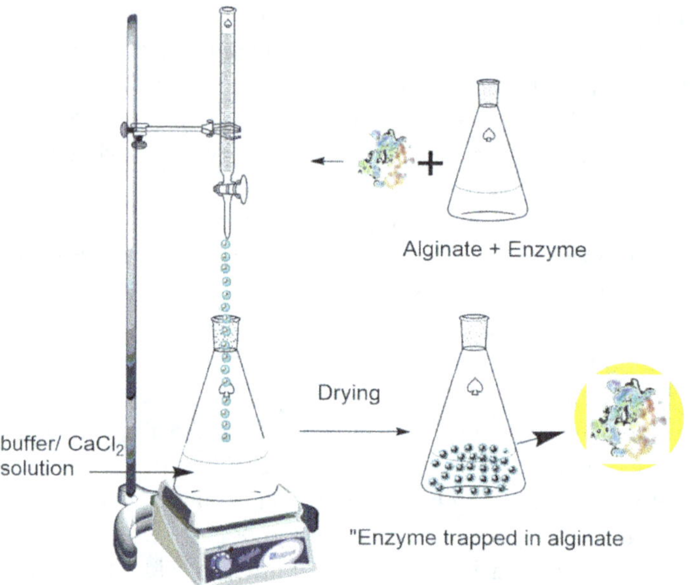


**Figure 3.19:** Illustration of alginate beads with entrapped enzyme.

**Immobilization sol-gel**: The formation of an inorganic framework progresses through a colloidal suspension (sol), which undergoes gelation to create the framework within a continuous liquid phase (gel). The precursors required for this process typically involve a metal (or metalloid) surrounded by various reactive ligands.

Metal alkoxides are the most commonly used precursors due to their ease of reaction with water. Among them, alkoxysilanes such as tetramethoxysilane or tetraethoxysilane are the most prevalent, although others like aluminates, titanates, or borates can also be utilized.

Sol-gel processes utilize acid- or base-catalyzed hydrolysis of tetraalkoxysilanes, $Si(OR)_4$.

Mechanistically, the silane precursor undergoes hydrolysis followed by cross-linking condensation, leading to the formation of a $SiO_2$ network that entraps the enzyme.

This immobilization approach has proven effective for a wide range of enzymes [41].

Sol-Gel process overview is as follows:

1. Sol-gel solution: The process begins with a liquid solution (sol) containing chemical precursors, such as metal alkoxides, which can undergo hydrolysis and condensation.
2. Hydrolysis: The precursors undergo a hydrolysis reaction, where hydroxyl groups (–OH) are added, generating more reactive species.
3. Condensation: The hydrolyzed species condense to create a solid, 3D network (gel). This gel can trap molecules or particles present in the solution.
4. Drying and hardening: The gel undergoes drying and hardening, transforming into a porous solid structure.

After the gel formation, various materials can be obtained depending on the drying method, such as aerogels, xerogels, and ambigels. Aerogels are produced through supercritical drying, xerogels through conventional drying, and ambigels through controlled thermal treatment, allowing for adjustments to their properties. All these processes originate from the initial formation of the gel using the sol-gel method.

Another type of compound used for gel formation is polyvinyl alcohol (PVA). Entrapping enzymes in PVA involves immobilizing them in a 3D structure formed by the polymer. During this process, enzymes are incorporated and trapped within the PVA matrix, ensuring their stability and preserving their biological activity. The matrix provides a stable, protective environment, which enhances enzyme stability.

Additionally, the process involves capturing enzymes in an aqueous phase and then solidifying that phase using PVA as a solidifying agent, forming a cohesive structure that retains the enzyme's activity. This method enables the use of enzymes in applications like immobilization or the creation of controlled-release systems. Both processes demonstrate the versatility of PVA in retaining and preserving enzymes for a wide range of biotechnological applications.

#### 3.6.2.3 Reverse Micelles Microencapsulation

In this approach, enzymes are encapsulated within semipermeable membranes that permit the diffusion of substrates and products while retaining the enzymes themselves. The membranes can be either permanent, formed through interfacial polymerization, or transient, generated by surfactants in structures commonly referred to as 'reverse micelles.' The resulting microcapsules are spherical, with diameters typically ranging from 1 to 100 µm (Figure 3.20).

Through this method, a wide variety of enzymes, cells, or biomolecules can be simultaneously encapsulated, allowing for multistep reactions to take place within the same system.

The association of surfactant molecules in nonpolar solvents (organic solvents) is primarily driven by dipole-dipole and ionic pair interactions between them. The colloids formed in these types of solvents are called reverse micelles and/or water-in-oil (w/o) microemulsions, and they are used as dispersing agents to solubilize water and/or water-soluble compounds in organic solvents. These structures consist of a microdroplet of water surrounded by a monolayer of surfactant molecules, with the hydrocarbon chains of the amphiphilic compound interacting directly with the organic solvent. These systems consist of three clearly distinguishable zones: an internal aqueous microphase ("aqueous pool"), the interface formed by the surfactant molecules, and the external organic phase.

Among the surfactants capable of forming reverse micelles are sodium 3,3-dimethyl-1-butylsulfosuccinate, sodium di(*n*-octyl) phosphinate, CB-Span 85®, and sodium di(2-ethylhexyl) sulfosuccinate (AOT), with hydrocarbon solvents being the most commonly used dispersing medium.

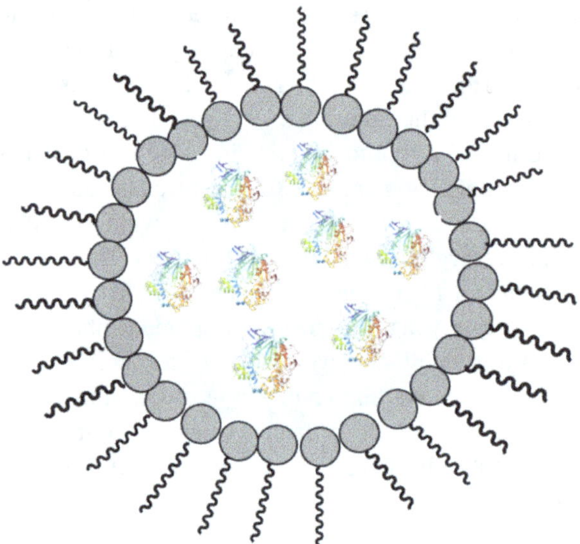

**Figure 3.20:** Schematic representation of a reverse micelle.

#### 3.6.2.4 Entrapment in Membranes

Another encapsulation strategy involves the use of liposomes, which form when water-insoluble polar lipids, such as phosphatidylethanolamine, phosphatidylcholine, cardiolipins, and phosphatidic acids, interact with water. This interaction leads to the formation of ordered structures, either as liposomes or smectic mesophases. These structures remain stable in the presence of excess water, which, due to unfavorable entropy, promotes further arrangement of the lipid molecules. The result is a system of concentric closed membranes, each constituting an intact bimolecular sheet of lipid molecules [42] (Figure 3.21).

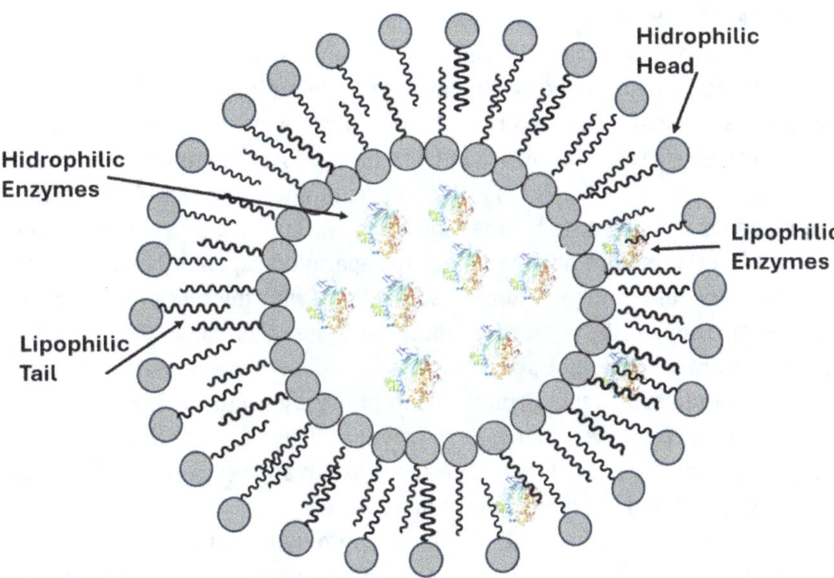


**Figure 3.21:** Schematic representation of a liposome.

#### 3.6.2.5 Entrapment in Hollow Fibers

One method of immobilizing enzymes involves the use of hollow fibers – asymmetric structures capable of entrapping enzymes between two barriers: an enzyme-impermeable inner layer and the aqueous or organic interface on the outer surface of the hollow fiber.

When selecting materials for hollow fibers, several factors should be considered, including biocompatibility, chemical resistance, mechanical strength, and the ability to create an environment conducive to enzyme activity. Additionally, the fabrication process of hollow fibers – particularly their pore size and structure can significantly impact the effectiveness of the enzyme immobilization process.

Various materials can be used to create hollow fibers, including:

- Polymers: Polyethylene and polypropylene
- Polymeric blends: Polyvinylidene fluoride blended with other polymers
- Cellulose-based materials: Cellulose acetate
- Silicon-based materials: Silicone rubber, offering flexibility and biocompatibility, ideal for certain enzyme immobilization applications
- Hydrogels: Polyacrylamide hydrogels, etc. [43].

The specific requirements of the enzyme, the intended application, and the operating conditions will guide the selection of the most suitable material for hollow fiber composition in enzyme immobilization.

#### 3.6.2.6 Entrapment in Membrane Reactors

The term "membrane reactor" generally refers to membrane-based devices designed to facilitate chemical conversion processes by leveraging the unique contact and separation characteristics of membranes. Specifically, it applies to systems where the membrane not only serves as a reactive surface but also functions as a matrix for immobilizing catalysts. A biocatalytic membrane reactor combines enzymatic or cellular biochemical reactions with selective permeation or mass transfer through a membrane.

Membrane bioreactors offer a valuable in situ separation capability that other bioreactor types lack. By combining membrane separations with the biocatalytic properties of enzymes, microbial cells, organelles, and even tissues from animals or plants, a wide range of membrane bioreactor systems can be developed. The membrane's capacity to retain cells while selectively removing inhibitory byproducts allows for high-density cell cultures, thereby maximizing the catalytic activity of enzymes and boosting bioreactor productivity. These reactors are particularly advantageous for enzyme reactions involving cofactors and the hydrolysis of macromolecules.

Membrane bioreactors can generally be divided into two main categories:

1. Cell retention systems: In these, the membrane retains cells, separating biocatalysts from the solution and creating differences in hydraulic and solid retention times. In this case, the biocatalysts remain suspended in the reactor.
2. Immobilized biocatalyst systems: Here, biocatalysts are immobilized on surfaces, within matrices, or between two membranes, rendering them nonmobile compared to those in suspension.

These two types of membrane bioreactors can further be distinguished by whether they operate in ultrafiltration or diffusion mode, or if they include units for product removal or aeration [44].

## 3.7 Entrapment in Ionic Liquids

Enzymes possess a key attribute that enhances their suitability for industrial applications: their potential for reuse. However, their activity often diminishes when stored in conventional aqueous or organic solvents. Ionic liquids (ILs) offer a solution to this challenge by promoting protein stability through the manipulation of various factors such as hydrophobicity, electrostatic properties, hydrogen bonding, and Brønsted acidity/basicity. This tunability creates an environment specifically tailored to a given enzyme, helping to maintain its structural integrity and catalytic activity.

As the protein-stabilizing properties of ILs have become increasingly evident, entrapment of enzymes within IL materials has emerged as a promising approach. The core principle of enzyme entrapment in IL materials lies in adjusting the solubility of the IL, thereby regulating the ingress and egress of chemicals. The IL provides a conducive medium for the enzyme to retain its active conformation, essential for catalysis. Entrapping enzymes in an IL gel helps encapsulate and preserve them in their active state. Achieving these benefits requires a deep understanding of enzyme dissolution, stability, and the structure-activity relationship within ILs.

**Manipulating surface charges** can make proteins more soluble in ILs. Protein modification can involve the covalent conjugation of charged or neutral molecules to the protein structure. Notably, polyether carboxylic acids and amine-based modifiers have been effectively applied to enhance the catalytic activity of myoglobin in pure anhydrous ILs, concurrently improving its thermal stability [45].

**Immobilization in ILs** can be accomplished through various techniques, such as encapsulating enzymes in IL gels or modifying surface properties to anchor enzymes to the IL matrix. This approach offers several advantages, including the ability to fine-tune the properties of the IL to match the specific enzyme, thereby optimizing its function.

## 3.8 Magnetic Immobilization

Magnetic immobilization of enzymes involves attaching them to magnetic particles, enabling easy separation from the reaction mixture by applying a magnetic field. This method offers several advantages over alternative immobilization techniques, including easy enzyme isolation, high enzyme loading capacity, and the potential for enzyme reuse (Figure 3.22).

In the field of biotechnology and bioprocess engineering, enzyme immobilization using magnetic supports plays a crucial role. This method binds enzymes to materials with magnetic properties, with iron oxides, particularly magnetite ($Fe_3O_4$) and maghemite ($\gamma$-$Fe_2O_3$), being the most commonly used MNPs These materials are favored for their biocompatibility, biodegradability, and ease of preparation and extraction from the reaction medium.

MNPs are synthesized through physical, wet chemical, and microbial methods, with co-precipitation being the simplest and most effective chemical method for producing particles typically smaller than 50 nm in diameter. In this process, ferric chloride and ferrous chloride are dissolved in deionized water, followed by the addition of ammonia, which leads to the formation of $Fe_3O_4$ nanoparticles with specific magnetic properties. Factors such as the stoichiometric ratio of ferric and ferrous salts, pH, ionic strength, temperature, and the type of alkaline medium influence the final properties of the iron oxide particles.

A common approach for enzyme immobilization involves combining polymeric supports such as PVA with magnetic materials like magnetite ($Fe_3O_4$). For instance, *Burkholderia cepacia* lipase has been immobilized on magnetic PVA supports and magnetized matrices of poly(styrene-co-divinylbenzene) with PVA. These methods have demonstrated high yields in the production of ethyl esters of fatty acids, allowing the reuse of the biocatalyst in successive reaction cycles without significant loss of catalytic activity, highlighting their potential for sustainable, long-term applications.

In another approach, chitosan modified with $Fe_3O_4$ nanoparticles was used as a support for immobilizing the α-glucosidase enzyme. Chitosan served as a stabilizer, enhancing the dispersion of MNPs. After the enzyme adsorbed onto the nanoparticles via electrostatic attraction, α-glucosidase was covalently immobilized through further modification with GA. The enzyme's activity was evaluated using capillary electrophoresis with diode array detection by monitoring the *p*-nitrophenol peak, demonstrating the versatility and effectiveness of this immobilization technique [46, 47].

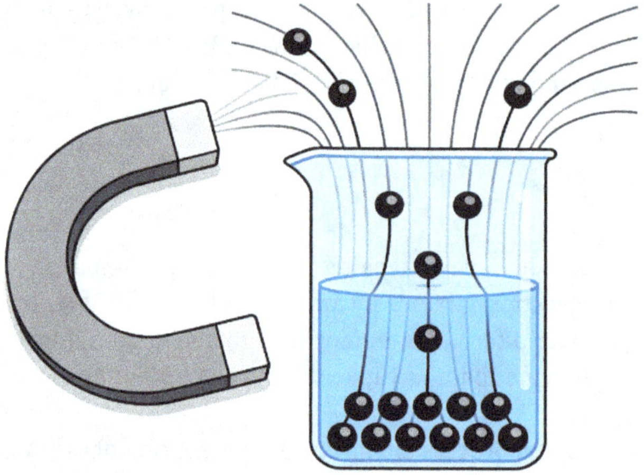

**Figure 3.22:** Schematic representation of a magnetically immobilized enzyme.

## 3.9 Modification of Enzymes

The chemical modification of enzymes lies at the intersection of biocatalysis and organic chemistry, playing a pivotal role in the advancement of industrial biotechnology. This process involves the controlled manipulation of enzymes by introducing specific chemical groups, enabling these biological molecules to adapt to distinct organic environments. This adaptation is crucial for their use in biocatalysis, where chemically modified enzymes can function effectively in organic media. Such modifications have diverse applications, from the synthesis of high-value organic compounds in the chemical industry to various other fields of biotechnology.

In the chemical modification of enzymes, chemical groups are added, forming specific covalent bonds at targeted sites. These bonds can alter the enzyme's 3D structure, influencing its activity and properties. This process is similar to unipoint cross-linking, where specific connections are made at a single point in the molecule. However, while the analogy is useful, the terms "chemical modification" and "cross-linking" are not always interchangeable. Cross-linking generally refers to joining two separate molecules, whereas chemical modification typically involves altering a single molecule (the enzyme itself).

Enzymes can be chemically modified in various ways, depending on the amino acid residues in their structure. Some common chemical modifications include:

- Phosphorylation on serine, threonine, and tyrosine residues
- Glycosylation on asparagine, serine, or threonine groups
- Acetylation, typically on lysine
- Methylation on lysine, arginine, or histidine residues
- Hydroxylation on proline, lysine, or asparagine residues
- Carboxylation on glutamate residues [48]

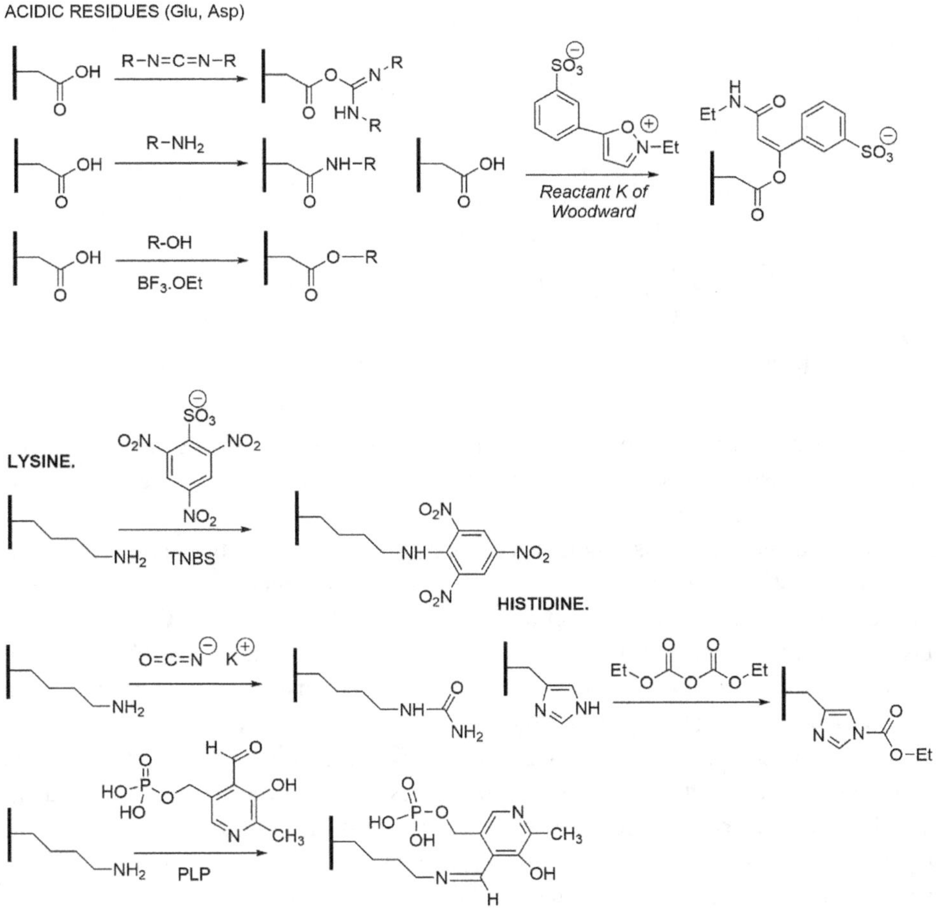

ACIDIC RESIDUES (Glu, Asp)

R-N=C=N-R

R-NH2

R-OH

BF3.OEt

Reactant K of Woodward

LYSINE.

TNBS

HISTIDINE.

PLP

Figure 3.23: Different strategies of modification of amino acid residues.

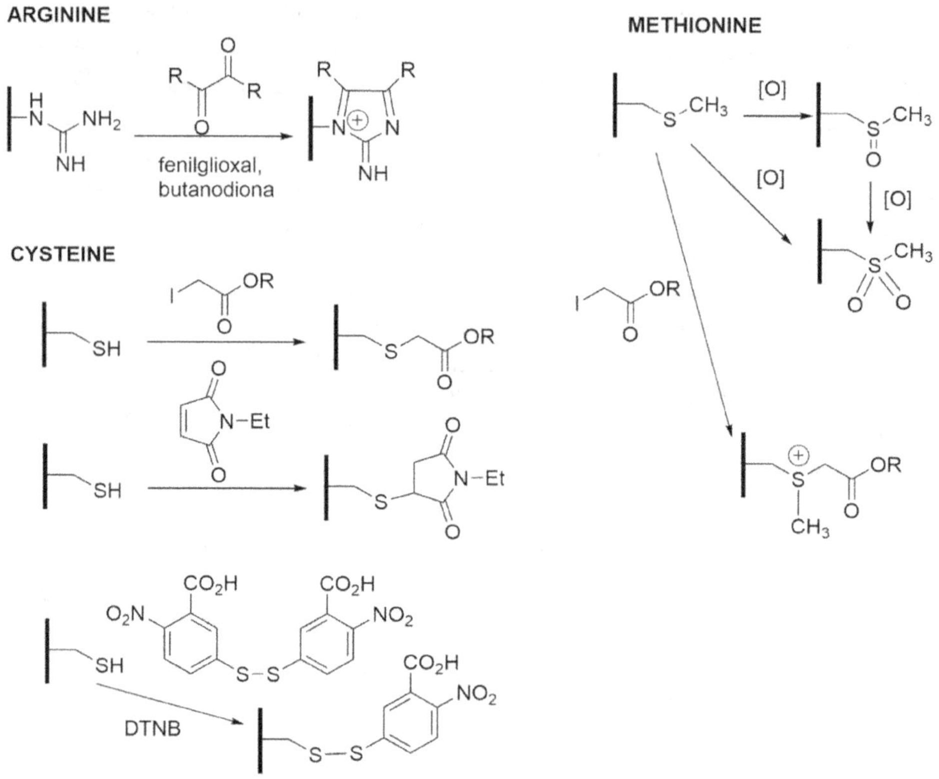

**Figure 3.23a:** (continued)

Site-specific chemical modification of enzymes can be achieved using a variety of chemical agents, allowing precise control over enzyme properties for specific applications (Figure 3.23 and 3.23a).

Unipoint modifications aimed at rendering enzymes insoluble in aqueous solutions are highly sought after. One of the most traditional methods to achieve this is through the covalent attachment of an amphiphilic polymer. Among the various amphiphilic polymers used, monomethoxypolyethylene glycol (mPEG) is the most common, with a variable molecular weight. The frequently used variant typically has a molecular weight exceeding 5 kDa.

Achieving complete insolubility in water depends on the degree of modification, which must be carefully controlled to avoid a loss of enzymatic activity. The covalent attachment is typically performed on lysine residues of the enzyme. As a result of these modifications, enzyme derivatives are obtained that are soluble in certain organic solvents, enabling catalysis to take place in a homogeneous manner [49].

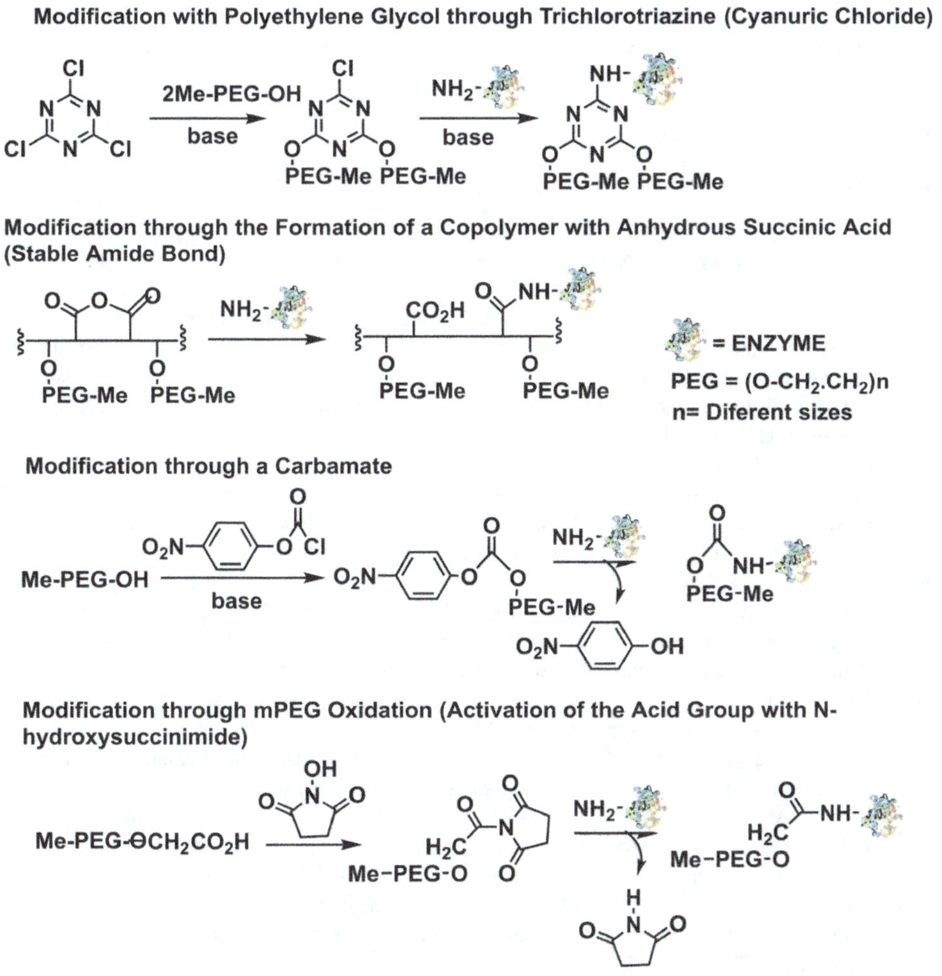
Modification with Polyethylene Glycol through Trichlorotriazine (Cyanuric Chloride)
2Me-PEG-OH
base
$NH_2$-
base
PEG-Me PEG-Me
PEG-Me PEG-Me
Modification through the Formation of a Copolymer with Anhydrous Succinic Acid (Stable Amide Bond)
$NH_2$-
$CO_2H$
NH-
PEG-Me PEG-Me
PEG-Me PEG-Me
= ENZYME
PEG = $(O\text{-}CH_2.CH_2)n$
n= Diferent sizes
Modification through a Carbamate
Me-PEG-OH
base
$O_2N$
PEG-Me
OH
Modification through mPEG Oxidation (Activation of the Acid Group with N-hydroxysuccinimide)
Me-PEG-$OCH_2CO_2H$
$H_2C$
Me-PEG-O
NH-

**Figure 3.24:** Different strategies of enzyme modification with PEG.

Covalent modification of proteins with the amphiphilic polymer polyethylene glycol (PEG) has been applied to improve solubility in organic solvents, simultaneously reducing antigenicity and enhancing in vivo stability. In this context, *C. rugosa* lipase modified with either para-nitrophenylchloroformate- or cyanuric chloride-activated PEG (Figure 3.24) exhibited increased stability in isooctane, with para-nitrophenylchloroformate-activated PEG providing superior stabilization. Notably, both modifications led to diminished lipase and esterase activities, whereas transesterification activity was significantly improved, highlighting a trade-off between stability and catalytic specificity [50].

## References

[1] Boudrant J, Woodley JM, Fernandez-Lafuente R Parameters necessary to define an immobilized enzyme preparation. Process Biochem 2020;90:66–80.

[2] Wingard LB. American Institute of Chemical E, Engineering Foundation Conference on Enzyme E. Enzyme engineering; papers in part from the Engineering Foundation Conference on Enzyme Engineering, August 9-13, 1971, Henniker, New Hampshire, and the 64th annual meeting of the American Institute of Chemical Engineers, December 1, 1971,San Francisco, California 1972, Editor. Wingard Jr. LB, New York: Interscience.

[3] Almeida FLC, Prata AS, Forte MBS Enzyme immobilization: What have we learned in the past five years?. Biofuels Bioprod Biorefin 2022;16:587–608.

[4] Hassan ME, Yang Q, Xiao Z, Liu L, Wang N, Cui X, et al, Impact of immobilization technology in industrial and pharmaceutical applications. Biotech 2019;3(9):440.

[5] Hettiarachchy NS, Feliz DJ, Edwards JS, Horax R The use of immobilized enzymes to improve functionality, In: Proteins in Food Processing 2018, Second Edition, p. 569–597.

[6] Nandini S, Anurag V, Ishman K, Soham C A Comprehensive review on the potential use of immobilized lipase as biosensor – present scenario and prospects. Res J Biotechnol 2022;17:145–150.

[7] Chapman J, Ismail AE, Dinu CZ Industrial applications of enzymes: Recent advances, techniques, and outlooks. Catalysts 2018;8. https://doi.org/10.3390/catal8060238.

[8] Kumar D, Nagar S, Bhushan I, Kumar L, Parshad R, Gupta VK Covalent immobilization of organic solvent tolerant lipase on aluminum oxide pellets and its potential application in esterification reaction. J Mol Catal B Enzym 2013;87:51–61.

[9] Ashkan Z, Hemmati R, Homaei A, Dinari A, Jamlidoost M, Tashakor A Immobilization of enzymes on nanoinorganic support materials: An update. Int J Biol Macromol 2021;168:708–721.

[10] Cao M, Li Z, Wang J, Ge W, Yue T, Li R, et al, Food related applications of magnetic iron oxide nanoparticles: Enzyme immobilization, protein purification, and food analysis. Trends Food Sci Technol 2012;27:47–56.

[11] Aber S, Mahmoudikia E, Karimi A, Mahdizadeh F Immobilization of Glucose Oxidase on Fe3O4 Magnetic Nanoparticles and its application in the removal of Acid Yellow 12. Water Air Soil Pollut 2016;227:93.

[12] Moritz M, Geszke-Moritz M Mesoporous materials as multifunctional tools in biosciences: Principles and applications. Mater Sci Eng C 2015;49:114–151.

[13] Costantini A, Califano V. Lipase Immobilization in Mesoporous Silica Nanoparticles for biofuel production. Catalysts 2021;11:629.

[14] Zdarta J, Meyer AS, Jesionowski T, Pinelo M A General overview of support materials for enzyme immobilization: Characteristics, properties, practical utility. Catalysts 2018;8:92.

[15] Jia H, Zhu G, Wang P Catalytic behaviors of enzymes attached to nanoparticles: The effect of particle mobility. Biotechnol Bioeng 2003;84:406–414.

[16] Thakur K, Attri C, Seth A Nanocarriers-based immobilization of enzymes for industrial application. Biotech 2021;3(11):427.

[17] Sirisha VL, Jain A, Jain A Chapter Nine – enzyme immobilization: An overview on methods, support material, and applications of immobilized enzymes, In Kim S-K, Toldrá F, editors. Advances in Food and Nutrition Research 2016, Academic Press; p. 179–211.

[18] Szekalska M, Puciłowska A, Szymańska E, Ciosek P, Winnicka K Alginate: Current use and future perspectives in pharmaceutical and biomedical applications. Int J Polym Sci 2016;2016:7697031.
[19] Jiang F, Xu XW, Chen FQ, Weng HF, Chen J, Ru Y, et al, Extraction, modification and biomedical application of agarose hydrogels: A review. Mar Drugs 2023;21.
[20] Maksym P, Tarnacka M, Dzienia A, Matuszek K, Chrobok A, Kaminski K, et al, Enhanced polymerization rate and conductivity of ionic liquid-based Epoxy Resin. Macromolecules 2017;50:3262–3272.
[21] Pundir CS, Chauhan NS, Bhambi M Activation of polyvinyl chloride sheet surface for covalent immobilization of oxalate oxidase and its evaluation as inert support in urinary oxalate determination. Anal Biochem 2008;374:272–277.
[22] Datta S, Christena LR, Rajaram YR Enzyme immobilization: An overview on techniques and support materials. Biotech 2013;3: 3 1–9.
[23] Zahirinejad S, Hemmati R, Homaei A, Dinari A, Hosseinkhani S, Mohammadi S, et al, Nano-organic supports for enzyme immobilization: Scopes and perspectives. Colloids Surf B Biointerfaces 2021;204:111774.
[24] Feng X, Ding X, Jiang D Covalent organic frameworks. Chem Soc Rev 2012;41:6010–6022.
[25] Cavalcante FTT, Cavalcante ALG, De Sousa IG, Neto FS, Dos Santos JCS Current status and future perspectives of supports and protocols for enzyme immobilization. Catalysts 2021;11:1222.
[26] Côté AP, Benin AI, Ockwig NW, O'Keeffe M, Matzger AJ, Yaghi OM Porous, crystalline, covalent organic frameworks. Science 2005;310:1166–1170.
[27] Huang S, Kou X, Shen J, Chen G, Ouyang G "Armor-Plating" Enzymes with Metal–Organic Frameworks (MOFs). Angew Chem Int Ed 2020;59:8786–8798.
[28] Liu F, Shi Z, Su W, Wu J State of the art and applications in nanostructured biocatalysis. Biotechnol Biotechnol Equip 2022;36:118–134.
[29] Li N, Qiao D, Zhao S, Lin Q, Zhang B, Xie F 3D printing to innovate biopolymer materials for demanding applications: A review. Mater Today Chem 2021;20:100459.
[30] Liu N, Li D, Wang W, Hollmann F, Xu L, Ma Y, et al, Production and immobilization of lipase PCL and its application in synthesis of α-linolenic acid-rich diacylglycerol. J Food Biochem 2018;42:e12574.
[31] Tang Q, Zhang L, Tan X, Jiao L, Wei Q, Li H Bioinspired synthesis of organic–inorganic hybrid nanoflowers for robust enzyme-free electrochemical immunoassay. Biosens Bioelectron 2019;133:94–99.
[32] Sun T, Dong Z, Wang J, Huang F-H, Zheng -M-M Ultrasound-assisted interfacial immobilization of lipase on hollow mesoporous silica spheres in a pickering emulsion system: A hyperactive and sustainable biocatalyst. ACS Sustain Chem Eng 2020;8:17280–17290.
[33] Sepahvand H, Heravi MM, Saber M, Hooshmand SE Techniques and support materials for enzyme immobilization using Ugi multicomponent reaction: An overview. J Iran Chem Soc 2022;19:2115–2130.
[34] Zhang J, Zhang F, Yang H, Huang X, Liu H, Zhang J, et al, Graphene Oxide as a matrix for enzyme immobilization. Langmuir 2010;26:6083–6085.
[35] Lyu XY, Gonzalez R, Horton A, Li T Immobilization of enzymes by polymeric materials. CATALYSTS 2021;11.
[36] Permana D, Putra HE, Djaenudin D Designed protein multimerization and polymerization for functionalization of proteins. Biotechnol Lett 2022;44:341–365.
[37] Khiari O, Bouzemi N, Sánchez-Montero JM, Alcántara AR Easy and versatile technique for the preparation of stable and active lipase-based CLEA-like Copolymers by using Two Homofunctional cross-linking agents: Application to the preparation of Enantiopure Ibuprofen. Int J Mol Sci 2023;24. https://doi.org/10.3390/ijms241713664.

[38] Smith MEB, Schumacher FF, Ryan CP, Tedaldi LM, Papaioannou D, Waksman G, et al, Protein modification, bioconjugation, and disulfide bridging using Bromomaleimides. J Am Chem Soc 2010;132:1960–1965.
[39] Soni S, Desai JD, Devi S *In situ* entrapment of α-chymotrypsin in the network of acrylamide and 2-hydroxyethyl methacrylate copolymers. J Appl Polym Sci 2000;77:2996–3002.
[40] Weng Y, Yang G, Li Y, Xu L, Chen X, Song H, et al, Alginate-based materials for enzyme encapsulation. Adv Colloid Interface Sci 2023;318:102957.
[41] Tielmann P, Kierkels H, Zonta A, Ilie A, Reetz MT Increasing the activity and enantioselectivity of lipases by sol-gel immobilization: Further advancements of practical interest. Nanoscale 2014;6:6220–6228.
[42] Yoshimoto M Stabilization of enzymes through encapsulation in Liposomes. Methods Mol Biol 2017;1504:9–18.
[43] Pietta PG, Agnellini D, Mazzola G, Vecchio G, Colombi S, Bianchi G Immobilization and characterization of enzymes on hollow fibers for a possible use in the biomedical field, In Weetall HH, Royer GP, editors. Enzyme Engineering 1980, Boston, MA: Springer US; Vol. 5, p. 235–237.
[44] Vladisavljević GT Biocatalytic membrane reactors (BMR). Phys Sci Rev 2016;1.
[45] Imam HT, PC M, AC M. Enzyme entrapment, biocatalyst immobilization without covalent attachment. Green Chem 2021;23:4980–5005.
[46] Guajardo N Immobilization of Lipases Using Poly(vinyl) Alcohol. Polymers (Basel) 2023;15.
[47] Vaghari H, Jafarizadeh-Malmiri H, Mohammadlou M, Berenjian A, Anarjan N, Jafari N, et al, Application of magnetic nanoparticles in smart enzyme immobilization. Biotechnol Lett 2016;38:223–233.
[48] Giri P, Pagar AD, Patil MD, Yun H Chemical modification of enzymes to improve biocatalytic performance. Biotechnol Adv 2021;53:107868.
[49] DeSantis G, Jones JB Chemical modification of enzymes for enhanced functionality. Curr Opin Biotechnol 1999;10:324–330.
[50] Hernáiz MJ, Sánchez-Montero JM, Sinisterra JV Influence of the nature of modifier in the enzymatic activity of chemical modified semipurified lipase from Candida rugosa. Biotechnol Bioeng 1997;55:252–260.

# Chapter 4
# Medium Engineering

**Abstract:** This chapter provides a comprehensive overview of medium engineering in biocatalysis, focusing on how the selection and design of solvent environments impact enzymatic performance in industrial processes. Traditionally, enzymes have been utilized in aqueous solutions due to their natural compatibility with water, but the inherent limitations, such as poor solubility of hydrophobic substrates, product inhibition, and stability issues, have spurred the development of alternative solvent systems for enzymatic reactions. Recent advancements in medium engineering have introduced the use of nonconventional solvents, including organic solvents and neoteric solvents, such as room-temperature ionic liquids, supercritical fluids, and deep eutectic solvents, which open new possibilities for improving reaction rates, selectivity, and scalability in biocatalysis. This chapter explores the underlying principles that govern enzyme stability and activity in these diverse media, emphasizing critical factors such as water activity, enzyme memory and the physicochemical properties of solvent systems. It highlights how tuning solvent polarity, viscosity, and water content can enhance enzymatic reactions, particularly for the synthesis of pharmaceuticals, fine chemicals, and biofuels. Detailed case studies are provided to illustrate industrial applications, including the synthesis of active pharmaceutical ingredients and the use of biphasic and neoteric solvents to overcome challenges of solubility and selectivity. Additionally, this chapter discusses the environmental and sustainability implications of medium engineering, including the adoption of green and bio-based solvents, processes for solvent recovery, and alignment with circular chemistry principles. Cutting-edge research in protein engineering, molecular dynamics simulations, and process optimization are reviewed to provide insight into future developments in the field.

**Keywords:** Non-aqueous enzymology, non-conventional media, water activity, biphasic and multiphase reaction systems, neoteric solvents

## 4.1 Introduction

Enzymes play a pivotal role in numerous industrial applications, including pharmaceuticals, food processing, biofuels, and bioremediation [1–6]. Their remarkable catalytic efficiency, specificity, and environmentally benign nature have positioned them as invaluable tools in modern biotechnology [7, 8]. Conventionally, enzymatic reactions have been conducted in aqueous environments due to the inherent compatibility of enzymes with water [9]. However, limitations such as poor solubility of substrates, product inhibition, and thermal instability have prompted exploration into alternative solvent systems [10], performing what has been termed “medium engi-

 | https://doi.org/10.1515/9783111122991-004

neering" [11–13]. In fact, nonconventional solvents (solvents other than pure aqueous media) offer unique advantages, such as:

i Enhanced solubility and reactivity: Many substrates and products in biocatalytic processes are poorly soluble in aqueous environments. Medium engineering allows the use of nonconventional solvents such as organic solvents, room-temperature ionic liquids (RTILs), supercritical fluids (SCFs) or deep eutectic solvents (DESs) that can significantly improve substrate solubility, increasing reaction rates and enabling conversions that would be difficult or impossible in water.

ii Stability of biocatalysts: The choice of medium can have a profound effect on the stability of enzymes or whole cells. Engineering the reaction medium to include solvents like ionic liquids, SCFs, or solvent-free systems can protect enzymes from denaturation, thermal deactivation, and inhibition by substrates or products, thus extending the operational life of the biocatalyst.

iii Improved selectivity and enzyme activity: Medium engineering can be used to modulate the selectivity of biocatalysts, particularly in enantioselective reactions. By carefully tuning factors such as solvent polarity, pH, and water content, the specificity of the enzyme can be enhanced, resulting in higher enantioselectivity or regioselectivity, which is especially important in the production of pharmaceuticals and fine chemicals.

iv Facilitation of multiphase systems: Many biocatalytic reactions involve multiple phases (e.g., organic-aqueous biphasic systems). Medium engineering allows the selection or creation of solvents that can integrate these phases more effectively, improving mass transfer and reducing inhibitory effects by isolating products or by-products from the biocatalyst.

v Minimization of product inhibition: Accumulation of products can often inhibit enzyme activity, leading to reduced reaction efficiency. Through medium engineering, inhibitory products can be selectively partitioned into a separate phase (e.g., organic solvent in a biphasic system) or be continuously removed, thus maintaining high enzyme activity throughout the reaction.

vi Sustainability and green chemistry: Medium engineering aligns with the principles of green chemistry by enabling the use of more sustainable solvents, such as bio-based solvents, supercritical $CO_2$, and water, or by reducing the need for harmful organic solvents. The development of solvent systems that can be easily recovered and reused (e.g., RTILs and SCFs) also reduces waste and environmental impact.

vii Control of reaction parameters: Medium engineering allows precise control over reaction parameters such as temperature, pH, ionic strength, and solvent properties (e.g., polarity, viscosity, and diffusivity). This fine-tuning can improve overall process efficiency and lead to better control over reaction kinetics and outcomes.

viii Integration of downstream processing: In some cases, the reaction medium can be engineered to facilitate easy product recovery. For example, SCFs can be used to integrate reaction and separation processes by allowing product extraction through phase changes (e.g., depressurization). This simplifies downstream processing and reduces energy consumption.

This chapter will critically examine the use of enzymes in several nonaqueous and mixed solvent systems, focusing on the opportunities, challenges, and future prospects in this emerging field.

## 4.2 Enzymes in Pure Organic Solvents: The Critical Role of Water

In nature, (cytosolic) enzymes are evolved to function optimally in aqueous environments, and aqueous buffers have been thus traditionally the first choice of reaction medium when characterizing a biocatalytic reaction. Similarly, a significant proportion of the natural substrates for enzymes, the cell metabolites, have been biochemically decorated to become highly water soluble [9]. Then, the classical enzymology was developed for aqueous media, in which the enzyme (E) and both the substrate (S) and the product (P) are soluble (homogeneous catalysis), as shown in Figure 4.1. According to conventional notion, enzymes are active only in water, because in other solvents (organic) they would denature by losing their native structure and their catalytic activity.

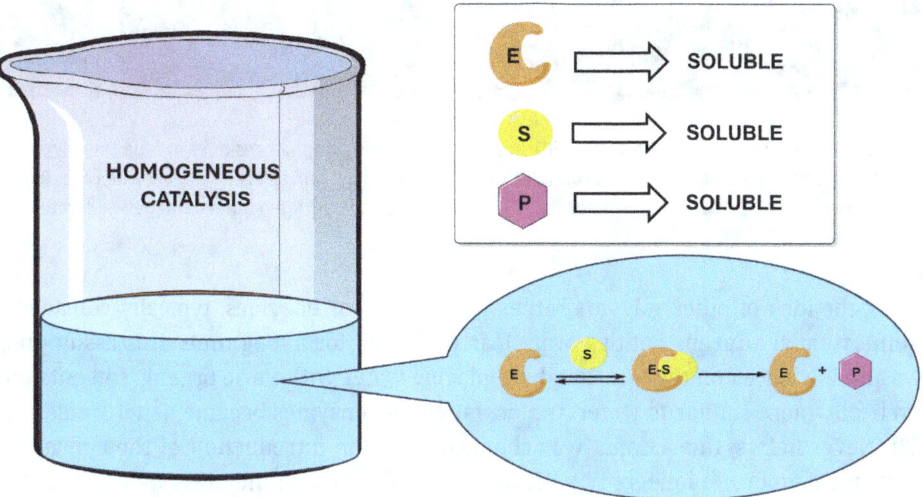


**Figure 4.1:** Homogeneous enzymatic catalysis in aqueous catalysis media.

In fact, the active 3D structure of an enzyme depends on a great variety of factors (recently, the development of artificial intelligence-based methods such as AlphaFold [14–17] has led to an almost magical way to predict these 3D structures), being the –S–S– bonds between cysteine residues [18], electrostatic bonds and van der Waals forces [19], hydrogen bonds [20], and hydrophobic bonds [21] those interactions that maintain a delicate equilibrium permitting the active conformation [22, 23]. For instance, in aqueous media, the balance between the location of lateral chains of hydrophilic aminoacidic residues (mainly) exposed to aqueous solvent (thus involved in a network of H-bonds with water molecules) and the hydrophobic ones, (mainly) buried inside the enzyme core (as illustrated in Figure 4.2 for the 3D structure of the lipase of *Rhizomucor miehei*, the first lipase to be crystallized more than 30 years ago (PDB Code 1TGL [24]) is vital for the establishment of the active conformation.

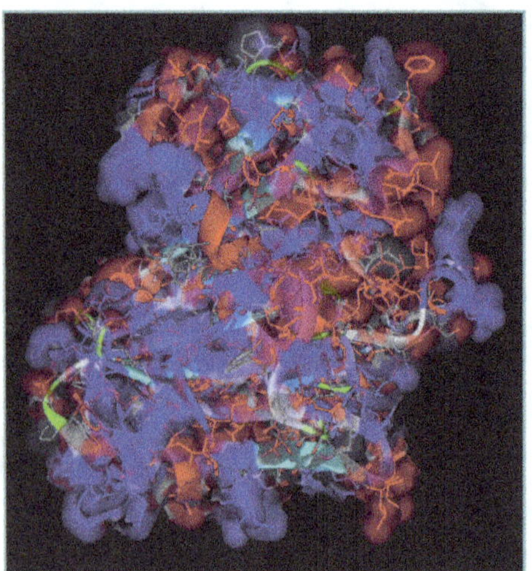
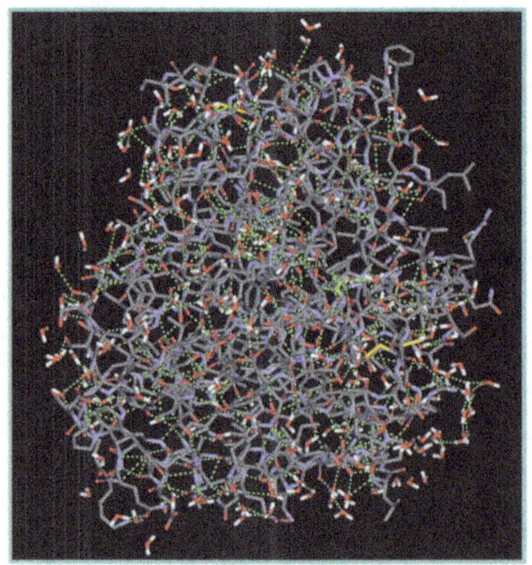

**Figure 4.2:** Left: *Rhizomucor miehei* lipase 3D structure showing hydrophilic residues in blue (mainly) exposed to solvent (aqueous) and hydrophobic ones in red (mainly) buried inside the enzyme core. Right: Same lipase showing water molecules of the crystal structure and hydrogen bonds. Images generated with Pymol.

Thus, the idea of other solvents rather than water for enzymes, typically considered quintessential aqueous entities, was clearly rejected for a long time. This assumption was mainly based on the fact that, by replacing water with those organic solvents theoretically more similar to water, that is, methanol, enzymes became denaturated [25–28]. Nevertheless, this axioma was changed after the introduction of the concept of "enzyme memory," pioneered by Alexander M. Klibanov in the mid-1980s, which revolutionized our understanding of enzyme behavior and stability in nonaqueous environments. Klibanov's groundbreaking work challenged the conventional notion that enzymes lose their biological activity outside of water due to denaturation. In his seminal paper published in 1984 [29], Klibanov demonstrated that certain enzymes exhibit remarkable stability and catalytic activity in organic solvents, even at temperatures as high as 100 °C. This unexpected resilience, he argued, was attributed to the enzyme's ability to maintain a "memory" of its original aqueous environment, preserving its native conformation and activity in nonaqueous media. In fact, Klibanov proposed that enzymes possess a "memory" of their native aqueous environment, retaining structural and functional characteristics even after prolonged exposure to nonaqueous solvents, as depicted in Figure 4.3.

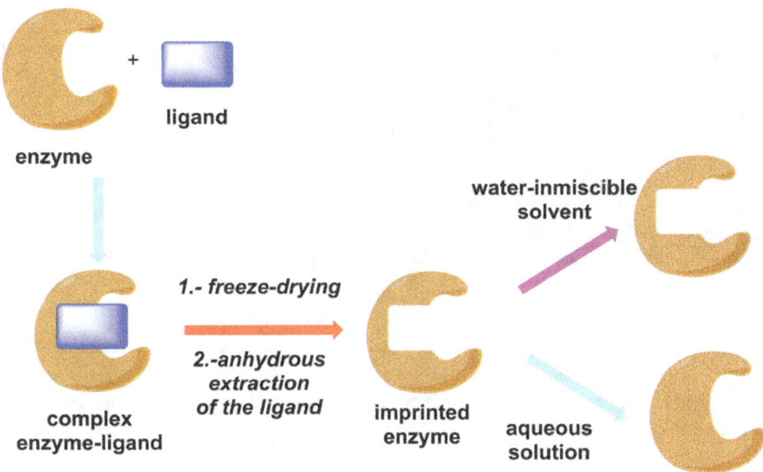


**Figure 4.3:** Left: *Rhizomucor miehei* lipase 3D structure, showing hydrophilic residues in blue (mainly) exposed to solvent (aqueous) and hydrophobic ones in red (mainly) buried inside the enzyme core. Right: Same lipase showing water molecules of the crystal structure and hydrogen bonds. Images generated with Pymol.

On binding of the ligand (blue rectangle, e.g., a substrate analogue or a reversible inhibitor) to the enzyme active site in water (first light blue arrow), this ligand induces a conformational change in the enzyme, which remains after a freeze-drying and a subsequent extraction of the ligand using an anhydrous solvent (red arrow). This altered conformation (memory) continues (purple arrow) if the enzyme is suspended in an anhydrous water-immiscible solvent, because the enzyme cannot return to its original aqueous conformation as a consequence of its rigidity in this "nonnatural" media. Contrarily, if the imprinted enzyme is redissolved in water (second light blue arrow), its intrinsic flexibility in aqueous media would make it to return to its original conformation, so that the imprint will disappear.

Furthermore, Klibanov proposed that enzyme memory could be harnessed for practical applications in biocatalysis, offering new opportunities for enzymatic reactions in nonaqueous solvents [30–44]. His research laid the foundation for the development of enzymatic processes in organic synthesis, pharmaceuticals, and other industries where traditional aqueous conditions are impractical.

As previously noted, water plays an essential role in biomolecular operations, influencing aspects like protein conformation, structural integrity, substrate identification, ligand association or dissociation, and thus overall catalytic performance [45–47]. In living systems, water molecules are classified into three categories based on their proximity to the protein surface, illustrated in Figure 4.4.

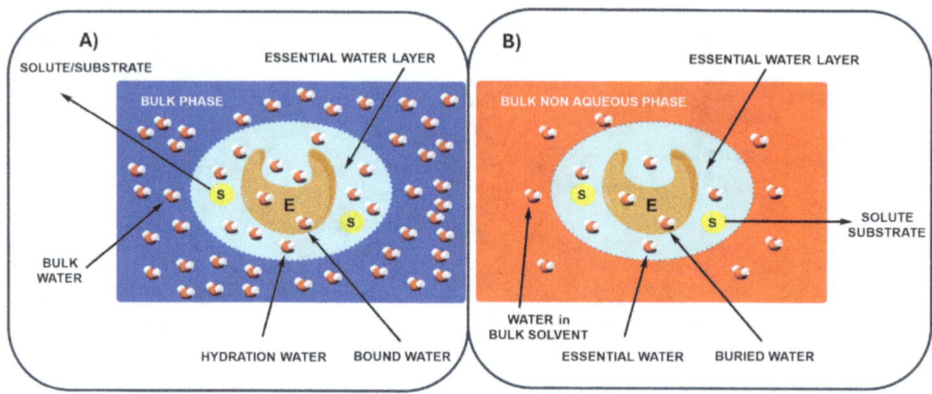


**Figure 4.4:** Graphical representation of the types of water molecules in a protein system: (A) in aqueous solution and (B) in pure immiscible organic solvents.

More than 98% of this water qualifies as "bulk water," given that these molecules reside farther than the van der Waals radius from the protein exterior, serving as an authentic solvent to promote protein mobility amid surrounding entities. In contrast, "hydration water" describes molecules engaging directly with the protein alongside bulk water, establishing a hydrogen-bond lattice across the protein exterior critical for maintaining protein architecture, operations, and motility. Lastly, "bound water" (frequently detected by diverse analytical approaches [48]) establishes numerous hydrogen bonds involving amino acid side chains, ligands, and nearby water within or adjacent to sequestered pockets and active sites, adhering firmly owing to the protein surface's dual solvophobic-hydrophilic character. Such bound water instances, alternatively termed "ordered water" or "structural water," induce kinetic stabilization of the functional protein form via intensified hydrogen bonding among protein constituents. Distinctions among these water categories become more evident in neat organic solvents, where the bulk phase holds minimal water dictated by solvent properties (as elaborated shortly), confining most molecules to the vital hydration or bound layers.

For instance, if we could remove the 231 bound water molecules present in the crystallographic structure of *Rh. miehei* lipase, we will have a picture of a "naked" enzyme, while the structural water molecules removed from the original structure would be the "dress," as shown in Figure 4.5. As it is shown, the appearance of the dress closely resembles the 3D structure of the enzyme. Consequently, if the dressed enzyme is submerged in a water-insoluble organic solvent (hexane, toluene, etc.), the solvent molecules will not replace the water molecules kinetically trapped by the enzyme; thus, the enzyme remains dressed, in a shape that resembles the original conformation. This means that water is acting in a similar way than the ligand represented in Figure 4.3, thus imprinting the enzyme molecules. Of course, if the dressed enzyme is submerged in water-soluble organic solvents, such as EtOH (ethyl alcohol) or MeOH (methyl alcohol), these molecules will compete with the water dressing the enzyme, and this would lead to enzyme denaturation. Figuratively, we can envision the enzyme as a sort of sea snail, whose shell protects the animal from the external medium while allowing it enough flexibility to move and live inside the shell.

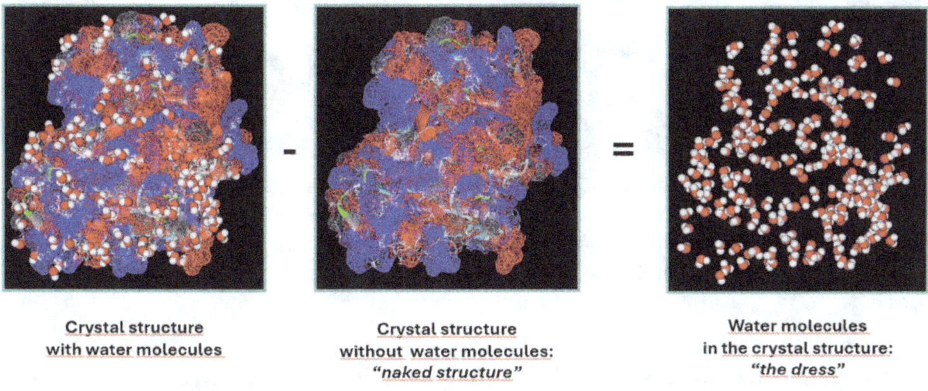


**Figure 4.5:** Graphical representation of the removal of the water molecules from an enzymatic 3D structure.

Nevertheless, the fact that water molecules present in a lyophilized enzyme are shielded from removing by external hydrophobic media does not mean that the enzyme is fully active in those media. In this sense, it is necessary to calculate the exact amount of water molecules that are required for allowing the enzyme to have the required flexibility to be able to recognize the substrate.

For this calculation, the physical chemistry to be assessed is not properly water amount, but rather water activity ($a_W$), which is defined as the ratio between the amount of water vapor present in a substance or material relative to the amount of water vapor that would be present if the substance were in equilibrium with pure water at the same temperature and pressure [49]. It is a dimensionless quantity, typically expressed on a scale from 0 to 1, where 0 represents completely dry conditions (no available water) and 1 represents pure water. One of the major advantages of using $a_W$ is that it is equal in all phases at equilibrium and hence can be measured in any phase.

To envision $a_w$, let us imagine a closed system in which we have a dry (*naked*) enzyme suspended in a water-immiscible organic solvent, and add some water in this solvent. Water molecules would tend to distribute among the several phases present, which are the enzyme, the solvent, and the vapor phase (Figure 4.6, left). Once the equilibrium is reached, the value of $a_w$ is the same in all phases, but we cannot say the same about the amount of water, as the partial solubility of water in the organic solvent will depend on the nature of this one.

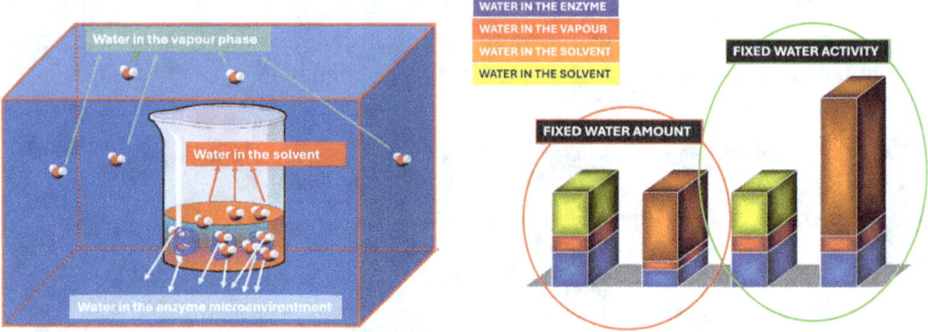


**Figure 4.6:** Left: Distribution of water molecules in a closed system formed by an enzyme suspended in a water-immiscible solvent. Right: Comparison between working in two solvents at fixed water amount and fixed $a_w$.

When comparing two organic solvents (depicted in orange and yellow colors, Figure 4.6, right) with different water solubilities, water is more soluble in the orange, and we work with the same amount of water added to the system; when comparing the two systems, the amount of water retained by the enzyme (presented in blue color) will be smaller when using the orange solvent, as more water molecules will go into this solvent. On the contrary, if $a_w$ is fixed, even if there are more water molecules dissolved in the orange solvent, the enzyme will have the same $a_w$ in both solvents. Therefore, water activity is unaffected by the properties of the organic solvent, which is not true for the total water content. When all the components reach equilibrium (Figure 4.6, left), $a_w$ can be measured in any of the components, with the vapor phase being the most accessible one. Additionally, if the water amount is low, those water molecules in the gas phase would behave as an ideal gas, so that $a_w$ is simply the ratio of the partial pressure of water in the vapor to the vapor pressure of pure water at the same temperature, and can be estimated by measuring the relative humidity (RH) of the vapor phase using a hygrometer.

Thus, if we are able to quantify the relationship between the water added (concentration) and the water which is really in the enzyme microenvironment we would be able to compare the real effect of different organic solvents on the enzymatic performance [50]. The way of representing this relationship is using water-sorption isotherms [50], by plotting the water amount added to the reaction medium (ordinate axis) versus the water activity (abscissa axis). To carry out a water adsorption isotherm for an enzyme (as depicted in Figure 4.7), we must begin by accurately weighing the enzyme sample, and dry it down to an almost anhydrous state, by using a strong water absorbent such as $P_2O_5$. Then, it must be placed in a sealed chamber equipped with a water-activity sensor and add a very small amount of water (in the μL range); after that, we must allow the enzyme to reach equilibrium and record the water activity of the enzyme. This process must be repeated to obtain a series of equilibrium data points, which are plotted to construct the adsorption isotherm.

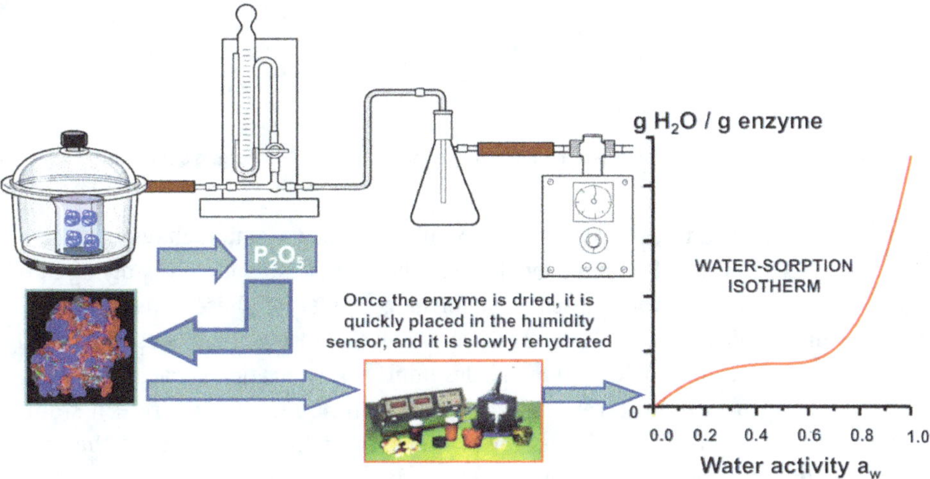


**Figure 4.7:** Protocol to plot water adsorption isotherms.

For enzymes dispersed in organic solvents, the water adsorption isotherm reveals distinct regions based on water activity levels, as depicted in Figure 4.8 (noting that bound water quantities mirror those in air at equivalent $a_W$, particularly at lower values [50]). In the initial phase, spanning from 0 (fully dehydrated "bare" enzyme) to 0.07 g water per g protein (commonly denoted as *h*), proton reorganization takes place (i.e., ionizable sites become charged) through water molecule engagement with those groups. This equates to a molar proportion of roughly 0–55 water molecules per enzyme (generally estimated using lysozyme as the reference enzyme [51]). Assuming the lysozyme surface accommodates approximately 400 water molecules for full coverage, this phase equates to 0–15% of the enzyme's surface hydration.

During phase 2 (0.07–0.30 h, equating to a 55–250 molar ratio or 15–65% coverage), water attaches to charged and polar regions. As hydration advances into phase 3 (0.30–0.50 h), less tightly bound areas (likely mostly nonpolar) get enveloped, culminating in a complete water monolayer at 0.50 h. Beyond this in phase 4 (>0.50 h), the

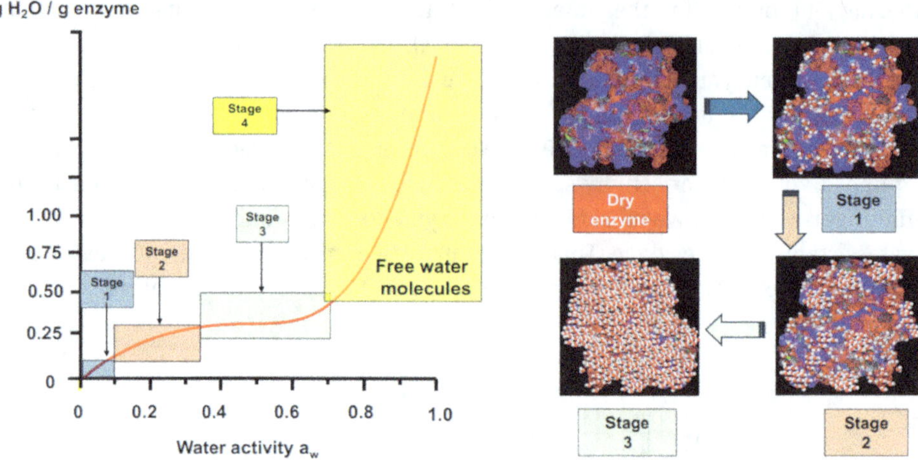


**Figure 4.8:** Typical water adsorption isotherm of an enzyme suspended in an organic solvent.

hydration layer expands, imparting solution-like thermodynamic behavior to the enzyme, with excess water becoming available in the organic medium – dissolving into it up to solubility limits if water-miscible, or forming a separate phase otherwise. Hence, enzymes in organic solvents effectively "dissolve water" (preventing its escape into the bulk solvent), as proteins hold onto molecules until their saturation capacity is met.

Thus, as proved by different experimental techniques and theoretical studies, water molecules serve as lubricants for proteins, providing the necessary flexibility for catalyzing reactions [46, 52]. Consequently, excessive drying can render enzymes inactive. Conversely, a high water content can also impede reactions in which water is produced while the enzymatic catalysis, such as esterification processes, driving the thermodynamic equilibrium toward hydrolysis. Furthermore, it has been shown that water molecules may cluster on the surface of lipase, thereby obstructing substrate molecules from accessing the active site [53, 54]; this fact implies that it is crucial to determine whether this clustering effect occurs when designing (by any kind of genetic engineering) more effective biocatalysts in low-water-containing media [55, 56].

So, the critical question to answer when working with enzymes in organic solvents is not if enzymes do need water to work (the answer is YES, of course), but rather how much water is needed to retain their activity and to reach the best catalytic behavior. This question cannot be universally answered, as it depends on the type of enzymes, but, as a rule of thumb, a threshold value of about 0.2 g $H_2O$/g protein is accepted to keep the required enzymatic flexibility for activity [57]. As previously noted, the behavior of different enzymes is very diverse: for instance, alpha-chymotrypsin needs only 50 molecules of water per molecule of protein to remain catalytically active (far less than a monolayer), and subtilisin and lipases are more or less similar [36]. On the other hand, polyphenol oxidase requires $3.5 \times 10^7$ molecules of water [35].

Therefore, for determining the optimal $a_w$ for enzymes in organic solvents, the use of water-sorption isotherms is highly recommended [49, 58–60], so that they can used to set the optimal $a_w$. In this sense, two primary methods for managing water activity have been proposed, either setting the initial water activity before starting the catalytic reaction or rather maintaining control of the water activity throughout the reaction's duration. If the final yield is critical, it is often necessary to regulate water activity during the reaction, particularly as esterification reactions produce water. However, if the initial reaction rate is the primary concern, a preliminary equilibration of water activity may suffice. In any case, to fix this $a_{w,}$ different techniques for the continuous control of saturated salt solutions have been proposed. One method entails directing saturated salt solutions or air balanced with them *via* silicone tubing or hollow fiber membranes interfacing the reaction mixture (see ref. [50]. for a detailed explanation). Nonetheless, these techniques are impractical for larger-scale applications due to the slow rate of water mass transfer. Instead, a more viable method for large-scale operations is sparging the substrate with either dry or humid air or nitrogen gas [61, 62], which allows high water transfer rates resulting from the extensive contact area between the gas bubbles and the substrate, and minimal impact on the enzyme.

In any case, it must be also considered that, when working with immobilized enzymes, a very practical and common strategy to render stable and reusable biocatalysts [63–69], the effect of the water activity of the support on the overall behavior is pivotal, as support usually constitutes the main component of the catalyst [70, 71]. In such scenarios, water distribution between the carrier and solvent can be quantified via the support's aquaphilicity (Aq), defined as the proportion of water retained on the support relative to that in the solvent under standardized conditions – a straightforward metric for assessing the material's water-retention ability [11].

As previously noted, water activity is independent of the water solubility of organic solvents. This parameter has been often quantified using log *P*, being P the partition coefficient of a certain substance (either a solute or a solvent) in the biphasic system *n*-octanol/water, according to the Nernst distribution law [11]. Thus, solvents can be roughly divided in *completely miscible* with water (log *P* ranging between and 2.5 and 0), partially miscible (between 0 and 1.5), *low miscible* (1.5–2.0), and *completely immiscible* (log $P > 2$) [72]. As a rule of thumb, these last types of solvents (toluene, cyclohexane, *n*-hexane, *n*- or *iso*-octane, amongst others) cause minor enzyme distortion and ensure high retention of activity for almost all enzymes (especially for lipases [73]). On the other hand, completely and partially miscible hydrophilic solvents (dimethylformamide, dimethyl sulfoxide, acetone, MeOH, EtOH, etc.) should not be used as "neat" organic solvents, as they cause serious enzyme distortion at elevated concentrations [72] (only very stable enzymes such as subtilisin [74, 75] or lipase B from *Candida antarctica* [56, 76] can work in those solvents in a monophasic system); usually, they are used in order to help solubilizing lipophilic substrates when working in aqueous systems, in a concentration generally not higher than 20% to avoid enzyme deactivation. Additionally, these polar solvents (especially acetone and lower alcohols) are commonly used for partial purification of protein by precipitation [77]; for some enzymes, adjusting $a_w$ after this treatment leads to enhanced activity [59].

In addition, some other parameters such as dielectric constant [78, 79], dipole moment [80], or solvent polarity [81, 82] have been tested for assessing their influence on the activity and stability of enzymes. However, the choice of solvents and solvent mixtures to tune enzymatic activity based on these well-established solvent descriptors is still a challenge, normally based on a trial-and-error methodology. The scenario is even worse if we try to explain [83, 84] or predict the effects of solvents on enzymatic stereoselectivity [85].

In recent years, there has been growing interest in the use of green solvents in all chemical processes [86–90]. Green solvents contribute significantly to enhancing both environmental sustainability and process efficiency. They are often derived from renewable resources, biodegradable, and exhibit lower toxicity compared to conventional solvents, making them more environmentally friendly. Green solvents further support green chemistry tenets by lowering the total carbon emissions from chemical production processes [91–94]. They often require less energy for temperature control and solvent recovery, further decreasing the environmental impact. Finally, adopting green solvents can improve workplace safety by reducing the exposure of workers to hazardous chemicals, fostering a safer and more sustainable industry. Moreover, the nontoxic nature of green solvents enables easier downstream processing and purification, potentially lowering costs. For sure, green solvents can improve the performance of biocatalysts, by providing a more suitable reaction environment, leading to increased reaction rates, higher yields, and reduced formation of unwanted by-products [95–99].

One of the key issues arising from the selection of greener solvents is how to assess the "greenness" of a solvent. The pursuit of methods to identify and choose solvents with reduced environmental, health, and safety impacts has notable examples within the pharmaceutical sector [100]. This is not surprising, as around 60% of the total mass used in producing active pharmaceutical ingredients (APIs) typically consists of organic solvents, which account for between 80% and 90% of the nonaqueous materials used [101]: indeed, many of the most commonly employed solvents are also recognized as chemicals of concern [102]. Some well-known solvent guides include those produced by GlaxoSmithKline [103–105], AstraZeneca [100], Pfizer [106], Sanofi [107], the American Chemical Society Green Chemistry Institute Pharmaceutical Roundtable (ACS GCI Pharmaceutical Roundtable) [108], or the CHEM21 consortium [109]. In addition to these guides, other tools and techniques have been developed through computational modeling approaches, enabling the evaluation of solvents under the specific conditions of a given process [110–113].

In any case, when using an enzyme in any organic solvent, we are facing a heterogeneous catalysis, as the enzyme is not soluble, but the substrate and the products are

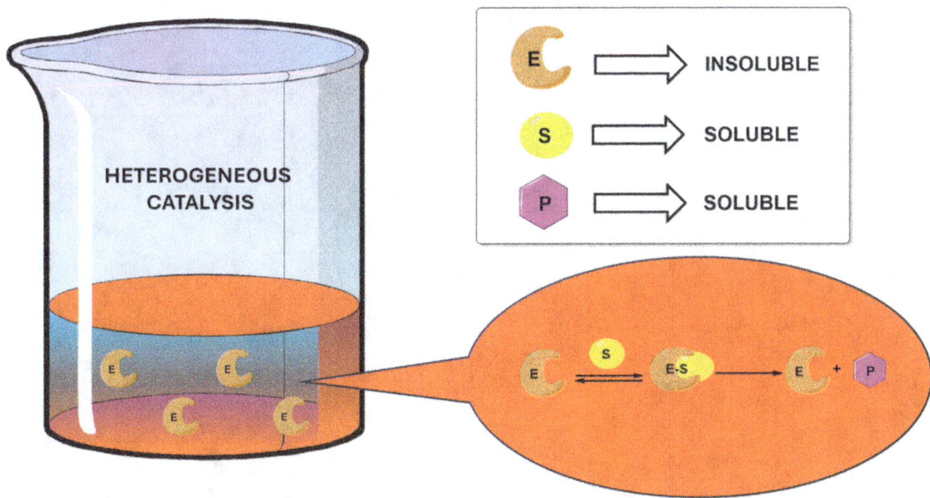


**Figure 4.9:** Heterogeneous enzymatic catalysis in organic solvents.

soluble, as shown in Figure 4.9, in a very simplistic manner (as water molecules attached to the enzyme are not depicted).

In modern literature, there are plenty of examples of enzymatic catalysis, mostly with lipases, in pure organic solvents [114–118]. Some examples focused on the preparation of APIs are shown in Figure 4.10 and discussed below.

Treprostinil (marketed as Remodulin®, Orenitram®, and Tyvaso®), as shown in Figure 4.10a, is a synthetic analogue of prostacyclin (PGI2) and functions as a vasodilator to treat pulmonary arterial hypertension [119]. Its synthesis is notably intricate and time-intensive [120]. In a 2016 patent awarded to Everlight Chemical Industrial Corporation [121], a stereo- and regioselective acylation of diol **1** with vinyl acetate was carried out. Employing vinyl esters as acylating agents proves highly advantageous, since the transesterification proceeds nearly irreversibly – the vinyl alcohol intermediate tautomerizes into acetaldehyde, thereby driving the equilibrium toward the target acylation [122]. This reaction, which occurred in pure *n*-hexane at 22 °C using Amano's lipase AK, shows the enzymatic precision of the biocatalyst, which selectively acylates the secondary cyclic hydroxyl group, leaving the exocyclic hydroxyl group unaffected, additionally achieving the desired stereochemistry for Treprostinil precursor **2**.

A 2018 patent from Wuxi Fortune Pharmaceutical Co. Ltd. [121] describes, as depicted in Figure 4.10b, a gram-scale kinetic resolution of the racemic hydroxynitrile (*rac*)-**4**, obtained from racemic epichlorohydrin 3, via an enzymatic transesterification employing perchlorophenyl acetate and *C. antarctica* lipase B immobilized on an acidic resin. This resolution affords the acetate (*R*)-**5**, which, after hydrolysis, furnishes L-carnitine, an essential cofactor in fatty acid metabolism and cellular energy production [123].

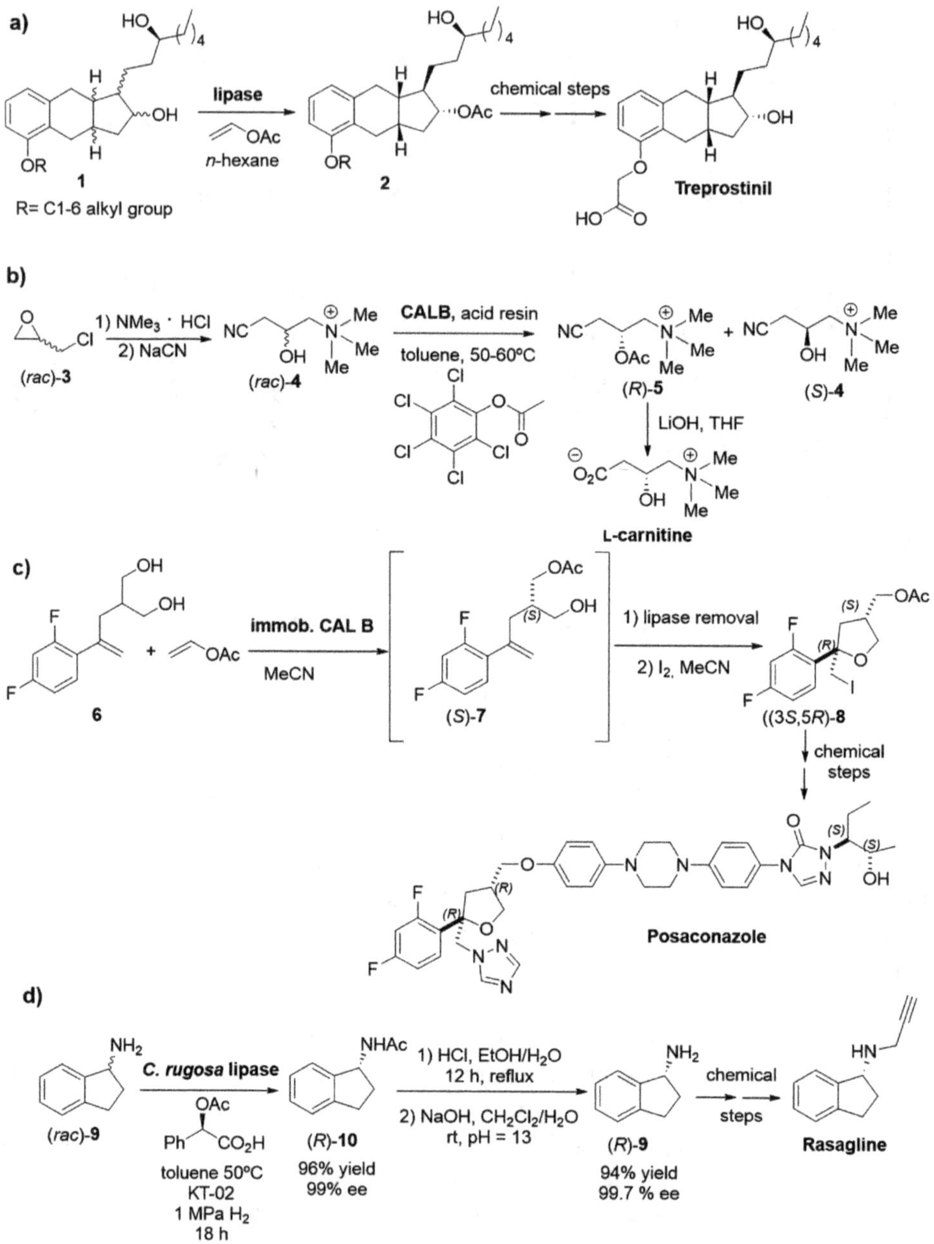

Figure 4.10: Preparation of APIs through biotransformations in organic solvents.

Figure 4.10c outlines a compelling biocatalytic route to a chiral synthon for Posaconazole production (sold as Noxafil® or Posanol®), an oral antifungal agent mainly deployed in clinical settings against *Aspergillus* spp., *Candida* spp., *Fonsecaea pedrosoi*, and *Fusarium* spp. During its manufacture [124], a key enzymatic stage installs two of the compound's four chiral centers by desymmetrizing prochiral diol **6** (Figure 4.10c). Merck's process (initially from Schering-Plough) relies on immobilized *C. antarctica* lipase B for the stereospecific acetylation of the pro-*S* hydroxy group in the diol, employing vinyl acetate as acyl donor. The transformation excels in efficiency, delivering strong conversion to monoacetate (*S*)-7 (with the residual 26% as diacetate) alongside superb stereocontrol. Acetonitrile (MeCN) serves as the vital medium, compatible not just for this enzymatic phase but also for the ensuing iodination and cyclization to yield tetrahydrofuran (3*S*,5*R*)-**8**; thus, lipase removal via filtration enables direct iodine introduction sans interim isolation. Full depletion of diol **6** remains imperative, given its reactivity with iodine – unlike the inert diacetate.

Resolving chiral amines represents a compelling research domain owing to the broad utility of these enantiomerically pure entities [125]. Figure 4.10d shows a pertinent case: the dynamic kinetic resolution of (*rac*)-**9**, yielding an enantiopure precursor for Rasagiline (Azilect®), a permanent monoamine oxidase-B blocker employed either as standalone therapy for initial Parkinson's symptoms or as supplementary treatment in progressive stages [126]. Though a distinct lipase drives this transformation, outcomes prove outstanding – a 0.5-L reaction scale delivers (*R*)-**9** with superior yield and enantiomeric purity.

## 4.3 Enzymes in Biphasic Systems

In another possible approach, the use of biphasic systems for biocatalysis (schematized in Figure 4.11) can be advantageous, in order to solve some issues associated with aqueous media, such as the limited solubility and stability of reagents, the occurrence of side reactions triggered by water, possible enzyme inhibition by either substrates or products, and the lengthy and expensive nature of downstream processing [9].

In these systems, the water-soluble enzymes remain within the aqueous phase or at the interface between the two liquids, while hydrophobic reactants are dissolved in the organic phase, serving either as a reservoir for substrates or as a means to extract products. This allows for higher substrate concentrations, potentially boosting productivity. Furthermore, reactions that are thermodynamically unfavorable in water, such as ester formation, could be facilitated in biphasic setups by preventing hydrolysis. Nevertheless, biphasic systems are still not widely employed, largely because most enzymes (lipases are an exception, as commented before) lose their activity when exposed to organic solvents, probably caused by two main factors, which have been termed "interfacial toxicity" (where enzymes are destabilized at the interface), and "molecular toxicity," caused by solvent molecules penetrating the aqueous phase and

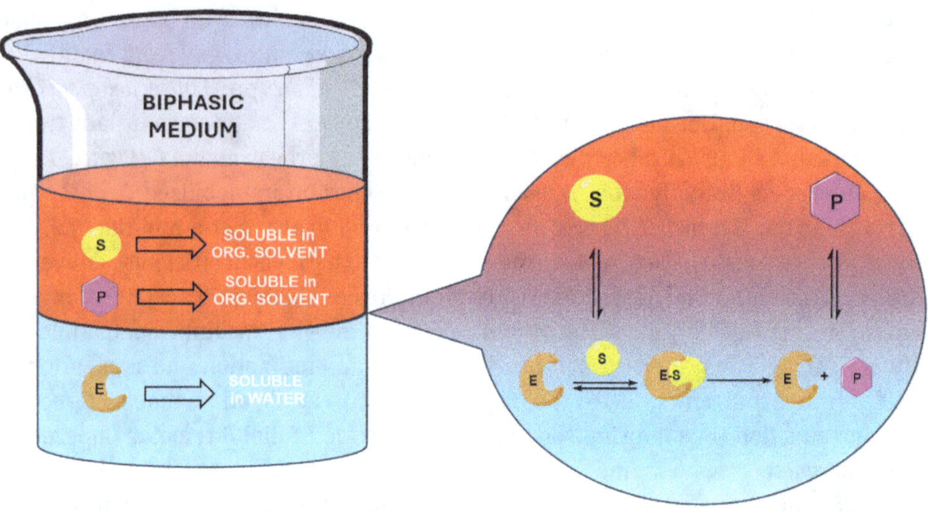


**Figure 4.11:** Enzymatic catalysis in biphasic aqueous-organic solvents.

interacting with proteins [127]. Ultimately, how an enzyme responds to organic solvents is determined by the specific properties of the solvents used, mainly log *P*, as noted before.

Most examples of biotransformations in biphasic media have been reported using hydrolases, as depicted in Figure 4.12.

An archetypal example was developed by the Swiss company Hoffmann-La Roche in the mid-1990s (Figure 4.12a), employing the protease subtilisin in stirred-tank reactors and biphasic toluene-buffer systems to carry out stereoselective hydrolysis of racemic esters such as **11** with excellent enantioselectivity [128]. In this process, the unreacted ester enantiomer (*S*)-**11** remains in the organic phase, while the acid is easily separated as a carboxylate (*R*)-**12** in the basic aqueous phase. This method enabled the large-scale synthesis of synthons used to produce compounds, such as Ro 31-9,790, a synthetic hydroxamic acid-based inhibitor used to inhibit metalloproteinases, particularly matrix metalloproteinases (MMPs) and other zinc-dependent enzymes, such as L-selectin and tumor necrosis factor-alpha from the surfaces of leukocytes [129]. Interestingly, the enzymatic resolution (45% conversion, 99% enantiomeric excess, 200 kg scale) can be improved, as the unreacted (*S*)-**11**, after separation of the toluene phase, was subsequently racemized (by heating with a catalytic amount of sodium ethoxide) and reintroduced into the reaction cycle, thus allowing an increase in the yield from 45% to 87%.

A 2019 patent by Daiichi Sankyo Co. Ltd. [121] presents a newer instance, as shown in Figure 4.12b, where kinetic resolution of (*rac*)-1-((((1,1,1,3,3,3-hexafluoropropan-2-yl) oxy)carbonyl)oxy)ethyl isobutyrate ((*rac*)-**14**) occurred through lipase-driven hydrolysis within a biphasic buffer/organic solvent setup. The method selectively liberated alcohol

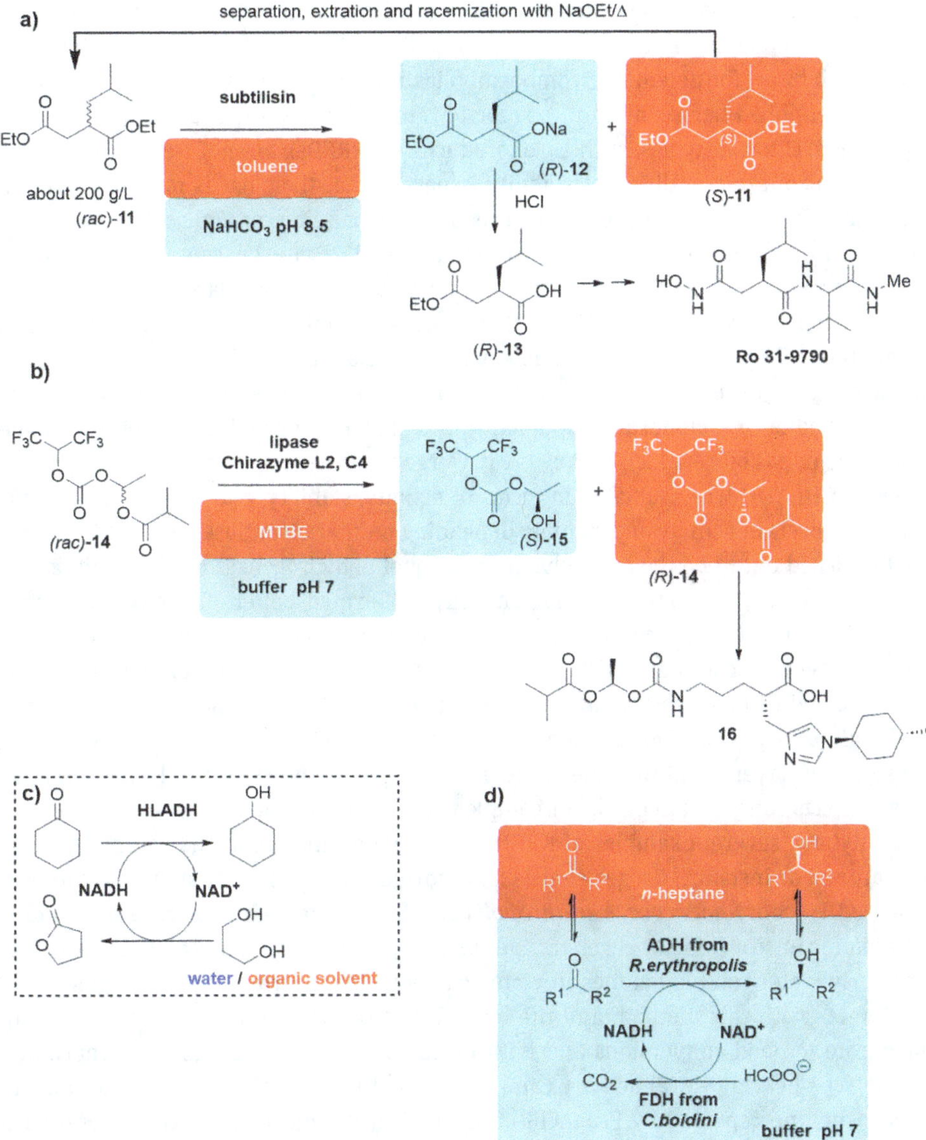


**Figure 4.12:** Some biotransformations in biphasic media.

(*S*)-**15**, isolating the target enantiopure ester (*R*)-**14**. Optimization involved screening commercial lipases (*C. antarctica*, *Candida rugosa*, and *Thermomyces lanuginosus*), cosolvents (acetonitrile, ethers, saturated/aromatic hydrocarbons, and alcohols), and conditions (0–80 °C temperatures, 1–120 h of durations). Optimal performance came from Chirazyme L2, C4 (*C. antarctica* lipase preparation) in a pH 7.0 buffer/methyl tert-butyl ether (MTBE) blend (5:1 v/v) at 20 °C, achieving ~40% conversions and >99% enantiomeric excess (ee). Ester (*R*)-**14** then reacted with a chiral amide to form carbamate **16**, a prodrug candidate inhibiting carboxypeptidase U (or TAFIa, thrombin-activatable fibrinolysis inhibitor) for potential thromboembolic disorder therapies [130].

As mentioned before, most of the reactions in biphasic systems are conducted with hydrolases. Other enzymes, such as ketoreductases (KREDs), are not frequently used in biphasic media due to different stability issues, so that their use is more common in aqueous buffer containing some amounts of water-miscible organic solvents, to help solubilizing hydrophobic carbonyl compounds [131, 132] . In this sense, a recent study has reported the effect of organic solvents on one of the most classical KREDs, horse liver alcohol dehydrogenase (HLADH) through a combination of experimental and computational approaches [127]. Thus, a variety of classic organic solvents with differing levels of hydrophobicity and functional properties were tested in the reduction of cyclohexanone to cyclohexanol, coupled to the oxidation of 1,4-butanediol to gamma-butyrolactone (Figure 4.12c), in 50:50 water/organic solvent (vol/vol). The catalytic behavior of HLADH was evaluated in terms of activity, stability, and selectivity, showing different levels of activity and stability depending on the solvent used. For this reaction, MTBE turned out to be the best solvent for alcohol dehydrogenase (ADH) catalysis and showed that hydrocarbon solvents generally performed better. The enzyme performance was further analyzed using molecular dynamics (MD) simulations to examine: (i) the enzyme's positioning within the biphasic system, (ii) the distribution of organic solvents, and (iii) changes in enzyme conformation. Both "molecular toxicity" and "interfacial toxicity" were found to induce conformational changes in the enzyme, disrupt its solvation layers, and in some cases, sterically block the substrate-binding tunnel. These effects can result in partial or complete loss of enzyme activity.

Anyhow, several examples of KREDs working on biphasic media can be found in literature. For instance, Gröger et al. [133] reported the use of fairly stable oxidoreductases (ADH from *Rhodococcus erythropolis* and formate dehydrogenase from *Candida boidinii* for in situ cofactor regeneration) useful for asymmetric biocatalytic reduction of poorly water-soluble ketones in water/*n*-hexane, at substrate concentrations of 10 −200 mM. With this method, and the optically active (*S*)-alcohols were formed with moderate to good conversions and with up to >99% ee (Figure 4.12d). Furthermore, whole cells bearing KRED activity can be also used in biphasic media; an example of this is the one reported by Fragnelli et al. [134] in the bioreduction of different 1,2-diaryl-1,2-ethanediones using cells from *Pichia glucozyma* yeast in water/*n*-hexane, leading to (*S*)-1,2-diaryl-2-hydroxyethanones (benzoins).

## 4.4 Biocatalysis in Neoteric Solvents

The term "neoteric solvents" was originally coined by K.R. Seddon in the late 1990s of the last century, to refer to innovative solvents with exceptional new properties that "pioneer new frontiers" and could offer significant potential for industrial applications [135, 136]. In that moment, this term was basically applied to RTILs [137] and SCFs [138], which were starting to be broadly used. Nowadays, this expression also includes DESs and fluorous solvents, which will be also commented.

### 4.4.1 Biocatalysis in Room-Temperature Ionic Liquids (RTILs)

RTILs are salts that exist in liquid form, typically consisting of an organic cation and an inorganic anion, with the defining characteristic of melting below 100 °C, so that they create a liquid phase at ambient temperature [137, 139]. Some of the most usual cations and anions used to furnish RTILs are shown in Figure 4.13.

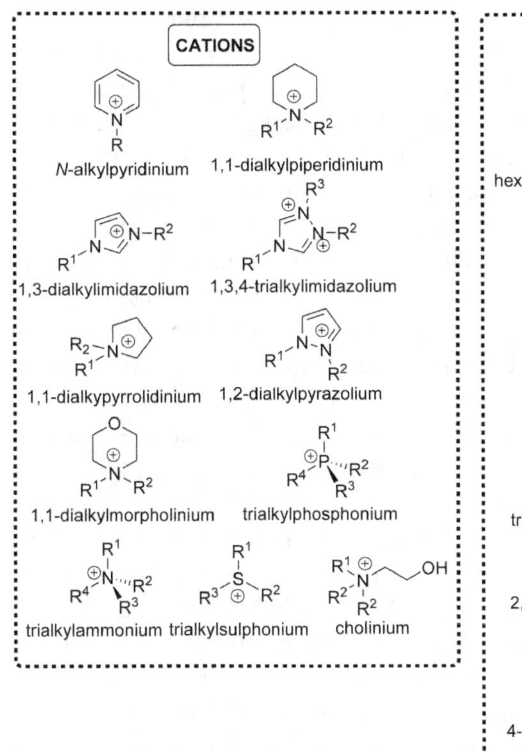


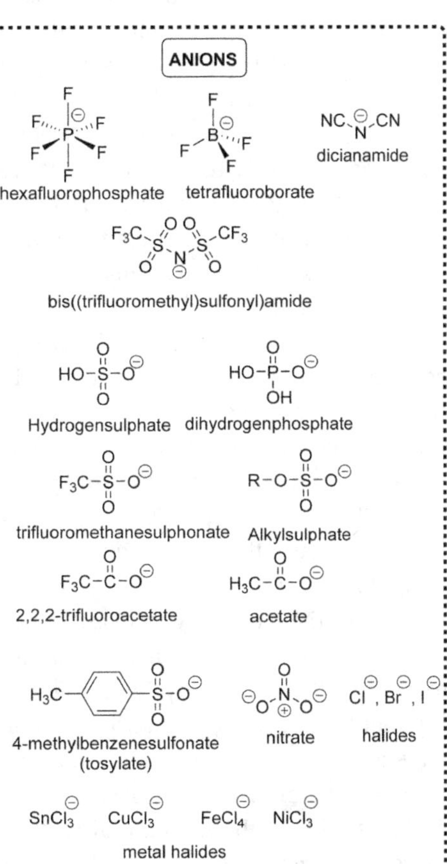


**Figure 4.13:** Some typical cations and anions used to generate RTILs.

As ionic rather than molecular compounds, RTILs possess a distinctive set of properties, such as low vapor pressure, high thermal stability, and significant solvent capacity, all of which contribute to their reputation as environmentally friendly solvents since they do not evaporate [139, 140], not leading to volatile organic compounds. Consequently, RTILs have usually been considered "green solvents" due to their ability to be recovered from the reaction medium and subsequently recycled through straightforward methods [141, 142]. Additionally, many of their physical and chemical properties, including polarity, solubility, miscibility, viscosity, and melting points, can be customized through the appropriate selection of cations and anions. Nevertheless, this cataloguing of RTILs as green solvents has been controversial since the very moment in which their first uses were reported [143]; in fact, RTILs pose several environmental challenges regarding their synthesis (mostly derived from oil), toxicity, and biodegradability, which cast doubt on their ecological benefits. A recent report by de Jesus and Maciel Filho [144] has assessed several processes involved in the synthesis, recovery, recycling, toxicity, and biodegradability of different RTILs, providing a comparative analysis with both fossil-based solvents and organic-based green solvents. By constructing synthesis trees, these authors found that all the RTILs assessed exhibit problematic stages, primarily due to the use of volatile compounds containing elements such as carbon, nitrogen, sulfur, and halogens. One of the most significant issues relates to the toxicity and biodegradability of RTILs, as the majority currently in use are either toxic or poorly/nonbiodegradable. Given the nearly limitless combinations of RTILs that can be synthesized, future progress in this field may depend on computational modeling combined with life cycle assessments. At the end, these authors literally stated: "Having analysed these four issues in this review, we can already answer the question: 'Are ionic liquids eco-friendly?' If we consider the data presented and mainly the 12 principles of green chemistry, current ionic liquids are not eco-friendly compounds" [144].

In any case, the first reports on enzyme catalysis in RTILs date back from 2000 [145, 146]. Since then, a wide variety of enzymes have been used in RTILs [147–149], because of their capacity to dissolve both hydrophilic and hydrophobic compounds, depending on the particular hydrophilic/hydrophobic nature of the ionic components of RTILs [150]. Particularly, hydrophobic RTILs have proven more effective in supporting biocatalytic activity than hydrophilic ones, as albeit being hydrophobic, RTILs have a hygroscopic nature that allows the co-immobilization of the essential water layer around enzymes [114, 150]. In addition, the strongly hydrophobic environment of RTILs inhibits the increase of water molecules' kinetic energy with temperature, thereby enhancing enzyme stability against thermal deactivation [151]. This allows for reactions to occur at temperatures higher than those optimal for enzymes in aqueous media, which in turn improves substrate solubility and diffusion; an impressive example was reported by Ramos-Martín et al. [152] (Figure 4.14), which showed how it was possible to work with thermophilic enzymes in RTILs at temperatures as high as 120 °C, in the ethanolysis of racemic *O*-butyryl mandelic acid rac-**17** catalyzed by the

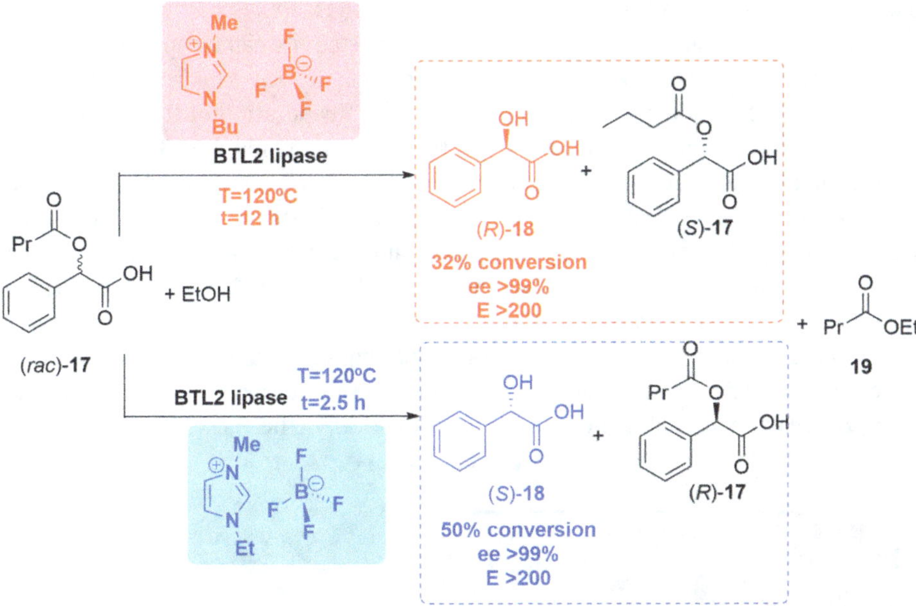


**Figure 4.14:** Kinetic resolution of racemic *O*-butyryl mandelic acid in RTILs at 120 °C, as reported by Ramos-Martín et al. [152].

lipase from *Geobacillus thermocathenulatus*, with excellent yield and enantioselectivity, showing also how the enzymatic stereobias leading to enantiopure compounds 17 and 18 was influenced by the nature of the cationic component of the RTIL.

It will be nearly impossible to cover all the examples of biocatalysis in RTILs, so we recommend the recent and excellent textbook edited by Prof. Pedro Lozano [97] for having a detailed and comprehensive panoramic view of the enormous possibilities of biocatalysis in RTILs, as well as the very recent review by Dupont et al. [149], which quote, respectively, 110 references (lipases), 26 references (proteases), and 23 references (redox enzymes) dealing with biocatalysis in RTILs.

### 4.4.2 Biocatalysis in Supercritical Fluids (SCFs)

An SCF is a substance that exists above its critical temperature ($T_c$) and pressure ($P_c$), although below the pressure required to condense to a solid, exhibiting properties of both liquids and gases [153]. These unique properties allow for greater solubility of organic compounds and faster diffusion than in conventional liquids [154]. Furthermore, as SCFs are nontoxic, cost-effective, and nonflammable, they have also been included in the category of green solvents [155–158]. One of the most notable properties of SCFs is the ability to adjust their density and, therefore, their solvent capacity by slightly modifying pressure and temperature conditions. This is particularly useful in biocatalysis, where substrate solubility and enzyme stability can be optimized by controlling the SCF environment [150, 159, 160]. Additionally, the use of SCFs allows the performance of simple downstream processes, via evaporation of fluid and the release of gas upon depressurization. Among the most commonly used SCFs, supercritical carbon dioxide ($scCO_2$) stands out for being nontoxic, economical, and easy to handle due to its low critical temperature (31 °C) and pressure (73.8 atm) [161, 162] (Figure 4.15, left). In fact, $scCO_2$ behaves like a nonpolar solvent, meaning it is ideal for solubilizing nonpolar substrates, something that is problematic in traditional aqueous media. In addition to their adjustable solubility, SCFs exhibit high diffusion rates and low viscosities compared to traditional liquids, facilitating the mass transport of substrates to the enzyme's catalytic site. These factors can accelerate enzyme-catalyzed reactions, making the process more kinetically efficient.

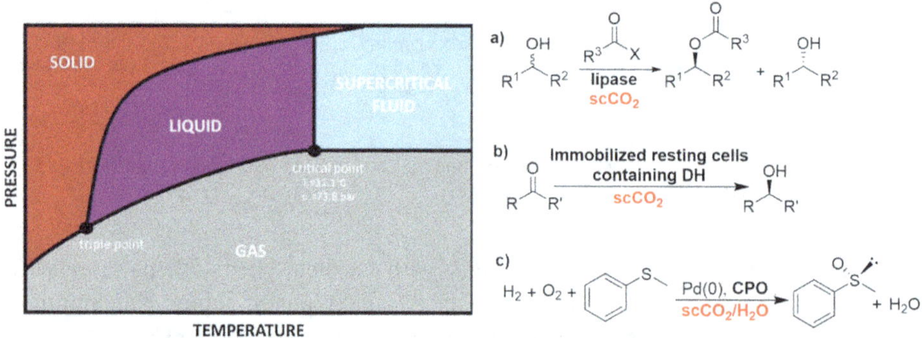


**Figure 4.15:** Left: Phase diagram of $CO_2$, reprinted from Knez et al. [158]. OPEN ACCESS, permission granted. Right: Schematic representation of some biocatalyzed reaction in $scCO_2$.

First reports of biocatalysis in SCFs date back to around 40 years ago [163–165]. Since then, many examples have been reported [118, 150, 159, 166], mainly using hydrolases (usually for acyl transfer processes such as esterifications or transesterifications, Figure 4.15a) or oxidoreductases (generally for ketone reductions, Figure 4.15b). When using hydrolases, as the enzyme stability can be affected by the SCF, most of them have been used as immobilized derivatives [118, 167]. In this area, many examples illustrating archetypal lipase-catalyzed processes in $scCO_2$ have been reported, including esterifications [168–173], transesterifications [174–178], interesterifications [179, 180], acidolysis [181], or polymer synthesis [182–185]. Matsuda and Hoanha have recently reported a review dealing with the use of lipase-catalyzed reaction in $scCO_2$, covering several examples of kinetic resolutions of secondary alcohols via enantioselective acylation (trans) procedures, both in batch and flow reactors [186]. Interestingly, these authors remarked how the enantioselectivity of the lipase-assisted discrimination between both enantiomers of a racemic substrate can be modulated by adjusting the pressure and temperature (and, consequently, the density) of $CO_2$; this effect was attributed to changes in the interaction of $CO_2$ with the enzyme surface (by formation of carbamates with superficial free amino groups), $CO_2$ adsorption on the enzyme, and/or incorporation of $CO_2$ in the substrate-binding pocket of the enzyme.

Additionally, other promiscuous reactions, such as aza-Michael additions [187], epoxidations [188], or even multicomponent reactions [189] have been reported with lipases in SCFs. Regarding the use of redox enzymes in SCFs, there are not many examples reported in literature, most of them describing bioreductions of ketones to furnish alcohols (Figure 4.15b). In this sense, immobilized resting cells from *Geotrichum candidum* were found to be active in $scCO_2$ at 10 MPa for the stereoselective bioreduction of prochiral ketones, leading to enantioenriched secondary alcohols [190–193]. Also whole cells from *Candida valida* [194] or even baker's yeast [195] have proven to be active in $scCO_2$ for catalyzing bioreductions. Also, HLADH was reported to use a synthetic fluorinated nicotinamide adenine dinucleotide (FNAD) as a soluble cofactor in fluorous solvents and liquid $CO_2$ [196]. On the other hand, stereoselective oxidation of prochiral thioethers to furnish enantioenriched sulfoxides was reported out by cascade reaction of Pd(0) catalyzed formation of $H_2O_2$ and enzymatic oxidation using chloroperoxidase from *Caldariomyces fumago* in $scCO_2/H_2O$ biphasic medium, as depicted in Figure 4.15c [197]. Finally, a very interesting and recent report from Queiroz et al. [198] has shown how the lyophilized whole cells and crude enzymatic extracts from the fungus *Stemphylium lycopersici*, immobilized within rigid polyurethane foam, were useful in the kinetic resolution of racemic 1-phenylethylamine via transamination in aqueous media: interestingly, when the biocatalysts were subjected to high-pressure $CO_2$ treatment, conversions of 49% (with 99% enantiomeric excess ketones for the (*R*)-enantiomer.) were obtained for the immobilized lyophilized fungus, demonstrating reuse for up to 20 cycles and over 50% recovery for the fungus. Furthermore, the immobilized lyophilized fungus was also able to transaminate some other secondary amines, leading to good conversions and stereoselectivity.

As a conclusion, and despite the many potential benefits of biocatalysis in SCFs, there are also significant limitations that must be overcome before this technology can be widely implemented at industrial scale. One of the main limitations is enzyme stability under supercritical conditions, which can vary significantly depending on the nature of the enzyme and the process conditions, demanding enzyme immobilization [167]. Moreover, while $scCO_2$ is a highly useful solvent, it is not suitable for all reactions due to its low polarity, which limits its ability to dissolve certain polar substrates. In these cases, the addition of cosolvents or modification of the SCF environment may be necessary [166]. Finally, the costs associated with producing and maintaining equipment to operate under supercritical conditions remain a barrier to the widespread adoption of this technology at industrial scale [159]. Therefore, optimizing operational costs and improving enzyme stability will be key to the future expansion of this technology.

### 4.4.3 Biocatalysis in Liquid $CO_2$ and $CO_2$ Expanded Solvents

Liquid $CO_2$ was employed for the first time in 2015 for immobilized CALB-catalyzed reactions, showing notable advantages, including lower pressure and temperature demands when compared to $scCO_2$, proving to be more efficient compared to traditional organic such as *n*-hexane and diisopropyl ether for transesterification reactions [199]. Same authors also reported the behavior of the immobilized lipase from *Burkholderia cepacia* (BCL) in liquid $CO_2$ as the reaction medium [200]. Typically, these two lipases face challenges when dealing with secondary alcohols that have bulky substituents flanking the chiral center, like 1-phenyl-1-dodecanol or 2-methyl-1-phenyl-1-propanol. However, these reactions were successfully carried out in liquid $CO_2$ (Figure 4.16).

OH
Ph R
**Novozym 435**
Vinyl acetate
*n*-hexane
low conversion for **R** > Et

**R** = Me, Et, Pr, Bu,pentyl, hexyl, undecyl, *iso*Pr, cyclopropyl

OH
Ph R
**Novozym 435**
Vinyl acetate
Liquid $CO_2$
OAc
Ph R
+
OH
Ph R
ee values up to > 99%

**Figure 4.16:** Kinetic resolution of racemic secondary alcohols in *n*-hexane and liquid $CO_2$, as reported by Hoang and Matsuda [199, 200].

Furthermore, flow processes incorporating liquid $CO_2$ with immobilized Novozym 435 have also been successfully reported [186].

On the other hand, $CO_2$-expanded liquids (CXLs) are solvents that have been modified by the addition of carbon dioxide under pressure; then, when $CO_2$ is dissolved in a liquid solvent, it changes the properties of that solvent, making it less viscous and more effective at dissolving certain substances [201]. This technique is often used in various applications, such as extraction processes, where it can help to selectively extract compounds from natural products or in the formulation of certain products [202]. CXLs are increasingly recognized for their versatility and sustainability, especially when the expanded organic solvent is derived from biomass (green bio-solvents [89, 90, 99]). A notable example is $CO_2$-expanded 2-MeTHF, which has been promoted as a more sustainable solvent with properties that can be fine-tuned [203–205]. By dissolving $CO_2$ (up to 10 bar) into 2-MeTHF, key solvent properties such as polarity and hydrophobicity can be regulated. This expanded solvent system has shown to enhance biocatalysis, particularly for the kinetic resolution of larger substrates that typically exhibit low or no activity when 2-MeTHF is used alone.

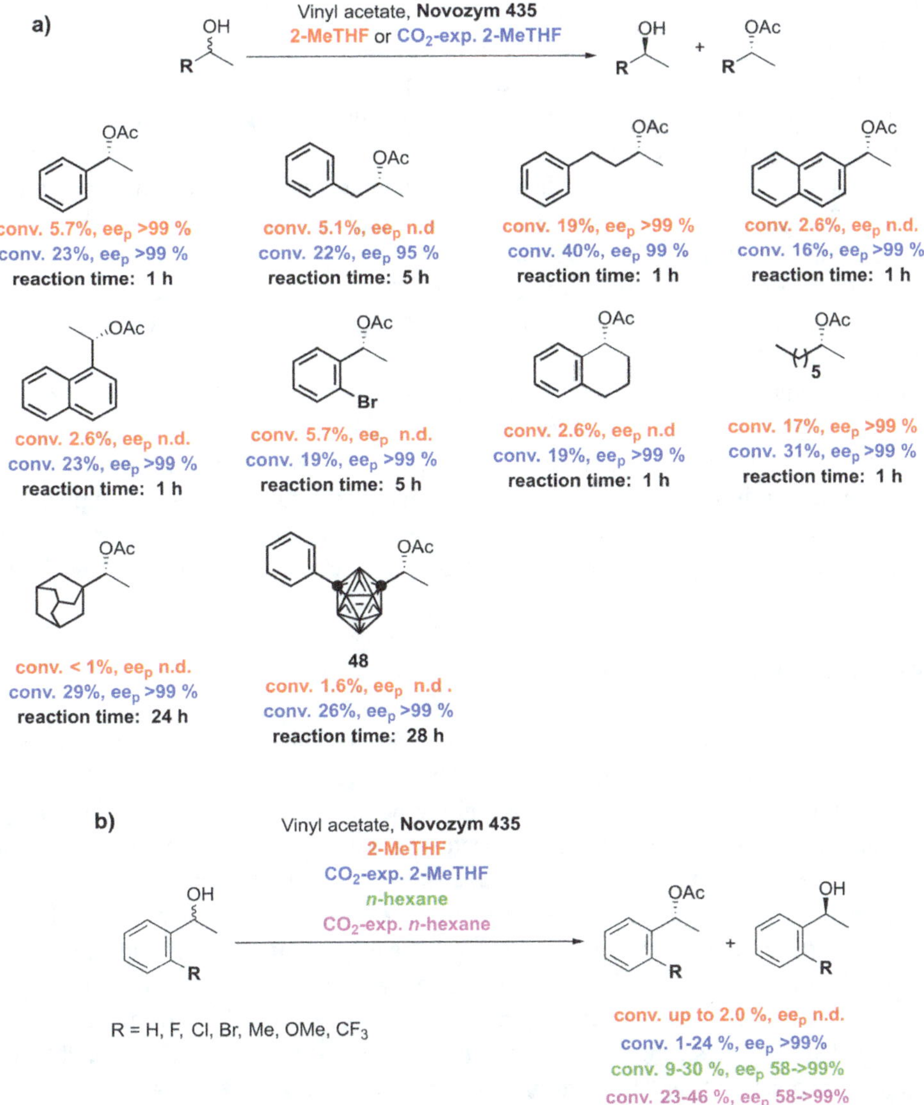


**Figure 4.17:** Kinetic resolution of secondary alcohols (methyl carbinols) in different $CO_2$-expanded solvents.

T. Matsuda's research team has investigated this dual-solvent system for lipase-mediated transesterification reactions involving diverse sterically hindered secondary alcohols, demonstrating outstanding enantioselectivity [206–209]. The presence of $CO_2$ in the solvent appears to make the enzyme's structure more flexible, enabling it to process larger substrates more efficiently. As shown in Figure 17a, the $CO_2$-expanded

2-MeTHF significantly improved conversion rates compared to reactions in the pure biosolvent, while maintaining the enzyme's high selectivity.

These reactions are also scalable, they can proceed at gram scale [206] . These authors put forward three possible reasons to rationalize the significant conversion observed during the kinetic resolution of sterically hindered substrates using CALB, which could be caused by (i) formation of carbamates between $CO_2$ and free amine groups like lysine on the enzyme's surface, potentially leading to structural shifts that enhance activity; (ii) increased flexibility or reduced rigidity of CALB when immersed in a $CO_2$-rich environment, which may allow for improved interaction with larger substrates; and/or (iii) transport properties and physical behavior of $CO_2$-expanded 2-MeTHF could be more favorable than those of the solvent on its own, promoting higher reactivity. In another paper from this group [208], it was found that while $CO_2$-expanded 2-MeTHF worked well for the kinetic resolution of several ortho-substituted 1-phenylethanols by CALB, the use of $CO_2$-expanded hexane yielded even better results, as shown in Figure 17b. Recently, this same group has reported the application of other of $CO_2$-expanded biosolvents, such as diethyl carbonate, γ-valerolactone, *p*-cymene, and (+)- or (–)-limonene), in the kinetic resolution of racemic-1-tetralols [209].

### 4.4.4 Biocatalysis in Deep Eutectic Solvents

DESs are a class of neoteric solvents formed by the combination of (at least) two compounds, a hydrogen bond donor (HBD) and a hydrogen bond acceptor (HBA), usually resulting in a eutectic mixture with a melting point significantly lower than that of the individual components [210, 211]. Strictly speaking, from a thermodynamic point of view, the most accurate definition of DES was reported by Abranches and Coutinho [210], which literally stated: "A eutectic solvent is a eutectic-type system that is a liquid at a given desired temperature where at least one of its components would, otherwise, be a solid unfit to be applied as a solvent. A deep eutectic solvent is a eutectic solvent whose components present enthalpic-driven negative deviations from thermodynamic ideality." We strongly recommend this seminal paper, which, as denoted by its rather provocative title ("Everything You Wanted to Know about Deep Eutectic Solvents but Were Afraid to Be Told"), clearly established the thermodynamic bases about the real nature of DESs, and which components should be chosen as HBDs and HBAs.

From an historical point of view, the term DES was originally reported by Abbot et al. in 2003, describing the mixture of urea and several quaternary ammonium salts [212]. Although they share several similarities with RTILs, such as high thermostability, high viscosity, low volatility, low vapor pressure, nonflammability, and tunable polarity (hydrophobic or hydrophilic), there are substantial differences between RTILs and DESs, the pivotal one being that RTILs are pure compounds, whereas DESs are mixtures of two components, an HBD and an HBA [213]. Depending on their compositions, there are five types of DESs [210]:

- Type I mixtures arise from combining a quaternary ammonium salt with a metal chloride, such as choline chloride (ChCl) paired with $ZnCl_2$. Within these type I DESs, surplus chloride ions from the ammonium salt form complexes with metal cations, yielding metal-chloride stoichiometries absent in the neat metal salt. This chloride relocation phenomenon accounts for the distinctive characteristics of such blends, which find common use in metal recovery and reuse, electrodeposition, and catalytic processes [214, 215].
- Type II, comprising a quaternary ammonium salt and a metal chloride hydrate (e.g., ChCl + $ZnCl_2.6H_2O$). Type II DESs are similar to type I but differ in that they incorporate hydrated metal chlorides instead of anhydrous ones. These solvents are often used for metal extraction and separation processes. The water content in type II DESs plays a crucial role in determining the solubility of the metal species [216].
- Type III, formed by a quaternary ammonium salt (HBA) and a molecular HBD, such as urea, glycerol, or carboxylic acids acting as HBD, as shown in Figure 18. Type III DESs are the most commonly studied and widely used [215] . These DESs have found applications in organic synthesis, biocatalysis, and pharmaceutical production, as they can solubilize a plethora of polar and nonpolar compounds [217, 218].
- Type IV, made of a metal chloride hydrate, is another compound acting as HBD ($ZnCl_2$ + urea). This type of DES is often used in the processing of inorganic materials, metal refining, and catalysis. The presence of the metal salt imparts unique reactivity to the solvent system, which is useful in specialized industrial processes [215, 216].
- Type V, the last reported type, also termed hydrophobic DESs, are formed by the interaction of only nonionic molecular HBAs and HBDs (Figure 18). Type V DESs are unique because, while types I–IV are based on the interaction of ionic species. Type V DESs are neutral in composition but still form highly functional liquid systems through hydrogen bonding. This neutral characteristic differentiates them from the conventional ion-based DESs and expands the scope of possible applications, particularly where ionic behavior might be undesirable [219–222].

Outside this classical categorization of DESs, the term NADESs (natural deep eutectic solvents) is used when DESs are formed from entirely natural components, using components often derived from plant-based sources such as sugars, amino acids, and organic acids [223]. NADESs are particularly appealing for pharmaceutical and food industries due to their biocompatibility and nontoxicity [224]. NADESs are also ideal for applications involving sensitive biomolecules, such as enzymes and DNA, because they can maintain enzyme activity and provide a protective medium [225].

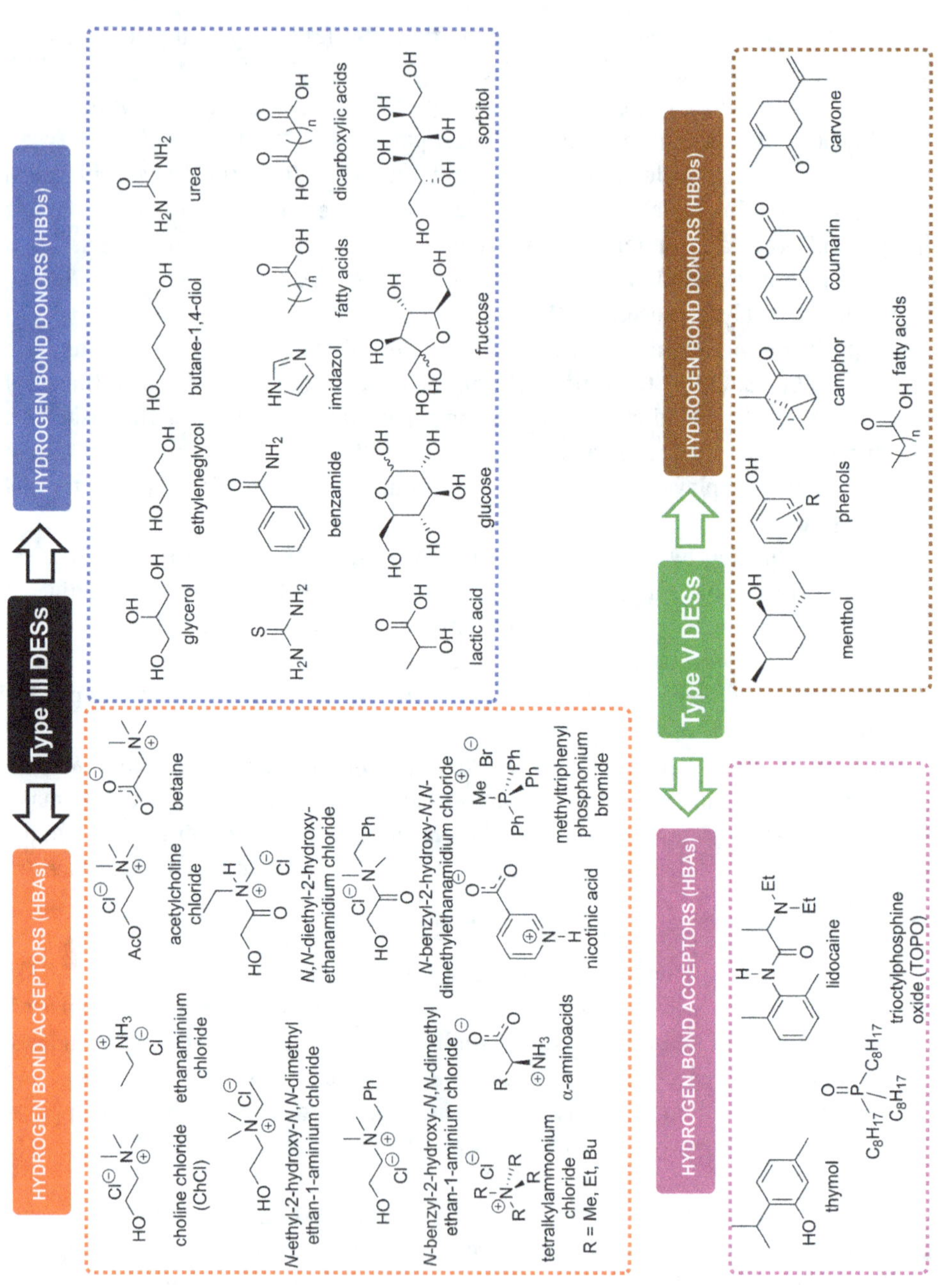


**Figure 4.18:** Components of the most useful DES types III and V.

DESs are considered an environmentally friendly alternative to traditional organic solvents due to their low toxicity, biodegradability, and potential for being synthesized from renewable resources, so that they have been considered "the solvents for the twenty-first century" [223, 226, 227]. Despite their reputation as environmentally friendly media, concerns have been raised about the actual "greenness" of DESs, stemming from the sustainability and natural origin of commonly used precursors, as well as from disputed aspects of their stability, biodegradability, and toxicity [210, 228].

Regardless, applying DESs as biocatalytic reaction environments offers multiple benefits, arising from their noted superior thermal resilience, elevated viscosity, negligible volatility, minimal vapor pressure, noninflammable nature, adjustable polarity (ranging from hydrophobic to hydrophilic), and simpler synthesis relative to RTILs. Consequently, enzymatic processes in DESs have surged in popularity in recent years [118, 225, 229–233]. As previously commented for pure organic solvents, the water activity ($a_W$) of the reaction environment plays a crucial role, as it reflects the extent of structural water associated with the enzyme's surface; consequently, adding water to create DES-water systems is a common practice to maintain adequate enzyme hydration and to reduce the viscosity of these solvents [234–238]. Moreover, DESs polarity plays a pivotal role and can be fine-tuned through alterations in constituent makeup or stoichiometric proportions, thereby enhancing the dissolution of nonpolar substrates and products.

In fact, there are plenty of publications dealing with the application of these solvents in biocatalysis, since the pioneering paper from Gorke et al. in 2008 reporting the use of an immobilized lipase for transesterifications in several DESs [239]. Once again, initial studies were carried out using hydrolases, based on their stability and versatility [239, 240], but nowadays it is broadly recognized that DESs are indeed very useful solvents for many other enzyme types, including redox enzymes [241, 242], lyases [243], and transferases [244]. It would be nearly impossible to cover all the published biotransformations in DESs, very recently, some detailed book chapters dealing with biocatalysis in DESs have been published [118, 245], illustrating how these different types of enzymes are suitable for being employed in DESs.

On the other hand, a wide variety of outcomes has emerged when trying to rationalize on a molecular scale the behavior of enzymes in DESs, due to differences in biocatalysts, reaction conditions, and system setups. Thus, if the relationship between DES characteristics and enzyme structure-function was fully understood, it would be possible to design optimized enzyme-specific solvents for each biocatalytic process. Computational approaches, notably MD simulations, serve as potent adjunct methods for probing molecular interactions within DESs and enzyme-solvent dynamics. Notably, MD analyses have yielded vital understanding – especially for lipases – on protein stabilization and activation mechanisms in DES environments. Early investigations again targeted established lipases, with Monhemi et al. [246] delivering the inaugural MD study on a lipase in 2014. This study, as well as some subsequent ones [247, 248], showed that lipase resilience in DESs stems from favorable hydrogen bonds established between DES constituents and the enzyme's exterior. Furthermore, the presence of some water molecules in the DESs allowed for enhanced enzyme activation by increasing mobility of the lid covering the active site [249]. To further examine water's role in enzyme activity, Qiao et al. [250] assessed lipase solvation in DESs revealing that water plays a critical role in enzyme active site behavior. Similarly, MD simulations have been also extended to explore the effects of DESs on redox biocatalysis using purified ADH in water-containing DESs, underscoring the influence of water content on activity [251–253], related to enzyme flexibility. Additional MD-based studies have addressed other enzymes, such as laccase and horseradish peroxidase, demonstrating enhanced enzyme stability in DESs through hydrogen bonding with enzyme residues [254].

## 4.5 Conclusions and Future Outlook

The aim of this chapter has been to prove that using enzymes in nonconventional media, such as pure organic solvents or neoteric solvents (RTILs, SCFs, ionic liquids, and DESs), offers multiple advantages that enhance both the efficiency and versatility of biocatalytic processes. First, these media can improve enzyme stability by reducing the denaturation risks that water-sensitive enzymes face in aqueous environments. Additionally, nonconventional solvents allow for greater control over reaction selectivity, enabling enzymes to catalyze reactions that may be less feasible in water, such as those involving hydrophobic substrates. This is especially valuable for producing complex or high-value compounds in pharmaceuticals, fine chemicals, and biofuels.

Another advantage is the ability to tailor the properties of these media to improve the solubility of specific substrates, particularly hydrophobic ones that have limited solubility in water. By adjusting the polarity, viscosity, and other physicochemical properties of solvents like DESs or RTILs, it's possible to create highly customized environments that optimize enzyme performance for desired reactions. Moreover, nonconventional media can shift reaction equilibria in favor of product formation, reducing side reactions and enhancing yields, particularly in synthetic processes like esterification or transesterification.

Finally, many nonaqueous media offer the benefit of recycling and reusability, making them more sustainable and reducing overall operational costs. This aligns well with the principles of green chemistry, where minimizing waste and maximizing resource efficiency are crucial goals. Overall, the use of enzymes in nonconventional media expands the range of feasible biocatalytic applications and enables more sustainable, high-performance industrial processes.

Future advances in medium engineering will be driven by integrating computational modeling, protein engineering, and green chemistry to design tailor-made solvent environments that maximize enzyme performance and sustainability. The development of novel bio-based and neoteric solvents, coupled with molecular simulations to predict enzyme-solvent interactions, will enable highly selective, efficient, and robust biocatalytic processes. Addressing challenges in enzyme stability and scalability in nonconventional media will expand industrial applications, fostering more sustainable manufacturing in pharmaceuticals, chemicals, and bioenergy, and paving the way for next-generation bioprocesses with reduced environmental impact.

## References


[1] Lewis RD, France SP, Martinez CA. Emerging technologies for biocatalysis in the pharmaceutical industry. ACS Catal;2023:5571–5577.

[2] Gallou F, Gröger H, Lipshutz BH. Status check: Biocatalysis; it's use with and without chemocatalysis. How does the fine chemicals industry view this area?. Green Chem 2023.

[3] Buller R, Lutz S, Kazlauskas RJ, Snajdrova R, Moore JC, Bornscheuer UT. From nature to industry: Harnessing enzymes for biocatalysis. Science 2023;382:eadh8615.

[4] Dean SRA. Green chemistry, biocatalysis, and the chemical industry of the future. ChemSusChem 2022;e202102628.

[5] Alcántara AR, Domínguez de María P, Littlechild JA, Schürmann M, Sheldon RA, Wohlgemuth R. Biocatalysis as key to sustainable industrial chemistry. ChemSusChem 2022.

[6] Kleinbeck F, Mahut M, Schlama T. Industrial processes using Biocatalysts, In de Gonzalo G, Lavandera I, editors. Biocatalysis for Practitioners: Techniques, Reactions and Applications 2021, Weinheim, Germany: WILEY-VCH, GmbH; p. 427–465.

[7] Lozano P, García-Verdugo E. From green to circular chemistry paved by biocatalysis. Green Chem 2023;25:7041–7057.

[8] Sheldon RA, Woodley JM. Role of biocatalysis in sustainable chemistry. Chem Rev 2018;118:801–838.

[9] Van Schie MMCH, Spöring JD, Bocola M, Domínguez de María P, Rother D. Applied biocatalysis beyond just buffers – From aqueous to unconventional media. Options and guidelines. Green Chem 2021;23:3191–3206.

[10] Holtmann D, Hollmann F. Is water the best solvent for biocatalysis?. Mol Cat 2022;517.

[11] Vermue MH, Tramper J. Biocatalysis in non-conventional media: Medium engineering aspects. Pure Appl Chem 1995;67:345–373.

[12] Carrea G, Riva S. Medium Engineering. Asymmetric Organic Synthesis with Enzymes 2008, John Wiley and Sons; p. 1–20.

[13] Katchalski-Katzir E. Medium and biocatalyst engineering. Prog Biotechnol;1992:3–9.

[14] Jumper J, Evans R, Pritzel A, Green T, Figurnov M, Ronneberger O, et al. Highly accurate protein structure prediction with AlphaFold. Nature 2021;596:583–589.

[15] Malhotra Y, John J, Yadav D, Sharma D, Rawal K, Mishra V, et al. Advancements in protein structure prediction: A comparative overview of AlphaFold and its derivatives. Comput Biol Med 2025:188.

[16] Graham F. Daily briefing: AlphaFold developers share Nobel Prize in Chemistry. Nature 2024.

[17] Abramson J, Adler J, Dunger J, Evans R, Green T, Pritzel A, et al. Accurate structure prediction of biomolecular interactions with AlphaFold 3. Nature 2024;630:493–500.

[18] Banach M, Kalinowska B, Konieczny L, Roterman I. Role of disulfide bonds in stabilizing the conformation of selected enzymes-An approach based on divergence entropy applied to the structure of hydrophobic core in proteins. Entropy 2016;18.

[19] Roth CM, Neal BL, Lenhoff AM. Van der Waals interactions involving proteins. Biophys J 1996;70:977–987.

[20] Pace CN, Fu H, Fryar KL, Landua J, Trevino SR, Schell D, et al. Contribution of hydrogen bonds to protein stability. Protein Science 2014;23:652–661.
[21] Camilloni C, Bonetti D, Morrone A, Giri R, Dobson CM, Brunori M, et al. Towards a structural biology of the hydrophobic effect in protein folding. Scientific Reports 2016:6.
[22] Godbey WT. Chapter 3 – Proteins, In Godbey WT, editor. Biotechnology and its Applications 2022, Academic Press; Second Edition. p. 47–72.
[23] Feher J. 2.3 – protein structure, In Feher J, editor. Quantitative Human Physiology 2017, Boston: Academic Press; Second Edition, p. 130–141.
[24] Brady L, Brzozowski AM, Derewenda ZS, Dodson E, Dodson G, Tolley S, et al. A serine protease triad forms the catalytic centre of a triacylglycerol lipase. Nature 1990;343:767–770.
[25] Fernández A, Sinanoğlu O. Denaturation of proteins in methanol/water mixtures. Biophysical Chemistry 1985;21:163–166.
[26] Yoshikawa H, Hirano A, Arakawa T, Shiraki K. Mechanistic insights into protein precipitation by alcohol. Int J Biol Macromol 2012;50:865–871.
[27] Herskovits TT, Gadegbeku B, Jaillet H. On the structural stability and solvent denaturation of proteins. I. Denaturation by the alcohols and glycols. J Biol Chem 1970;245:2588–2598.
[28] Bull HB, Breese K. Interaction of alcohols with proteins. Biopolymers 1978;17:2121–2131.
[29] Zaks A, Klibanov AM. Enzymatic catalysis in organic media at 100 °C. Science 1984;224:1249–1251.
[30] Kirchner G, Scollar MP, Klibanov AM. Resolution of racemic mixtures via lipase catalysis in organic-solvents. J Am Chem Soc 1985;107:7072–7076.
[31] Zaks A, Klibanov AM. Enzyme-catalyzed processes in organic-solvents. Proc Natl Acad Sci USA 1985;82:3192–3196.
[32] Zaks A, Klibanov AM. Substrate-specificity of enzymes in organic-solvents vs water is reversed. J Am Chem Soc 1986;108:2767–2768.
[33] Klibanov AM. Enzyme-catalyzed processes in organic-solvents. Ann NY Acad Sci 1987;501:129–29.
[34] Russell AJ, Klibanov AM. Inhibitor-induced enzyme activation in organic solvents. J Biol Chem 1988;263:11624–11626.
[35] Zaks A, Klibanov AM. The effect of water on enzyme action in organic media. J Biol Chem 1988;263:8017–8021.
[36] Zaks A, Klibanov AM. Enzymatic catalysis in nonaqueous solvents. J Biol Chem 1988;263:3194–3201.
[37] Zaks A, Klibanov AM. The effect of water on enzyme action in organic media. J Biol Chem 1988;263:8017–8021.
[38] Klibanov AM. Enzymatic catalysis in anhydrous organic-solvents. Trends BiochemSci 1989;14:141–144.
[39] Russell AJ, Klibanov AM. Enzymes in organic solvents. Biochem Soc Trans 1989;17:1145.
[40] Klibanov AM. Asymmetric transformations catalyzed by enzymes in organic-solvents. Accounts Chem Res 1990;23:114–120.
[41] Fitzpatrick PA, Klibanov AM. How can the solvent affect enzyme enantioselectivity?. J Am Chem Soc 1991;113:3166–3171.
[42] Fitzpatrick PA, Steinmetz AC, Ringe D, Klibanov AM. Enzyme crystal structure in a neat organic solvent. Proc Natl Acad Sci USA 1993;90:8653–8657.
[43] Narayan VS, Klibanov AM. Are water-immiscibility and apolarity of the solvent relevant to enzyme efficiency. Biotechnol Bioeng 1993;41:390–393.
[44] Wescott CR, Klibanov AM. The solvent dependence of enzyme specificity. Biochim Biophys Acta 1994;1206:1–9.
[45] Ball P. Water as an active constituent in cell biology. Chem Rev 2008;108:74–108.
[46] Bellissent-Funel MC, Hassanali A, Havenith M, Henchman R, Pohl P, Sterpone F, et al. Water determines the structure and dynamics of proteins. Chem Rev 2016;116:7673–7697.

[47] Laage D, Elsaesser T, Hynes JT. Water dynamics in the hydration shells of biomolecules. Chem Rev 2017;117:10694–10725.

[48] Verma R, Mitchell-Koch K. *In silico* studies of small molecule interactions with enzymes reveal aspects of catalytic function. Catalysts 2017;7.

[49] Labuza TP, Altunakar B. Water activity prediction and moisture Sorption Isotherms, In Water Activity in Foods: Fundamentals and Applications wiley; 2020, p. 161–205.

[50] Adlercreutz P. Fundamentals of Biocatalysis in neat organic solvents, In Carrea G, Riva S, editors. Organic Synthesis with Enzymes in Non-Aqueous Media 2008, Weinheim: WILEY-VCH Verlag GmbH & Co. KGaA; p. 3–24.

[51] Oleinikova A, Smolin N, Brovchenko I. Influence of water clustering on the dynamics of hydration water at the surface of a lysozyme. Biophysical Journal 2007;93:2986–3000.

[52] Zhang J, Chen J, Sha Y, Deng J, Wu J, Yang P, et al. Water-mediated active conformational transitions of lipase on organic solvent interfaces. Int J Biol Macromol 2024:277.

[53] Banik SD, Nordblad M, Woodley JM, Peters GH. Effect of water clustering on the activity of Candida antarctica lipase B in organic medium. Catalysts 2017;7.

[54] Tjørnelund HD, Brask J, Woodley JM, Peters GHJ. Optimised protocol for drying aqueous enzyme solutions in organic solvents – comparison of free and immobilised Candida antarctica Lipase B. ChemCatChem 2023;15.

[55] Cui H, Vedder M, Schwaneberg U, Davari MD. Using Molecular simulation to guide protein engineering for Biocatalysis in organic solvents, In Magnani F, Marabelli C, Paradisi F, editors. Enzyme Engineering: Methods and Protocols 2022, New York, NY: Springer US; p. 179–202.

[56] Chen M, Jin T, Nian B, Cheng W. Solvent tolerance improvement of lipases enhanced their applications: State of the art. Molecules 2024;29.

[57] Daniel RM, Dunn RV, Finney JL, Smith JC. The role of dynamics in enzyme activity. Annu Rev Biophys Biomol Struct 2003;32:69–92.

[58] López-Belmonte MT, Alcántara AR, Sinisterra JV. Enantioselective esterification of 2-arylpropionic acids catalyzed by immobilized *Rhizomucor miehei* lipase. J Org Chem 1997;62:1831–1840.

[59] Chamorro S, Alcántara AR, De la Casa RM, Sinisterra JV, Sánchez-Montero JM. Small water amounts increase the catalytic behaviour of polar organic solvents pre-treated *Candida rugosa* lipase. J Mol Catal B Enzym 2001;11:939–947.

[60] Domínguez de María P, Martínez-Alzamora F, Moreno SP, Valero F, Rúa ML, Sánchez-Montero JM, et al. Heptyl oleate synthesis as useful tool to discriminate between lipases, proteases and other hydrolases in crude preparations. Enzyme Microb, Technol 2002;31:283–288.

[61] Petersson AEV, Adlercreutz P, Mattiasson B. A water activity control system for enzymatic reactions in organic media. Biotechnol Bioeng 2007;97:235–241.

[62] Causevic A, Gladkauskas E, Olofsson K, Adlercreutz P, Grey C. Impact of critical parameters influencing enzymatic production of structured lipids using response surface methodology with water activity control. Biochem Eng J 2022;187.

[63] Riaz R, Ashraf M, Hussain N, Baqar Z, Bilal M, Iqbal HMN. Redesigning robust biocatalysts by engineering enzyme microenvironment and enzyme immobilization. Catalysis Letters 2022.

[64] Guisan JM, Fernandez-Lorente G, Rocha-Martin J, Moreno-Gamero D. Enzyme immobilization strategies for the design of robust and efficient biocatalysts. Curr Opin Green Sustain Chem 2022;35.

[65] Ferreira ML. Biocatalyst immobilization: Foundations and applications 2022, Amsterdam, The Netherlands: Academic Press, Elsevier

[66] Bolivar JM, Woodley JM, Fernandez-Lafuente R. Is enzyme immobilization a mature discipline? Some critical considerations to capitalize on the benefits of immobilization. Chem Soc Rev 2022.

[67] Almeida FLC, Prata AS, Forte MBS. Enzyme immobilization: What have we learned in the past five years?. Biofuels Bioprod Biorefin 2022;16:587–608.

[68] Sheldon RA, Basso A, Brady D. New frontiers in enzyme immobilisation: Robust biocatalysts for a circular bio-based economy. Chem Soc Rev 2021.
[69] Rocha-Martín J, Betancor L, López-Gallego F. Immobilization techniques for the preparation of supported Biocatalysts: Making better Biocatalysts through protein immobilization, In Biocatalysis for Practitioners: Techniques, Reactions and Applications wiley; 2021, p. 63–88.
[70] Karume I. Nature of support system and enzyme as key factors in immobilized biocatalyzed processes. Sustain Chem Environ 2023;4:100048.
[71] Adlercreutz P. On the importance of the support material for enzymatic synthesis in organic media. Support effects at controlled water activity. Progress in Biotechnology;1992:55–61.
[72] Faber K. Biotransformations in organic chemistry: A textbook 2018, Cham, Switzerland: Springer International Publishing AG.
[73] Ingenbosch KN, Vieyto-Nuñez JC, Ruiz-Blanco YB, Mayer C, Hoffmann-Jacobsen K, Sanchez-Garcia E. Effect of organic solvents on the structure and activity of a minimal Lipase. J Org Chem 2022;87:1669–1678.
[74] Klibanov AM. Improving enzymes by using them in organic solvents. Nature 2001;409:241–246.
[75] Evans DJ, Williams A, Alcántara AR, Pryce RJ. Specificity to leaving group in transesterification of substituted phenyl-esters in organic-solvents catalyzed by subtilisin microgel sols. J Mol Catal 1993;81:119–131.
[76] Park HJ, Joo JC, Park K, Kim YH, Yoo YJ. Prediction of the solvent affecting site and the computational design of stable *Candida antarctica* lipase B in a hydrophilic organic solvent. J Biotechnol 2013;163:346–352.
[77] Novák P, Havlíček V. Protein extraction and precipitation, In Proteomic Profiling and Analytical Chemistry: The Crossroads: Second Edition 2016, Elsevier Inc.; p. 52–62.
[78] Clark DS. Characteristics of nearly dry enzymes in organic solvents: Implications for biocatalysis in the absence of water. Philos Trans R Soc Lond B Biol Sci 2004 discussion 307, 323–8;359:1299–1307.
[79] Affleck R, Haynes CA, Clark DS. Solvent dielectric effects on protein dynamics. Proc Natl Acad Sci USA 1992;89:5167–5170.
[80] Trodler P, Pleiss J. Modeling structure and flexibility of *Candida antarctica* lipase B in organic solvents. BMC Structural Biology 2008;8.
[81] Katritzky AR, Fara DC, Yang H, Tämm K, Tamm T, Karelson M. Quantitative Measures of Solvent Polarity. Chem Rev 2004;104:175–198.
[82] Spange S, Weiß N, Schmidt CH, Schreiter K. Reappraisal of empirical solvent polarity scales for organic solvents. Chemistry Methods 2021;1:42–60.
[83] Wang Y, Li Q, Zhang Z, Ma J, Feng Y. Solvent effects on the enantioselectivity of the thermophilic lipase QLM in the resolution of (R, S)-2-octanol and (R, S)-2-pentanol. J Mol Catal B Enzym 2009;56:146–150.
[84] Mathpati AC, Bhanage BM. Combined docking and molecular dynamics study of lipase catalyzed kinetic resolution of 1-phenylethanol in organic solvents. J Mol Catal B Enzym 2016;133:S119–S27.
[85] Chatzikonstantinou AV, Norra GF, Stamatis H, Voutsas E. Prediction of solvent effect on enzyme enantioselectivity. Fluid Phase Equilibria 2017;450:126–132.
[86] Clarke CJ, Tu W-C, Levers O, Bröhl A, Hallett JP. Green and sustainable solvents in chemical processes. Chem Rev 2018;118:747–800.
[87] Wu XF, Yin Z, He L, Wang F. Green Solvents in Organic Synthesis 2024, Weinheim, Germany: WILEY-VCH GmbH
[88] Rani M, Shanker U. Green solvents in chemical reactions, In Inamuddin AMI, Asiri AM, editors. Industrial Applications of Green Solvents 2019, Millersville: Materials Research Forum Llc; I, p. 165–241.
[89] Miele M, Pillari V, Pace V, Alcántara AR, de Gonzalo G. Application of biobased solvents in asymmetric catalysis. Molecules 2022;27.

[90] Miele M, Ielo L, Pillari V, Fernández M, Alcántara AR, Pace V. Biomass-derived solvents, In Protti S, Palmieri A, editors. Sustainable Organic Synthesis: Tools and Strategies 2022, Croydon, UK: The Royal Society of Chemistry; p. 239–279.
[91] Winterton N. The green solvent: A critical perspective. Clean Technol Environ Policy 2021;23:2499–2522.
[92] Sheldon RA. The greening of solvents: Towards sustainable organic synthesis. Curr Opin Green Sustain Chem 2019;18:13–19.
[93] Häckl K, Kunz W. Some aspects of green solvents. Comptes Rendus Chimie 2018;21:572–580.
[94] Welton T. Solvents and sustainable chemistry. Proc Math Phys Eng Sci 2015;471.
[95] de María PD. Green solvents and biocatalysis: A bigger picture. EFB Bioecon J 2023;100056.
[96] Tiwari VK, Kumar A, Rajkhowa S, Tripathi G, Singh AK. Green solvents: Application in organic synthesis, In Tiwari VK et al. editors. Green Chemistry: Introduction, Application and Scope 2022, Singapore: Springer Nature Singapore; p. 79–112.
[97] Lozano P. Biocatalysis in green solvents, 2022 London, U.K: Academic Press, Elsevier.
[98] Hernáiz M, Alcántara AR, García JI, Sinisterra JV. Applied biotransformations in green solvents. Chem Eur J 2010;16:9422–9437.
[99] Miele M, Ielo L, Pace V, Alcántara AR. Biocatalysis in green biosolvents, In Nag A, editor. Greener synthesis of organic compounds 2022, Boca Raton: CRC Press.
[100] Prat D, Hayler J, Wells A. A survey of solvent selection guides. Green Chem 2014;16:4546–4551.
[101] Jimenez-Gonzalez C. Life cycle considerations of solvents. Curr Opin Green Sustain Chem 2019;18:66–71.
[102] Ashcroft CP, Dunn PJ, Hayler JD, Wells AS. Survey of solvent usage in papers published in organic process research & development 1997–2012. Org Process Res Dev 2015;19:740–747.
[103] Jiménez-González C, Curzons AD, Constable DJC, Cunningham VL. Expanding GSK's Solvent Selection Guide – Application of life cycle assessment to enhance solvent selections. Clean Technol Environ Policy 2005;7:42–50.
[104] Henderson RK, Jimenez-Gonzalez C, Constable DJC, Alston SR, Inglis GGA, Fisher G, et al. Expanding GSK's solvent selection guide – Embedding sustainability into solvent selection starting at medicinal chemistry. Green Chem 2011;13:854–862.
[105] Alder CM, Hayler JD, Henderson RK, Redman AM, Shukla L, Shuster LE, et al. Updating and further expanding GSK's solvent sustainability guide. Green Chem 2016;18:3879–3890.
[106] Alfonsi K, Colberg J, Dunn PJ, Fevig T, Jennings S, Johnson TA, et al. Green chemistry tools to influence a medicinal chemistry and research chemistry based organisation. Green Chem 2008;10:31–36.
[107] Prat D, Pardigon O, Flemming H-W, Letestu S, Ducandas V, Isnard P, et al. Sanofi's solvent selection guide: A step toward more sustainable processes. Org Process Res Dev 2013;17:1517–1525.
[108] Diorazio LJ, Hose DRJ, Adlington NK. Toward a more holistic framework for solvent selection. Org Proc Res Develop 2016;20:760–773.
[109] Prat D, Wells A, Hayler J, Sneddon H, McElroy CR, Abou-Shehada S, et al. CHEM21 selection guide of classical- and less classical-solvents. Green Chem 2016;18:288–296.
[110] Byrne FP, Jin S, Paggiola G, Petchey THM, Clark JH, Farmer TJ, et al. Tools and techniques for solvent selection: Green solvent selection guides. Sustain chem process 2016;4:7.
[111] Piccione PM, Baumeister J, Salvesen T, Grosjean C, Flores Y, Groelly E, et al. Solvent selection methods and tool. Org Process Res Dev 2019;23:998–1016.
[112] González-Miquel M, Díaz I. Green solvent screening using modeling and simulation. Curr Opin Green Sustain Chem 2021;29:8.
[113] Wen HQ, Nan SH, Wu D, Sun QH, Tong Y, Zhang J, et al. A Systematic review on intensifications of artificial intelligence assisted green solvent development. Ind Eng Chem Res 2023;62:20473–20491.
[114] Nieto S, Villa R, Donaire A, Lozano P. Nonconventional biocatalysis: From organic solvents to green solvents, In Biocatalysis in Green Solvents 2022, Elsevier; p. 23–55.

[115] Hinzmann A, Adebar N, Betke T, Leppin M, Gröger H. Biotransformations in pure organic medium: Organic solvent-labile enzymes in the batch and flow synthesis of Nitriles. Eur J Org Chem, 2019 2019;6911–6916.

[116] Hollmann F, Kara S. Enzymes in organic solvents: Biocatalysis in non-conventional reaction media. BioSpektrum 2014;20:100–102.

[117] Cao C, Matsuda T. Chapter 3 – Biocatalysis in organic solvents, supercritical fluids and ionic liquids A2 – Goswami, Animesh, In Stewart JD, editor. Organic Synthesis Using Biocatalysis 2016, Academic Press; 67–97.

[118] Zhang N, Kara S. Chapter 10 – Asymmetric biocatalysis in nonconventional media: Neat conditions, eutectic solvents, and supercritical conditions, In Gonzalo GD, Alcántara AR, editors. Biocatalysis in Asymmetric Synthesis 2024, Academic Press; 323–370.

[119] Cassady SJ, Almario JAN, Ramani GV. Therapeutic potential of treprostinil inhalation powder for patients with pulmonary arterial hypertension: Evidence to date. Drug Healthc Patient Saf 2024;16:51–59.

[120] Moriarty RM, Rani N, Enache LA, Rao MS, Batra H, Guo L, et al. The intramolecular asymmetric Pauson-Khand cyclization as a novel and general stereoselective route to benzindene prostacyclins: Synthesis of UT-15 (treprostinil). J Org Chem 2004;69:1890–1902.

[121] Domínguez de María P, de Gonzalo G, Alcántara AR. Biocatalysis as useful tool in asymmetric synthesis: An assessment of recently granted patents (2014–2019). Catalysts 2019;9:802.

[122] Kawasaki M, Goto M, Kawabata S, Kometani T. The effect of vinyl esters on the enantioselectivity of the lipase-catalysed transesterification of alcohols. Tetrahedron-Asymmetry 2001;12:585–596.

[123] Bloomer RJ, Butawan M, Farney TM, McAllister MJ. An Overview of the Dietary Ingredient Carnitine 2019, London: Academic Press Ltd-Elsevier Science Ltd

[124] Langner M, De Souza D, Pise AC, Bhuta S. Method for preparation and purification of posaconazole and posaconazole intermediates 2011, Sandoz AG, Switz: p. 71.

[125] Mutti FG, Knaus T. Enzymes applied to the synthesis of Amines, In de Gonzalo G, Lavandera I, editors. Biocatalysis for Practitioners: Techniques, Reactions and Applications 2021, Weinheim, Germany: WILEY-VCH, GmbH; p. 143–180.

[126] McCormack PL. Rasagiline: A review of its use in the treatment of Idiopathic Parkinson's Disease. CNS Drugs 2014;28:1083–1097.

[127] Zhang N, Bittner JP, Fiedler M, Beretta T, Domínguez de María P, Jakobtorweihen S, et al, Unraveling alcohol Dehydrogenase Catalysis in organic–aqueous biphasic systems combining experiments and molecular dynamics simulations. ACS Catal 2022:9171–9180.

[128] Doswald S, Estermann H, Kupfer E, Stadler H, Walther W, Weisbrod T, et al. Large scale preparation of chiral building blocks for the P3 site of renin inhibitors. Bioorg Med Chem 1994;2:403–410.

[129] Borland G, Murphy G, Ager A. Tissue inhibitor of metalloproteinases-3 inhibits shedding of L-selectin from leukocytes. J Biol Chem 1999;274:2810–2815.

[130] Ni R, Neves MAD, Wu C, Cerroni SE, Flick MJ, Ni H, et al. Activated thrombin-activatable fibrinolysis inhibitor (TAFIa) attenuates fibrin-dependent plasmin generation on thrombin-activated platelets. J Thromb Haemost 2020;18:2364–2376.

[131] Gröger H, Borchert S, Krauber M, Hummel W. Enzyme-catalyzed asymmetric reduction of ketones, In Flickinger MC, Editor. Encyclopedia of Industrial Biotechnology: Bioprocess, Bioseparation and Cell Technology 2010, New York: John Wiley & Sons; p. 2094–2110.

[132] Moore JC, Pollard DJ, Kosjek B, Devine PN. Advances in the enzymatic reduction of ketones. Acc Chem Res 2007;40:1412–1419.

[133] Gröger H, Hummel W, Buchholz S, Drauz K, Van Nguyen T, Rollmann C, et al. Practical asymmetric enzymatic reduction through discovery of a dehydrogenase-compatible biphasic reaction media. Org Lett 2003;5:173–176.

[134] Fragnelli MC, Hoyos P, Romano D, Gandolfi R, Alcántara AR, Molinari F. Enantioselective reduction and deracemisation using the non-conventional yeast Pichia glucozyma in water/organic solvent biphasic systems: Preparation of (S)-1,2-diaryl-2-hydroxyethanones (benzoins). Tetrahedron 2012;68:523–528.
[135] Seddon KR. Ionic liquids for clean technology, In Proceedings of J Chem Technol Biotechnol 1997, John Wiley & Sons Ltd.
[136] Seddon KR. Room-temperature ionic liquids: Neoteric solvents for clean catalysis. Kinetics and Catalysis 1996;37:693–697.
[137] Cull SG, Holbrey JD, Vargas-Mora V, Seddon KR, Lye GJ. Room-temperature ionic liquids as replacements for organic solvents in multiphase bioprocess operations. Biotechnol Bioeng 2000;69:227–233.
[138] Eckert CA, Knutson BL, Debenedetti PG. Supercritical fluids as solvents for chemical and materials processing. Nature 1996;383:313–318.
[139] Wasserscheid P, Welton T. Ionic Liquids in Synthesis 2008, John Wiley and Sons; Second Edition, 1.
[140] Hallett JP, Welton T. Room-temperature ionic liquids: Solvents for synthesis and catalysis. 2. Chemical Reviews 2011;111:3508–3576.
[141] Earle MJ, Seddon KR. Ionic liquids. Green solvents for the future, In Proceedings of Pure Appl Chem 2000, Walter de Gruyter GmbH.
[142] Earle MJ, Seddon KR. Ionic liquids: Green solvents for the future. ACS Symposium Series 2002;819:10–25.
[143] Nelson WM. are ionic liquids green solvents?, In ACS Symposium Series 2002, 818, p. 30–41.
[144] de Jesus SS, Maciel Filho R. Are ionic liquids eco-friendly?. Renew Sustain Energy Rev 2022;157:112039.
[145] Erbeldinger M, Mesiano AJ, Russell AJ. Enzymatic catalysis of formation of Z-aspartame in ionic liquid an alternative to enzymatic catalysis in organic solvents. Biotechnol Prog 2000;16:1129–1131.
[146] Madeira Lau R, Van Rantwijk F, Seddon KR, Sheldon RA. Lipase-catalyzed reactions in ionic liquids. Org Lett 2000;2:4189–4191.
[147] Marr AC, Imam HT, Marr PC. Biocatalysis in ionic liquids for a low carbon future, In Biocatalysis in Green Solvents Elsevier; 2022, p. 299–316.
[148] Imam HT, Krasňan V, Rebroš M, Marr AC. Applications of ionic liquids in whole-cell and isolated enzyme biocatalysis. Molecules 2021;26.
[149] Dupont J, Leal BC, Lozano P, Monteiro AL, Migowski P, Scholten JD. Ionic liquids in metal, photo-, electro-, and (Bio) catalysis. Chem Rev 2024.
[150] Villa R, Donaire A, Nieto S, García-Verdugo E, Lozano P. Biocatalytic processes in ionic liquids and supercritical carbon dioxide biphasic systems, In Biocatalysis in Green Solvents Elsevier: 2022, p. 403–433.
[151] Zhao H. Stability and stabilization of biocatalysts by ionic liquids, In Biocatalysis in Green Solvents Elsevier; 2022, p. 105–153.
[152] Ramos-Martin J, Khiari O, Alcantara AR, Sanchez-Montero JM. Biocatalysis at extreme temperatures: Enantioselective synthesis of both enantiomers of mandelic acid by transesterification catalyzed by a thermophilic lipase in ionic liquids at 120 degrees C. Catalysts 2020;10:1055.
[153] Knez Z, Markocic E, Leitgeb M, Primozic M, Hrncic MK, Skerget M. Industrial applications of supercritical fluids: A review. Energy 2014;77:235–243.
[154] Brunner G. Applications of supercritical fluids. Annu Rev Chem Biomol Eng 2010;1:321–342.
[155] Deepthi A, Sathi V. Supercritical fluids as green solvents, In Materials Horizons: From Nature to Nanomaterials 2021, Springer Nature; p. 323–338.
[156] Chaudhary A, Dwivedi A, Upadhyayula S. Supercritical fluids as green solvents, In Handbook of Greener Synthesis of Nanomaterials and Compounds: Volume 1: Fundamental Principles and Methods 2021, Elsevier; p. 891–916.

[157] Pattnaik S, Arun G, Swain K. Supercritical fluid technologies: A green solvent approach for pharmaceutical product development, In Nanotechnology in the Life Sciences 2020, Springer Science and Business Media B.V.; p. 1–14.

[158] Knez Ž, Pantić M, Cör D, Novak Z, Knez Hrnčič M. Are supercritical fluids solvents for the future?. Chem Eng Process: Process Intensif 2019;141.

[159] Knez Ž, Leitgeb M, Primožič M. Biocatalysis in subcritical and supercritical fluids, In Biocatalysis in Green Solvents Elsevier: 2022, p. 377–401.

[160] Matsuda T. Recent progress in biocatalysis using supercritical carbon dioxide. J Biosci Bioeng 2013;115:233–241.

[161] Peach J, Eastoe J. Supercritical carbon dioxide: A solvent like no other. Beilstein J Org Chem 2014;10:1878–1895.

[162] Asiri AM, Isloor AM. Green sustainable process for chemical and environmental engineering and science, In Supercritical carbon dioxide as green solvent 2019, Elsevier.

[163] Nakamura K, Chi YM, Yamada Y, Yano T. Lipase activity and stability in supercritical Carbon Dioxide. Chem Eng Commun 1986;45:207–212.

[164] Randolph TW, Blanch HW, Prausnitz JM, Wilke CR. Enzymatic catalysis in a supercritical fluid. Biotechnol Lett 1985;7:325–328.

[165] Hammond DA, Karel M, Klibanov AM, Krukonis VJ. Enzymatic-reactions in supercritical gases. Appl Biochem Biotechnol 1985;11:393–400.

[166] Bagheri S, Bagheri H, Sedghamiz MA, Rahimpour MR. Supercritical $CO_2$ for biocatalysis, In Green Sustainable Process for Chemical and Environmental Engineering and Science: Green Solvents for Biocatalysis Elsevier; 2020, p. 55–72.

[167] Kumar P, Kermanshahi-Pour A, Brar SK, He QS, Rainey JK. Influence of elevated pressure and pressurized fluids on microenvironment and activity of enzymes. Biotechnol Adv 2023;68.

[168] Nyari NLD, Zabot GL, Zamadei R, Paluzzi AR, Tres MV, Zeni J, et al. Activation of *Candida antarctica* lipase B in pressurized fluids for the synthesis of esters. J Chem Technol Biotechnol 2018;93:897–908.

[169] Dias ALB, Hatami T, Martínez J, Ciftci ON. Biocatalytic production of isoamyl acetate from fusel oil in supercritical $CO_2$. J Supercritical Fluids 2020;164.

[170] Quintana-Gómez L, Ladero M, Calvo L. Enzymatic production of biodiesel from alperujo oil in supercritical $CO_2$. J Supercritical Fluids 2021;171.

[171] Dias ALB, Ubeyitogullari A, Hatami T, Martínez J, Ciftci ON. Continuous production of isoamyl acetate from fusel oil under supercritical $CO_2$: A mass transfer approach. Chem Eng Res Des 2021;176:23–33.

[172] Bourkaib MC, Guiavarc'h Y, Chevalot I, Delaunay S, Gleize J, Ghanbaja J, et al. Non-covalent and covalent immobilization of *Candida antarctica* lipase B on chemically modified multiwalled carbon nanotubes for a green acylation process in supercritical $CO_2$. Catalysis Today 2020;348:26–36.

[173] Dos Santos P, Meireles MAA, Martínez J. Production of isoamyl acetate by enzymatic reactions in batch and packed bed reactors with supercritical $CO_2$. J Supercritical Fluids 2017;127:71–80.

[174] Liu KJ. Synthesis of lipophilic arbutin ester by enzymatic transesterification in high pressure carbon dioxide. Enzyme Microb Technol 2021;148.

[175] Villa R, Alvarez E, Nieto S, Donaire A, Garcia-Verdugo E, Luis SV, et al. Chemo-enzymatic production of omega-3 monoacylglycerides using sponge-like ionic liquids and supercritical carbon dioxide. Green Chem 2020;22:5701–5710.

[176] More SB, Gogate PR, Waghmare JS, Naik SN. Intensified synthesis of structured triacylglycerols from fish, flaxseed and rice bran oil using supercritical $CO_2$ or ultrasound. Chem Eng Process: Process Intensif 2019;144.

[177] Kmecz I, Varga Z, Székely E. One pot kinetic resolution and product separation with corn germ oil and supercritical carbon dioxide. J Supercritical Fluids 2018;141:218–223.

[178] Pollardo AA, Lee HS, Lee D, Kim S, Kim J. Effect of supercritical carbon dioxide on the enzymatic production of biodiesel from waste animal fat using immobilized *Candida antarctica* lipase B variant. BMC Biotechnol 2017;17.
[179] Vafaei N, Eskin MNA, Rempel CB, Jones PJH, Scanlon MG. Interesterification of Soybean Oil with Propylene Glycol in Supercritical Carbon Dioxide and Analysis by NMR Spectroscopy. Appl Biochem Biotech 2020;191:905–920.
[180] Yu D, Qi X, Ren Y, Wang W, Sun L, Xu D, et al. Thermal and crystal characteristics of enzymatically interesterified fats of fatty acid-balanced oil and fully hydrogenated soybean oil in supercritical $CO_2$ system. Int J Food Prop 2017;20:2675–2685.
[181] More S, Gogate P, Waghmare J, Naik SN. Intensified synthesis of structured lipids from oleic acid rich moringa oil in the presence of supercritical $CO_2$. Food Bioprod Process 2018;112:86–95.
[182] Baheti P, Gimello O, Bouilhac C, Lacroix-Desmazes P, Howdle SM. Sustainable synthesis and precise characterisation of bio-based star polycaprolactone synthesised with a metal catalyst and with lipase. Polymer Chemistry 2018;9:5594–5607.
[183] Guindani C, Jaramillo WAG, Candiotto G, Rebelatto EA, Tavares FW, Pinto JC, et al. Synthesis of polyglobalide by enzymatic ring opening polymerization using pressurized fluids. Journal of Supercritical Fluids 2022:186.
[184] Santos RD, Rebelatto EA, Guindani C, Madalosso HB, Scorsin L, Mayer DA, et al. Lipase-catalyzed synthesis of poly(ω-pentadecalactone-co-globalide) in supercritical carbon dioxide. J Supercritical Fluids 2025:215.
[185] Polloni AE, Veneral JG, Rebelatto EA, De Oliveira D, de Oliveira JV, Araújo PHH, et al. Enzymatic ring opening polymerization of ω-pentadecalactone using supercritical carbon dioxide. J Supercritical Fluids 2017;119:221–228.
[186] Matsuda T, Hoang HN. Medium engineering of lipase-catalyzed reaction using $CO_2$. JAOCS J Am Oil Chem Soc 2024.
[187] Zhang J, Wang C, Wang C, Shang W, Xiao B, Duan S, et al. Lipase-catalyzed aza-Michael addition of amines to acrylates in supercritical carbon dioxide. J Chem Technol Biotechnol 2019;94:3981–3986.
[188] Zhang J, Qian W, Wang C, Cao Z, Chen S, Zhang L, et al. Lipase-mediated epoxidation of alkenes in supercritical carbon dioxide. Green Chem Lett Rev 2018;11:508–512.
[189] Li F, Wang C, Xu Y, Gao X, Xu Y, Xie H, et al. Lipase-catalyzed synthesis of anthrone functionalized benzylic amines via a multicomponent reaction in supercritical Carbon Dioxide. ChemistrySelect 2022:7.
[190] Matsuda T, Harada T, Nakamura K. Enantioselective oxidation and reduction by Geotrichum candidum. J Synth Org Chem Jpn 2001;59:659–669.
[191] Nakamura K, Matsuda T, Harada T. Chiral synthesis of secondary alcohols using Geotrichum candidum. Chirality 2002;14:703–708.
[192] Matsuda T, Watanabe K, Kamitanaka T, Harada T, Nakamura K. Biocatalytic reduction of ketones by a semi-continuous flow process using supercritical carbon dioxide. Chem Commun 2003;3:1198–1199.
[193] Harada T, Kubota Y, Kamitanaka T, Nakamura K, Matsuda T. A novel method for enzymatic asymmetric reduction of ketones in a supercritical carbon dioxide/water biphasic system. Tetrahedron Lett 2009;50:4934–4936.
[194] Ying X, Yao S, Guan Y. Asymmetric reduction of acetophenone to α-phenylethanol by *Candida valida* under compressed $CO_2$. Chin J Catal 2006;27:631–635.
[195] Da Silva Serres JD, Taisline Bandeira P, Cabral Zappani P, Piovan L, Corazza ML. A greener bioreduction using baker's yeast cells in supercritical carbon dioxide and glycerol system. J Supercritical Fluids 2019;143:330–335.
[196] Panza JL, Russell AJ, Beckman EJ. Enzyme activity using a perfluoropolyether-modified NAD(H) in fluorous solvents and carbon dioxide. Clean Solvents 2002;819:64–81.

[197] Karmee SK, Roosen C, Kohlmann C, Lütz S, Greiner L, Leitner W. Chemo-enzymatic cascade oxidation in supercritical carbon dioxide/water biphasic media. Green Chem 2009;11:1052–1055.

[198] Queiroz MSR, Nyari NLD, Neves CLV, Pinheiro ADS, Mignoni ML, Dallago RM, et al. *Stemphylium lycopersici* immobilized in polyurethane (PU) treated in pressurized fluid CO2 as biocatalyst for ω-transamination reactions. Biocatal Agric Biotechnol 2024:58.

[199] Hoang HN, Matsuda T. Liquid carbon dioxide as an effective solvent for immobilized *Candida antarctica* lipase B catalyzed transesterification. Tetrahedron Lett 2015;56:639–641.

[200] Hoang HN, Matsuda T. Expanding substrate scope of lipase-catalyzed transesterification by the utilization of liquid carbon dioxide. Tetrahedron 2016;72:7229–7234.

[201] Jessop PG, Subramaniam B. Gas-expanded liquids. Chem Rev 2007;107:2666–2694.

[202] Weidener D, Klose H, Graf Von Westarp W, Jupke A, Leitner W, Domínguez De María P, et al. Selective lignin fractionation using $CO_2$-expanded 2-methyltetrahydrofuran (2-MTHF). Green Chem 2021;23:6330–6336.

[203] Pace V, Hoyos P, Castoldi L, Domínguez de María P, Alcántara AR. 2-Methyltetrahydrofuran (2-MeTHF): A biomass-derived solvent with broad application in organic chemistry. ChemSusChem 2012;5:1369–1379.

[204] Alcántara AR, Dominguez de Maria P. Recent advances on the use of 2-methyltetrahydrofuran (2-MeTHF) in biotransformations. Current Green Chemistry 2018;5:85–102.

[205] Pace V, Holzer W, Hoyos P, Hernaíz MJ, Alcántara AR 2-Methyltetrahydrofuran, in Encyclopedia of Reagents for Organic Synthesis [Online] 2014, John Wiley & Sons Ltd., http://onlinelibrary.wiley.com/book/10.1002/047084289X [accessed date].

[206] Hoang HN, Granero-Fernandez E, Yamada S, Mori S, Kagechika H, Medina-Gonzalez Y, et al. Modulating biocatalytic activity toward sterically bulky substrates in $CO_2$-expanded biobased liquids by tuning the physicochemical properties. ACS Sustain Chem Eng 2017;5:11051–11059.

[207] Hoang HN, Nagashima Y, Mori S, Kagechika H, Matsuda T. CO2-expanded bio-based liquids as novel solvents for enantioselective biocatalysis. Tetrahedron 2017;73:2984–2989.

[208] Otsu M, Suzuki Y, Koesoema AA, Hoang HN, Tamura M, Matsuda T. $CO_2$-expanded liquids as solvents to enhance activity of *Pseudozyma antarctica* lipase B towards ortho-substituted 1-phenylethanols. Tetrahedron Lett 2020;61.

[209] Suzuki Y, Taniguchi K, Hoang HN, Tamura M, Matsuda T. Rate enhancement of lipase-catalyzed reaction using CO2-expanded liquids as solvents for chiral tetralol synthesis. Tetrahedron Lett 2022;99.

[210] Abranches DO, Coutinho JAP. Everything you wanted to know about deep eutectic solvents but were afraid to be told. Annu Rev Chem Biomol Eng 2023;14:141–163.

[211] Martins MAR, Pinho SP, Coutinho JAP. Insights into the nature of eutectic and deep eutectic mixtures. J Solut Chem 2019;48:962–982.

[212] Abbott AP, Capper G, Davies DL, Rasheed RK, Tambyrajah V. Novel solvent properties of choline chloride/urea mixtures. Chem Commun 2003;70–71.

[213] Płotka-Wasylka J, De la Guardia M, Andruch V, Vilková M. Deep eutectic solvents vs ionic liquids: Similarities and differences. Microchem J 2020;159.

[214] Abranches DO, Schaeffer N, Silva LP, Martins MAR, Pinho SP, Coutinho JAP. The role of charge transfer in the formation of type i deep eutectic solvent-analogous ionic liquid mixtures. Molecules 2019;24.

[215] Hansen BB, Spittle S, Chen B, Poe D, Zhang Y, Klein JM, et al. Deep Eutectic solvents: A review of fundamentals and applications. Chem Rev 2021;121:1232–1285.

[216] Omar KA, Sadeghi R. Database of deep eutectic solvents and their physical properties: A review. J Mol Liq 2023;384.

[217] Patzold M, Siebenhaller S, Kara S, Liese A, Syldatk C, Holtmann D. Deep Eutectic solvents as efficient solvents in biocatalysis. Trends Biotechnol 2019;37:943–959.

[218] Domínguez de María P, Guajardo N, Kara S. Enzyme Catalysis: In DES, with DES, and in the presence of DES, In Ramón DJ, Guillena G, editors. Deep Eutectic Solvents 2019, Weinheim, Germany: Wiley-VCH Verlag GmbH & Co. KGaA.; p. 257–271.
[219] Van Osch DJGP, Dietz CHJT, Van Spronsen J, Kroon MC, Gallucci F, Van Sint Annaland M, et al. A search for natural hydrophobic deep eutectic solvents based on natural components. ACS Sustainable Chem Eng 2019;7:2933–2942.
[220] Van Osch DJGP, Dietz CHJT, Warrag SEE, Kroon MC. The curious case of hydrophobic deep eutectic solvents: A story on the discovery, design, and applications. ACS Sustainable Chem Eng 2020;8:10591–10612.
[221] Křížek T, Bursová M, Horsley R, Kuchař M, Tůma P, Čabala R, et al. Menthol-based hydrophobic deep eutectic solvents: Towards greener and efficient extraction of phytocannabinoids. J Clean Prod 2018;193:391–396.
[222] Abranches DO, Coutinho JAP. Type V deep eutectic solvents: Design and applications. Curr Opin Green Sustain Chem 2022;35.
[223] Paiva A, Craveiro R, Aroso I, Martins M, Reis RL, Duarte ARC. Natural deep eutectic solvents – Solvents for the 21st century. ACS Sustainable Chem Eng 2014;2:1063–1071.
[224] Liu Y, Friesen JB, McAlpine JB, Lankin DC, Chen SN, Pauli GF. Natural deep eutectic solvents: Properties, applications, and perspectives. Journal of Natural Products 2018;81:679–690.
[225] Carreiro EP, Federsel HJ, Hermann GJ, Burke AJ. Stereoselective Catalytic Synthesis of Bioactive Compounds in Natural Deep Eutectic Solvents (NADESs): A Survey across the Catalytic Spectrum. Catalysts 2024;14.
[226] Florindo C, Lima F, Ribeiro BD, Marrucho IM. Deep eutectic solvents: Overcoming 21st century challenges. Curr Opin Green Sustain Chem 2019;18:31–36.
[227] Prabhune A, Dey R. Green and sustainable solvents of the future: Deep eutectic solvents. J Mol Liq 2023;379.
[228] Zaib Q, Eckelman MJ, Yang Y, Kyung D. Are deep eutectic solvents really green?: A life-cycle perspective. Green Chem 2022;24:7924–7930.
[229] Zhang N, Domínguez de María P, Kara S. Biocatalysis for the synthesis of active pharmaceutical ingredients in deep eutectic solvents: State-of-the-art and prospects. Catalysts 2024;14:84.
[230] Jesus AR, Paiva A, Duarte ARC. Current developments and future perspectives on biotechnology applications of natural deep eutectic systems. Curr Opin Green Sustain Chem 2023;39.
[231] Yu D, Xue Z, Mu T. Deep eutectic solvents as a green toolbox for synthesis. Cell Rep Phys Sci 2022;3.
[232] Rente D, Cvjetko Bubalo M, Panić M, Paiva A, Caprin B, Radojčić Redovniković I, et al. Review of deep eutectic systems from laboratory to industry, taking the application in the cosmetics industry as an example. J Clean Prod 2022;380:135147.
[233] Panic M, Bubalo MC, Redovnikovic IR. Designing a biocatalytic process involving deep eutectic solvents. J Chem Technol Biotechnol 2021;96:14–30.
[234] Dai Y, Witkamp GJ, Verpoorte R, Choi YH. Tailoring properties of natural deep eutectic solvents with water to facilitate their applications. Food Chem 2015;187:14–19.
[235] Hammond OS, Bowron DT, Edler KJ. The effect of water upon deep eutectic solvent nanostructure: An unusual transition from ionic mixture to aqueous solution. Angew Chem Int Ed 2017;56:9782–9785.
[236] Ma C, Laaksonen A, Liu C, Lu X, Ji X. The peculiar effect of water on ionic liquids and deep eutectic solvents. Chem Soc Rev 2018;47:8685–8720.
[237] El Achkar T, Fourmentin S, Greige-Gerges H. Deep eutectic solvents: An overview on their interactions with water and biochemical compounds. J Mol Liq 2019;288.
[238] Zheng Q, Yang F, Tan H, Wang X. Influence of water on the properties of hydrophobic deep eutectic solvent. J Chem Thermodyn 2024;195.
[239] Gorke JT, Srienc F, Kazlauskas RJ. Hydrolase-catalyzed biotransformations in deep eutectic solvents. Chem Commun 2008;1235–1237.

[240] Tan JN, Dou Y. Deep eutectic solvents for biocatalytic transformations: Focused lipase-catalyzed organic reactions. Appl Microbiol Biotechnol 2020;104:1481–1496.

[241] Baby EK, Savitha R, Kinsella GK, Nolan K, Ryan BJ, Henehan GTM. Influence of deep eutectic solvents on redox biocatalysis involving alcohol dehydrogenases. Heliyon 2024;10.

[242] Gotor-Fernandez V, Paul CE. Deep eutectic solvents for redox biocatalysis. J Biotechnol 2019;293:24–35.

[243] Maugeri Z, Domínguez De María P. Benzaldehyde lyase (BAL)-catalyzed enantioselective CC bond formation in deep-eutectic-solvents-buffer mixtures. J Mol Catal B Enzym 2014;107:120–123.

[244] Paris J, Telzerow A, Ríos-Lombardía N, Steiner K, Schwab H, Morís F, et al. Enantioselective One-Pot Synthesis of Biaryl-Substituted Amines by Combining Palladium and Enzyme Catalysis in Deep Eutectic Solvents. ACS Sustainable Chem Eng 2019;7:5486–5493.

[245] Paul CE, Gotor-Fernández V. Applied biocatalysis in deep eutectic solvents. Biocatalysis in Green Solvents: Elsevier;2022:467–510.

[246] Monhemi H, Housaindokht MR, Moosavi-Movahedi AA, Bozorgmehr MR. How a protein can remain stable in a solvent with high content of urea: Insights from molecular dynamics simulation of *Candida antarctica* lipase B in urea: Choline chloride deep eutectic solvent, In Proceedings of Phys Chem Chem Phys. Royal Society of Chemistry 2014, p. 24930496.

[247] Nian B, Cao C, Liu Y. How *Candida antarctica* lipase B can be activated in natural deep eutectic solvents: Experimental and molecular dynamics studies. J Chem Technol Biotechnol 2020;95:86–93.

[248] Kovács A, Yusupov M, Cornet I, Billen P, Neyts EC. Effect of natural deep eutectic solvents of non-eutectic compositions on enzyme stability. J Mol Liq 2022;366.

[249] Shehata M, Unlu A, Sezerman U, Timucin E. Lipase and water in a deep Eutectic solvent: Molecular dynamics and experimental studies of the effects of water-in-deep Eutectic Solvents on Lipase stability. J Phys Chem B 2020;124:8801–8810.

[250] Qiao Q, Shi J, Shao Q. Effects of water on the solvation and structure of lipase in deep eutectic solvents containing a protein destabilizer and stabilizer. Phys Chem Chem Phys 2021;23:23372–23379.

[251] Huang LP, Bittner JP, Domínguez de María, Jakobtorweihen S, Kara S. Modeling alcohol Dehydrogenase catalysis in deep Eutectic solvent/water mixtures. ChemBiochem 2020;21:811–817.

[252] Bittner JP, Huang L, Zhang N, Kara S, Jakobtorweihen S. Comparison and validation of force fields for deep eutectic solvents in combination with water and Alcohol Dehydrogenase. J Chem Theory Comput 2021;17:5322–5341.

[253] Bittner JP, Zhang N, Huang L, Domínguez De María P, Jakobtorweihen S, Kara S. Impact of deep eutectic solvents (DESs) and individual des components on alcohol dehydrogenase catalysis: Connecting experimental data and molecular dynamics simulations. Green Chem 2022;24:1120–1131.

[254] Bittner JP, Smirnova I, Jakobtorweihen S. Investigating biomolecules in deep eutectic solvents with molecular dynamics simulations: Current state, challenges and future perspectives. Molecules 2024;29:703.

# Chapter 5
# Industrial Applications of Enzymes

**Abstract:** This chapter offers a comprehensive, state-of-the-art review of the industrial applications of enzymes as catalysts, charting their historical evolution and underscoring their transformative impact across multiple manufacturing sectors. Beginning with ancient uses in food processing and extending through to the modern biotechnological era, it details how enzymes – owing to their catalytic efficiency, high selectivity, and eco-friendly operation – have become central to sustainable and economically viable industry practices. The review traces pivotal developments such as the commercial extraction of rennet, the first enzyme-based detergents, and breakthrough applications in the food, textile, pharmaceutical, pulp and paper, biofuel, animal feed, and environmental remediation sectors.

A key conceptual framework is the "wave" model of biocatalysis, describing technological progress in distinct phases. This chapter distinguishes the first wave, reliant on wild-type enzymes and early immobilization strategies; the second wave, characterized by gene technology and rudimentary protein engineering; the third wave, defined by directed evolution and sophisticated recombinant approaches; and the emerging fourth wave, enabled by computational design, artificial intelligence, novel biosynthetic pathways, and cellular factories. Each wave is illustrated by landmark industrial processes, including the enzymatic synthesis of chiral drug intermediates, green manufacturing of active pharmaceutical ingredients such as montelukast, statins, sitagliptin, esomeprazole, and advanced enzyme cascades for complex pharmaceutical production. The review also highlights the impact of modern bioinformatics tools, patent literature, and market trends, with global enzyme sales valued at up to USD 14.6 billion in 2024–2025 and robust growth projections. Through rigorous academic synthesis and up-to-date referencing, we show how advances in enzyme engineering, process intensification, and computational modeling have positioned biocatalysis at the heart of the green and circular economy transition. The review concludes by outlining ongoing challenges and the future potential for de novo protein design, flow biocatalysis, and cell factory integration, establishing enzymes as indispensable tools for next-generation industrial biotechnology.

**Keywords:** Industrial biocatalysis, enzyme applications, biocatalysis waves, green chemistry, pharmaceutical manufacturing

## 5.1 Introduction

Enzymes, the highly proficient catalysts evolved in nature, have profoundly reshaped industrial production by providing effective routes to address key technolog-

 | https://doi.org/10.1515/9783111122991-005

ical and societal challenges. Their extensive use across manufacturing sectors arises from several decisive advantages. To begin with, enzymes can accelerate chemical transformations by many orders of magnitude compared with the uncatalyzed reactions, delivering exceptionally high reaction rates. In addition, their pronounced preference for particular substrates and reactions greatly limits the formation of side-products, thereby simplifying downstream purification and enhancing overall product purity. A further benefit is that enzymatic processes usually proceed under relatively gentle conditions of temperature and pH, which helps to cut energy demand, protects labile substrates, and lessens problems such as corrosion and associated maintenance in processing equipment. Taken together, this combination of outstanding catalytic performance, selectivity, and inherently more sustainable operating conditions has established enzymes as central and indispensable components of contemporary biotechnology [1–6]. Crucially, enzymes possess intrinsic biodegradability and originate from renewable biological sources, making them ideally suited to green chemistry doctrines and the worldwide transition to circular economic frameworks. Employing them empowers sectors to diminish dependence on perilous substances, curtail waste output, and cut down on greenhouse gas discharges, while sustaining – or even boosting – operational effectiveness and commercial viability [7, 8]. In fact, the urgency of addressing climate change, resource depletion, and environmental pollution has placed sustainability at the forefront of industrial innovation. Enzymes, with their unique catalytic capabilities and green credentials, are poised to play a central role in the transition to more sustainable manufacturing paradigms. The convergence of biotechnology, computational science, and process engineering is unlocking new possibilities for enzyme discovery, design, and application, promising to reshape industries and deliver significant environmental and societal benefits.

Consequently, enzymes' capacity to expedite chemical transformations with exceptional precision and in benign conditions has rendered them indispensable across diverse industrial sectors, as depicted in Figure 5.1.

- **Food and beverages:** Enzymes are used to improve dough quality in baking, clarify juices, enhance cheese ripening, and increase yields in brewing and starch processing. Their application leads to better product consistency, reduced processing times, and lower environmental footprints [9–11].
- **Textiles:** Enzyme-based processes have supplanted aggressive chemical agents in textile operations like desizing, scouring, and bio-polishing, yielding lower water and energy usage, diminished chemical effluent, and superior fabric attributes [12–16].
- **Biofuels:** Cellulases and hemicellulases play pivotal roles in breaking down lignocellulosic biomass into fermentable sugars, facilitating the manufacture of second-generation biofuels that exert a reduced ecological footprint [17, 18].

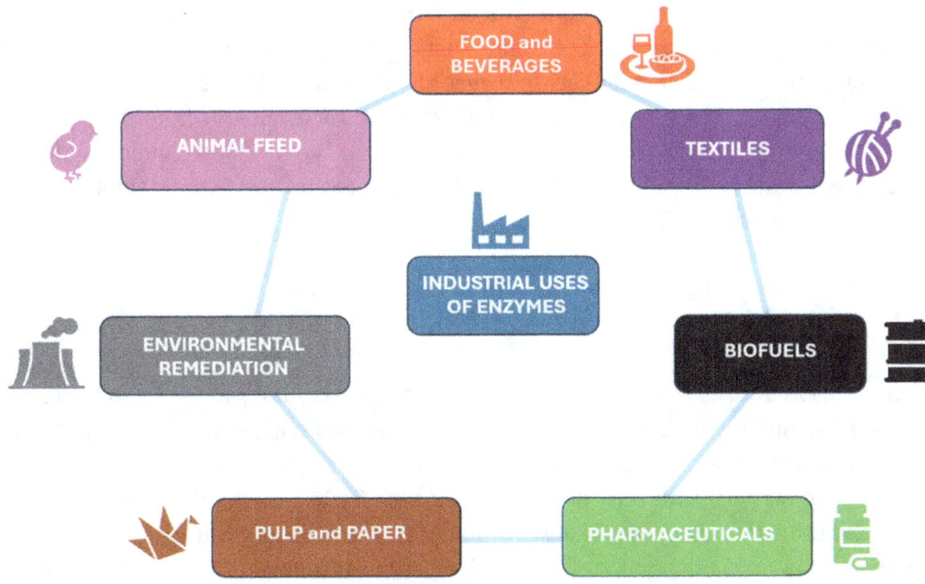


**Figure 5.1:** Schematic representation of industrial uses of enzymes.

- **Pharmaceuticals:** Biocatalysis enables the synthesis of complex, chiral drug molecules with high selectivity and fewer steps, reducing waste and the need for toxic pharmaceutical reagents [19–21].
- **Pulp and paper:** Enzymatic bleaching and fiber modification processes minimize chlorine use, reduce energy consumption, and improve paper quality [22–24].
- **Animal feed:** Enzymes are added to animal feed to break down complex nutrients like fiber, starch, and protein, making them more digestible and improving nutrient absorption in livestock. Overall, enzymes in animal feed promote animal health, support sustainable livestock production, and enhance profitability for producers [25–28].
- **Environmental remediation:** Enzymes are being developed for the degradation of persistent pollutants, plastics, and dyes, offering new tools for environmental cleanup and waste management [29–31].

As the global community grapples with the dual imperatives of economic growth and environmental stewardship, enzymes have emerged as critical enablers of sustainable, resource-efficient, and cleaner production processes [32–36]. Recent data highlight the enzyme sector's robust expansion and financial significance, pegging its worldwide valuation at USD 11.42–14.63 billion for 2024–2025 across various reports [37–40]. Projections anticipate vigorous uplift, with values projected to climb to USD 21.9–27.55 billion by 2033–2034, reflecting a compound annual growth rate (CAGR) of roughly 4–6% [37–40]. This trajectory is propelled by escalating needs for eco-conscious production methods, refined food handling, and greener substitutes in domains like animal nutrition, textiles, and bioenergy. While North America commands the biggest portion today, Asia-Pacific surges ahead with the briskest advance, fueled by broadening industrial deployment and heightened requirements in nutrition, medical care, and renewable fuels. Principal catalysts encompass biotechnology innovations, green imperatives, and heightened adoption within drugs, alimentation, and clean power arenas [37–40].

## 5.2 Historical Perspective and Industry Evolution

Humans have utilized enzymes since ancient times, particularly in food and beverage production, long before understanding their biochemical nature. Processes like brewing, baking, cheesemaking, and fermentation in ancient civilizations relied on natural enzymatic reactions, with records dating back over 6,000 years to the Sumerians and Babylonians [41]. Early scientific observations in the eighteenth century, such as those by Réaumur and Spallanzani, noted digestive processes involving mysterious substances in gastric juices, laying the groundwork for enzyme science [42].

A very recent article has reviewed the pioneering scientists at the origin of the field of biocatalysis and biotransformations from 1833 until the 1950s [43]. French chemists Payen and Persoz published in 1833 an article [44] describing a cost-effective method for extracting the agent responsible for breaking down the starch grain's outer layer, converting starch into sugar and gum using water, heat, and sprouted barley. Their primary aim was to isolate the substance that transformed dextrin into sugar. They identified this agent as a white, shapeless solid that did not dissolve in alcohol but was soluble in water and diluted alcohol. The solution it formed was neutral, nearly tasteless, and was not precipitated by lead subacetate; however, it quickly decomposed at room temperature, becoming acidic. When heated with starch at temperatures between 65 and 75 °C, this substance demonstrated an extraordinary ability to instantly rupture starch granules, releasing dextrin, which dissolved readily in water, while the insoluble husks either floated or sank depending on the liquid's density. This straightforward separation process inspired Payen and Persoz to name the substance "diastase," from the Greek word διάστασις (diastasis, meaning "separation"). Later, this enzyme was termed amylase, and nowadays "diastase" refers to any α-, β-, or γ-amylase capable to break down carbohydrates. Notably, the convention of using the suffix "-ase" to name enzymes is credited to the French scientist Émile Duclaux, who introduced this practice in 1899 as a tribute to the discoverers of diastase [45].

Another hallmark is the development of a method to extract and standardize rennet from the stomachs of calves or lambs, containing rennin (chymosin, EC 3.4.23.4, an aspartic protease) for cheese production by Christian Hansen in 1874 [46], which revolutionized the dairy industry, being considered one of the earliest examples of commercial enzyme use. The term "enzyme" (Greek ἔνζυμος, leavened, from ἐν, in, and ζύμη, leaven) was introduced by Wilhelm Kühne in 1878 [47], and the protein nature of enzymes was confirmed in 1926 by James B. Sumner, who crystallized urease [48].

The early developments in the industry can be traced back to 1914, when German chemist Otto Röhm pioneered the creation of the first commercial enzyme product, utilizing trypsin extracted from animal sources to enhance detergent performance [49]. Röhm was initially successful in applying enzymes for processing leather, which prompted him to shift his attention to the laundry sector. He found inspiration after witnessing Swabian butchers clean bloodstained garments with water infused with animal pancreases – a practice he personally observed. Motivated by this traditional technique, Röhm formulated an enzyme-powered laundry agent called BURNUS®, a name that references the distinctive white garments worn by North African Bedouins. Launched in 1914, this product was so effective that consumers in Germany were dubious due to the modest size of its packaging, leading the company to switch to larger boxes to satisfy demand. The detergent industry experienced a surge in enzyme applications with the development of microbial proteases. In 1956, the first laundry detergent containing bacterial enzymes, marketed as Bio-40, was introduced by the Swiss firm Schweizerische Ferment AG [50]. Since early enzymes were not sufficiently effective in alkaline washing environments, efforts intensified to identify or engineer more resilient enzyme variants. Thus, by 1958, Novo Industry (precursor of Novozymes) in Denmark launched an innovative detergent component containing a bacterial protease known as Alcalase®, which maintained strong activity and stability in alkaline conditions, specifically at pH levels between 8 and 10 [51]. This advancement paved the way for broader market acceptance. The breakthrough took on greater significance when Novozymes began mass-producing these enzymes, launching its first detergent enzyme plant in 1960, so that, by 1963, Alcalase® was being produced at scale for industrial application in detergents. This product was rapidly taken up by the detergent industry due to its superior efficacy compared to earlier animal-derived enzymes. Thus, by around 1965, major detergent manufacturers worldwide had embraced this technology [51].

Turning back to the food industry, several noteworthy milestones should be highlighted. For instance, the 1950s marked a turning point in breadmaking with the widespread adoption of fungal alpha-amylase by the baking industry [52]. This enzyme significantly enhanced bread quality and extended shelf life, addressing key limitations of traditional bread production techniques and responding to industrial demands for more consistent, high-volume, and prolonged-freshness baked products. In fact, bread quality is shaped by the transformation of starches during dough fermentation and baking. While traditional sources of enzymes for starch conversion included malt and cereal amylases, the introduction of fungal alpha-amylase, notably from species such as *Aspergillus oryzae* and *Aspergillus niger*, provided more predictable and effective hydrolysis of flour starches into fermentable sugars. Modern enzyme technology enabled large-scale fermentation and efficient purification of fungal amylases during this decade, making industrial applications feasible and economical [53].

The advent of glucose isomerase during the late 1950s marked a pivotal milestone in food biotechnology, facilitating cost-effective, industrial-scale manufacture of high-fructose corn syrup and fundamentally altering worldwide sweetener commerce [54]. Glucose isomerase (alternatively termed xylose isomerase, EC 5.3.1.5) functions as the catalyst for reversible conversion between D-glucose and D-fructose [55, 56]. Its commercial application addressed a key technological limitation: while glucose syrups derived by hydrolyzing cornstarch were easily produced, their utility was restricted by glucose's lower sweetness and inferior solubility compared to sucrose. Fructose, in contrast, possesses a sweetness comparable to or greater than sucrose and is highly soluble, making it desirable for the food and beverage industries.

As noted earlier, rennet – sourced from the gastric mucosa of immature ruminants – is used in cheesemaking, yet this practice is increasingly rejected by growing vegan and vegetarian populations, who may accept products made with microbial rennet, an enzymatic blend generated by select fungi, chiefly *Rhizomucor miehei* and *Rhizomucor pusillus* [57]. Unlike traditional animal rennet, microbial rennet is generated through fermentation processes in controlled environments, making it suitable for vegetarians and often classified as vegan if no animal-derived substances are involved in its production [58, 59]. During the 1960s and 1970s, shortages and increasing costs of calf rennet – caused in part by a declining veal industry and growing animal welfare concerns – prompted the search for alternative milk coagulants for cheese production. Microbial rennet, particularly from fungi such as *R. miehei*, began to enter the market in the 1960s. By the 1980s, microbial rennet represented about half of the rennet used in commercial cheese production in the United States, and by the 1990s, it was the dominant form as animal rennet supplies became increasingly inconsistent and expensive [60]. The introduction of recombinant chymosin (fermentation-produced chymosin) in the late 1980s and early 1990s further accelerated this shift [61], but the first wide commercial use of strictly microbial (non-GMO, from fungi) rennet was established from the 1960s onward.

The emergence of enzyme immobilization, predominantly refined during the 1970s, represents yet another cornerstone in industrial biotechnology, delivering substantial gains for producing pharmaceuticals, foodstuffs, specialty chemicals, and remediation technologies [62–64], as already commented in Chapter 3. The notion of enzyme immobilization originated in 1916, with Nelson and Griffin proving that invertase preserved its reactivity when adsorbed on substrates such as charcoal or aluminum hydroxide [65]. The inaugural commercial deployment came in 1967 from Chibata's group, who engineered an immobilized variant of *A. oryzae* aminoacylase to separate racemic synthetic amino acids [66]. This breakthrough rendered biocatalytic methods viable at production scale, spurring broad integration of fixed enzymes across pharmaceuticals, nutrition, and chemical industries. By the 1970s, such techniques had matured sufficiently for routine industrial uptake, thanks to refinements in methodologies and substrates that supported expansive operations. Essentially, attaching enzymes to rigid carriers or encapsulating them in gels bolsters their robustness – they withstand severe regimes like high heat or aberrant pH levels prevalent in manufacturing. Such fortification prolongs service life, simplifies retrieval and recycling, curtails expenses, and slashes residues. Moreover, it supports uninterrupted flows, streamlines product isolation, heightens selectivity, and unlocks nonaqueous reactions, thus widening enzymatic utility. The escalating embrace of immobilization underscores its fiscal and environmental merits, cementing its status as essential in scenarios including alimentation, medicinals, fabrics, cleaning agents, and effluent management [58, 64, 67, 68].

The last decade has witnessed remarkable advances in enzyme discovery, engineering, and application, driven by breakthroughs in molecular biology, protein engineering, synthetic biology, and computational modeling (see Chapter 6). As commented, protein-engineering techniques, including rational design, directed evolution, and machine learning-guided approaches, have enabled the fine-tuning of enzyme properties such as substrate specificity, catalytic efficiency, and operational stability [4, 69].

## 5.3 Biocatalysts in Pharma Industry

### 5.3.1 Introduction

Although, as clearly stated, the use of biocatalysts is a powerful tool in different types of industries, we will focus on the application of these versatile catalysts in the pharma industry, more specifically in the preparation of chiral drugs or intermediates [5, 6, 70]. This is mainly because of the extraordinary precision of enzymes, in terms of chemo-, regio-, and stereoselectivity, together with the inherent sustainability associated with their use, make them an outstanding tool in the industrial and lab-scale preparation of drugs or intermediates, generally requiring asymmetric synthetic protocols [20, 33, 71–75].

Chemoselectivity, according to IUPAC definition [76], denotes a reagent's tendency to interact preferentially with a single functional group amid several alternatives (Figure 5.2). A reagent qualifies as highly chemoselective when it targets just a narrow range of distinct functional groups. Sodium tetrahydroborate, for instance, exhibits greater chemoselectivity as a reductant compared to lithium tetrahydroaluminate. Although commonplace, the term lacks a formal quantitative formulation from IUPAC. It may also describe molecules or intermediates displaying preference for disparate reagents. Certain researchers employ "chemospecificity" to signify absolute chemoselectivity, though IUPAC advises against this. Nonetheless, enzymes demonstrate extraordinary accuracy in distinguishing functional groups of comparable reactivity, as illustrated in Figure 5.2 with the selective hydrolysis of nitriles alongside hydrolyzable amides [77].

Site selectivity, classified as a subtype of constitutional selectivity, refers to the ability to distinguish between nonidentical functional groups within a molecule that are susceptible to identical reaction types [78, 79], exemplified in Figure 5.3, is carried out very efficiently using enzymes. As an archetypal example, we will mention the

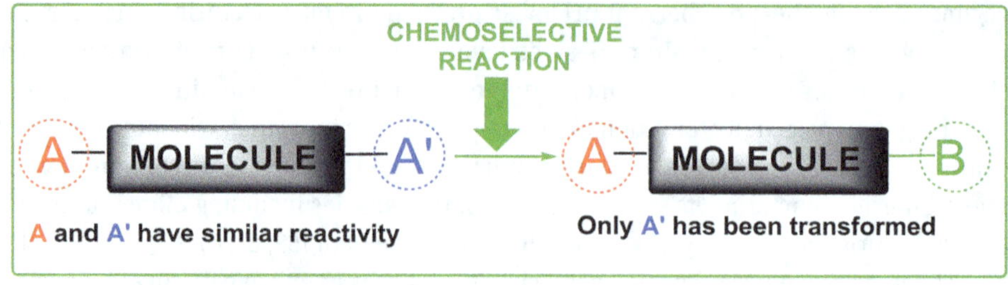


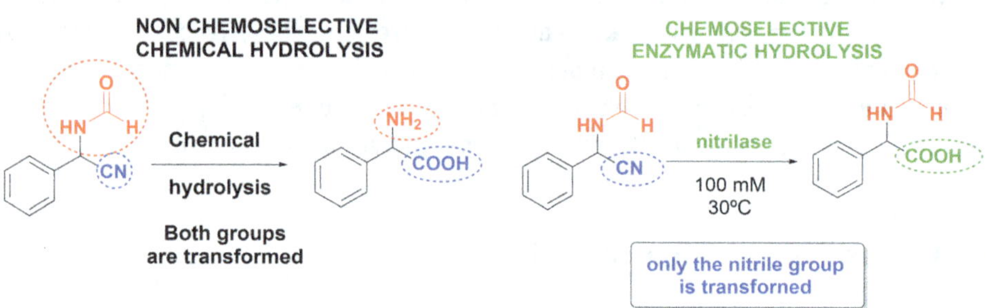


**Figure 5.2:** Schematic representation of enzymatic precision: chemoselectivity, as exemplified in the chemoselective hydrolysis of a molecule containing an amide and a nitrile.

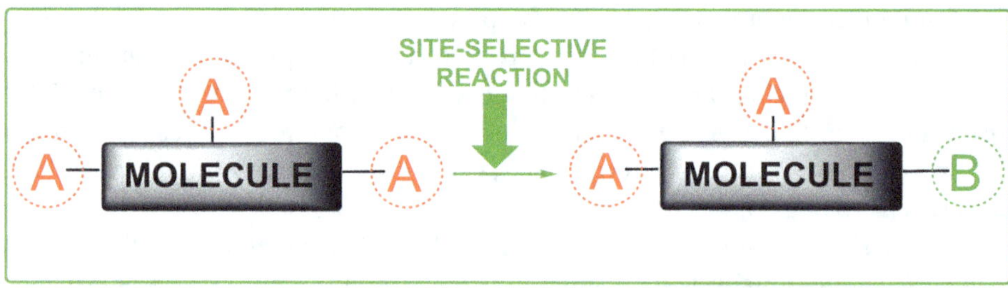


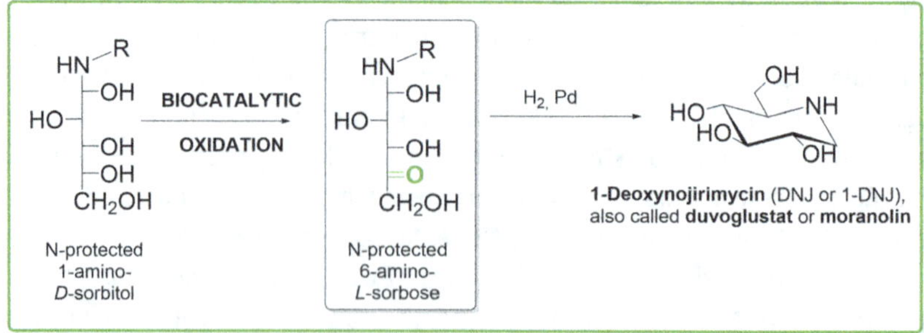


**Figure 5.3:** Schematic representation of enzymatic precision: site-selectivity, as exemplified in the selective oxidation of a single OH functional group in a polyol.

site-selective oxidation of only the secondary OH group at C5 of (*N*-protected) sorbitol to render the corresponding (*N*-protected) 6-aminosorbose, catalyzed by whole cells from *Gluconobacter oxydans*, was reported by scientists from Bayer AG in the 1980s as an efficient way en route to the antidiabetic 1-deoxynojirimycin [80].

Alongside chemo- and site-selectivity, biocatalysts are mostly used because they possess an extraordinary ability to recognize and enforce stereoselectivity, a property that is foundational to biological systems and chiral synthesis. Stereoselectivity refers to the preference for a catalyst (either chemical or biological) for transforming a particular stereoisomer or for producing a specific enantiomer in a chemical transformation. This phenomenon arises from the intricate, three-dimensional arrangement of amino acid residues within enzyme active sites, which tightly and specifically interact with substrates [81, 82]. The stereoselective recognition depends on the complementarity between the enzyme's binding pocket and the configuration of the substrate's stereocenters. A widely accepted model, the stereocenter-recognition model, posits that for a protein or enzyme to distinguish between enantiomers, it must make at least three essential simultaneous contacts with different "locations" on the substrate for a single chiral center, and even more for substrates with multiple stereocenters,

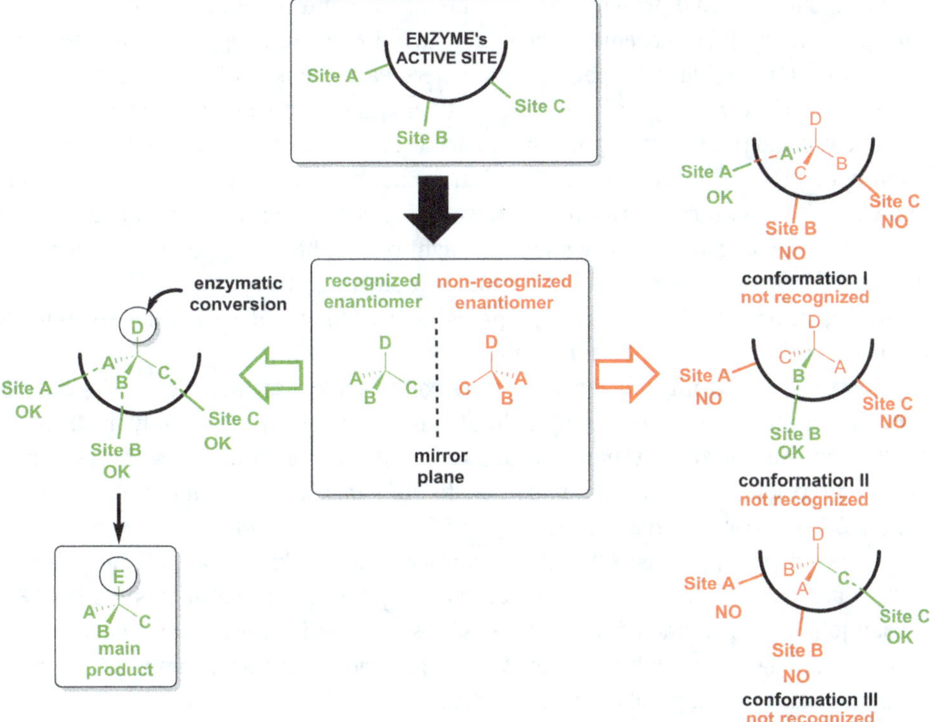


**Figure 5.4:** Schematic representation of enzymatic precision: stereoselectivity, as the capability of discriminating between enantiomers based on the three interactions in the active site.

as shown in Figure 5.4. Such molecular "fitting" ensures selectivity and excludes other isomers through steric, electronic, and hydrophobic effects [83].

### 5.3.2 Historical Perspective: The First Examples

Regarding the recognition of enzymes' stereodiscrimination capability, going back in time to the initial findings, we must mention that Louis Pasteur achieved a landmark in scientific discovery in 1848 when he painstakingly separated the two mirror-image crystal forms of sodium ammonium tartrate by hand [84, 85], as shown in Figure 5.5. These crystals exhibited macroscopic asymmetry (dissymmetry, as defined by Louis Pasteur) and differed in their influence on polarized light, rotating it either to the left or right. In addition to Pasteur's remarkable experimental expertise and powers of observation, the utilization of sodium-ammonium tartrate proved extraordinarily fortuitous in his studies [86]. In reality, fewer than 10% of racemic compounds are able to undergo spontaneous resolution by crystallization as conglomerates – physical mixtures comprising two distinct enantiomorphic crystals, each containing solely one enantiomer – rather than as racemates, in which both enantiomers coexist within individual crystals in equal proportions. Both hemihedry and spontaneous resolution are rare phenomena [87]. Moreover, achieving such a crystalline aggregate hinges critically on exacting laboratory parameters, where Pasteur's keen insights proved vital once more. He carefully documented that "crystallization ought to commence at dawn, since daytime warming prompts partial redissolution of the crystals, erasing the tiny hemihedral facets." At 20 °C, sodium-ammonium tartrate yields a conglomerate, yet at 30 °C, it forms a racemate, triggered by conglomerate dehydration around 28 °C. This outcome, fortuitously attained, might readily have escaped notice. Remarkably, over four decades passed before the next such discovery, reported by Arnaldo Piutti (Figure 5.5) in 1886 concerning asparagine [88, 89], underscoring the exceptional nature of the conditions underpinning Pasteur's discovery.

Pasteur subsequently advanced his investigations into dissymmetry, making significant contributions to the emerging field of stereochemistry [90]; notably, in 1858, he showcased the microbial separation of racemic tartaric acid (2,3-dihydroxysuccinic acid) via fermentation of its ammonium salt with *Penicillium glaucum* mold, isolating the (–) (*S,S*)-tartaric acid enantiomer (Figure 5.6A) [91]. In 1862, while probing ethanol-to -vinegar conversion, he discerned that a surface pellicle – known as "mother of vinegar" – facilitated oxygen delivery to sundry compounds [92]. Subsequently, in 1886, Adrian John Brown validated these observations, pinpointing *Bacterium xylinum* as the vinegar-forming agent; he further revealed its proficiency in transforming propanol to propionic acid and mannitol to fructose (Figure 5.6B) [93].

Despite these early and pioneering revelations in the field of chirality and biocatalysis, a comprehensive understanding of asymmetric synthesis took time to develop, and practical methods for such transformations advanced only gradually. Emil Fisch-

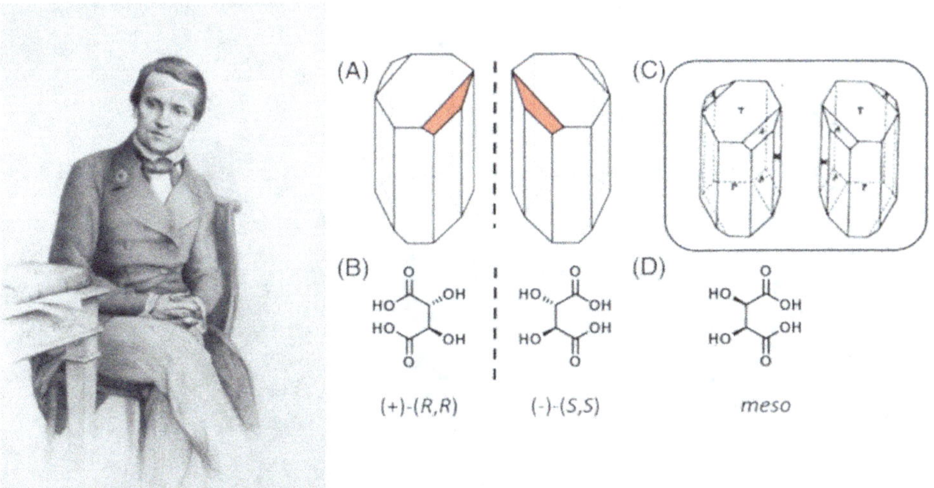


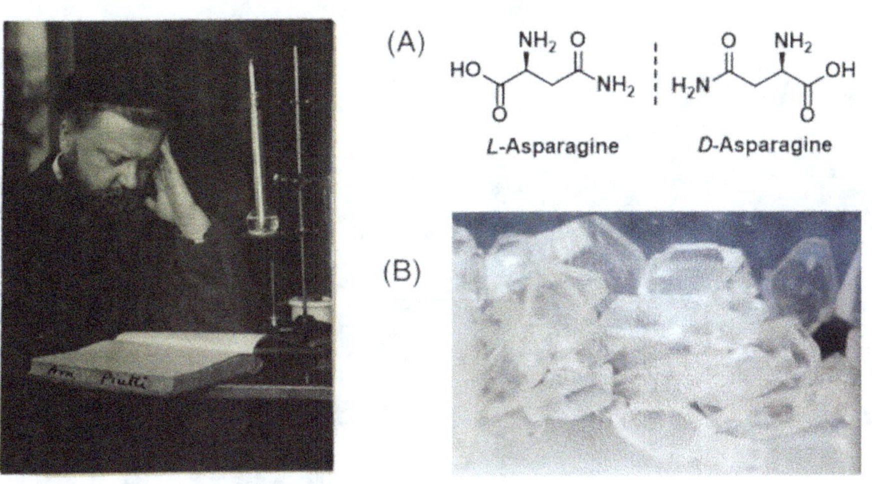


**Figure 5.5:** *Top Left*, Louis Pasteur; *Top right*: (A) hemihedral crystals of double sodium ammonium tartrates; (B) chemical structures of enantiomeric tartaric acids showing the facet (in red) that allows to visualize the crystals mirror images; (c) Hemihedral crystals as drawn by Pasteur; (D) achiral meso stereoisomer of tartaric acid. Reprinted with permission from Vantomme and Crassous [86] . *Bottom left*, Arnaldo Piutti (Image © Università di Padova, PHAIDRA repository, CC BY_NC_SA 4.0.). Bottom right: (A) chemical structures of enantiomers of asparagine; (B) crystals of L-asparagine. Adapted from L. Colli, A. Guarna [89].

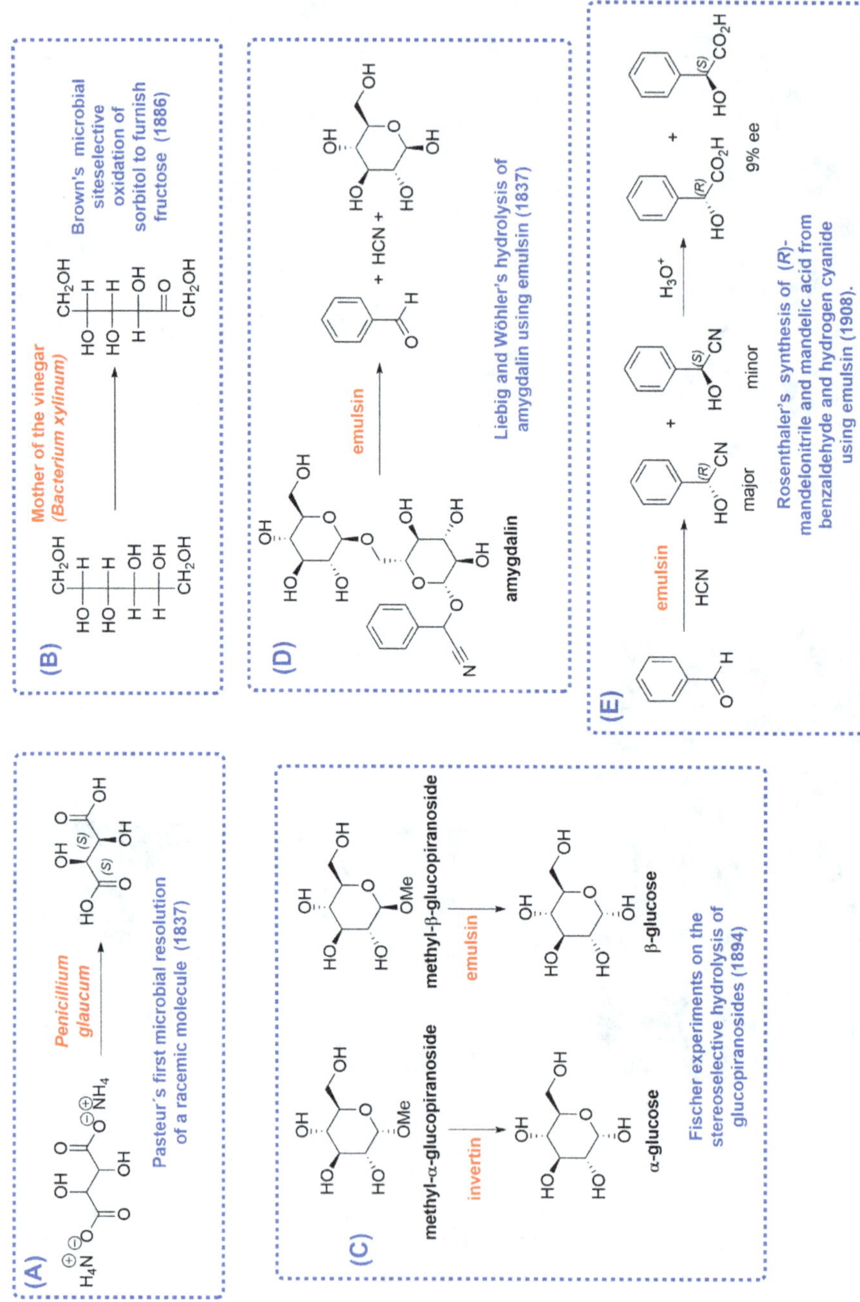


**Figure 5.6:** Some early discoveries (1800–1900) related to the enzymatic site- or stereoselective production of pure compounds.

er's work with the enzymes invertin (now known as invertase) and emulsin (β-D-glucosidase) stands as a foundational achievement in the field of enzymology and carbohydrate chemistry. In the early 1890s, Fischer was deeply involved in the study of sugars and their transformations, both chemically and through microbial (yeast) and enzymatic processes. By systematically synthesizing a variety of sugars and glucosides, Fischer set the stage for pioneering experiments to probe enzyme specificity and stereochemistry. Fischer employed invertin (from yeast) and emulsin (from bitter almonds) to study how enzymes catalyze the hydrolysis of glycosidic bonds in sugars and glucosides. Through a comparative approach, Fischer found that invertin specifically hydrolyzes α-glucosidic bonds (e.g., those in sucrose), while emulsin targets β-glucosidic bonds (e.g., in cellobiose and certain synthetic glucosides). This clear distinction highlighted the remarkable stereospecificity of enzymes – even when acting on seemingly similar substrates – only a precise molecular configuration was recognized and acted upon [94]. Fischer synthesized a suite of artificial methyl and ethyl glucosides of various configurations and demonstrated experimentally that these enzyme preparations were highly selective: invertin would only hydrolyze glucosides with α-configuration, while emulsin was specific to those with β-configuration (Figure 5.6C). These findings illustrated, for the first time with robust experimental evidence that **enzyme action depends critically on the three-dimensional fit (stereochemistry) between enzyme and substrate.**

This line of research culminated in Fischer's famous "lock-and-key" analogy ("Schlussel-Schloss-Prinzip," in German), published in 1894, literally stating: "The restricted action of the enzymes on glucosides may therefore be explained by the assumption that only in the case of similar geometrical structure can the molecules so closely approach each other as to initiate a chemical action. To use a picture I would like to say that enzyme and glucoside have to fit together like lock and key in order to exert a chemical effect on each other." [94, 95]. Fischer's hypothesis – that enzymes and substrates must possess complementary structures to interact – became the intellectual cornerstone of modern enzymology, influencing not only chemistry but also the life sciences broadly. Eduard Buchner's discovery of cell-free fermentation in 1897 further expanded this foundation [96].

Another milestone had been already set decades before when an almond-derived protein was utilized in asymmetric synthesis. As early as 1837, Liebig and Wöhler had already identified emulsin, β-D-glucosidase present in sweet and bitter almonds, as the key agent breaking down amygdalin into bitter almond oil (benzaldehyde), hydrocyanic acid, and a sugar (Figure 5.6D) [97]. The use of such cell-free enzymes, and the conceptual framework they enabled, was first put into practice in 1908, when Leopold Rosenthaler reported the first example of an enzyme-catalyzed asymmetric synthesis [98]. In this instance, benzaldehyde was reacted with hydrogen cyanide in the presence of emulsin. This reaction yielded an optically active cyanohydrin. Subsequent hydrolysis produced (–)(*R*)-mandelic acid with an optical purity of 9%, corresponding to an enantiomeric ratio of approximately 55:45 (Figure 5.6E).

### 5.3.3 Historical Perspective: The First Wave

These pioneer experiments have been included in what has been described as the "first wave" of biocatalysis [99–102], which goes from the dawn of time (fermentations to produce bread, cheese, or wine) until roughly the 1980s [100]. Despite their potential, enzymes did not assume a central role within the discipline of Chemistry at that time. Notably, as early as 1936, Otto Warburg demonstrated that specific pyridine molecules – similar in nature to nicotinamide, which could be obtained through the hydrolysis of a factor he called "co-zymase" – were capable of reversibly transferring hydride ions, so that he can be considered the "father" of oxidative enzymes [103], as this was one of the first biochemical mechanisms that were deciphered for enzymes; nevertheless, this hallmark was overshadowed by the prestige of the Nobel Prize in Physiology or Medicine that Warburg received in 1931 for his pioneering research on the role of iron in cellular respiration [104]. Concurrently, the earliest known industrial application of enantioselective synthesis (Figure 5.7A) dates from 1921, when Neuberg and Hirsch discovered that the condensation of benzaldehyde with acetaldehyde in the presence of yeast whole cells produces an optically active compound: 1-hydroxy -1-phenyl-2-propanone (also known as (*R*)-phenylacetylcarbinol) [105], which was subsequently converted into L-(–)-ephedrine through various chemical steps reported in 1930 in a patent from Knoll AG [106]. However, this milestone has been largely overlooked in standard chemistry education and textbooks [102].

Several pioneering biocatalytic applications from the initial phase merit attention. Notably, *Acetobacter suboxydans*, isolated in 1923, was harnessed for its precise oxidation prowess to convert D-sorbitol into L-sorbose (Figure 5.7B) [107], which by the mid-1930s fueled the Reichstein-Grüssner route to L-ascorbic acid (vitamin C) [108]. A further emblematic case involves the selective steroid hydroxylation introduced by Peterson et al., who in 1953 showed that *Rhizopus arrhizus* cultures could enact the stereospecific hydroxylation of progesterone to 11α-hydroxyprogesterone, a precursor to cortisone (Figure 5.7C). This microbial method drastically streamlined corticosteroid production; contrasted with Merck's prior all-chemical pathway from deoxycholic acid [109], demanding 31 stages to yield 1 kg cortisone acetate from 615 kg starting material – it offered a major leap forward, slashing cortisone's cost from $200 to $6 per gram. Later refinements dropped it below $1 per gram [110]. Amongst later first-wave triumphs, the late-1950s deployment of glucose isomerase for high-fructose corn syrup (Figure 5.7D) – which upended sweetener trade [54] (debuted commercially in 1967 by Clinton Corn Processing in Iowa, USA [111]), or the expanded use of proteases in laundry detergents in the 1970s–1980s [112].

The main obstacles of this first wave of biocatalysis were [100, 102]:

- Limited enzyme availability and diversity: Only wild-type enzymes were used, sourced from plants or animals. The natural repertoire of enzymes did not match many desired synthetic transformations.

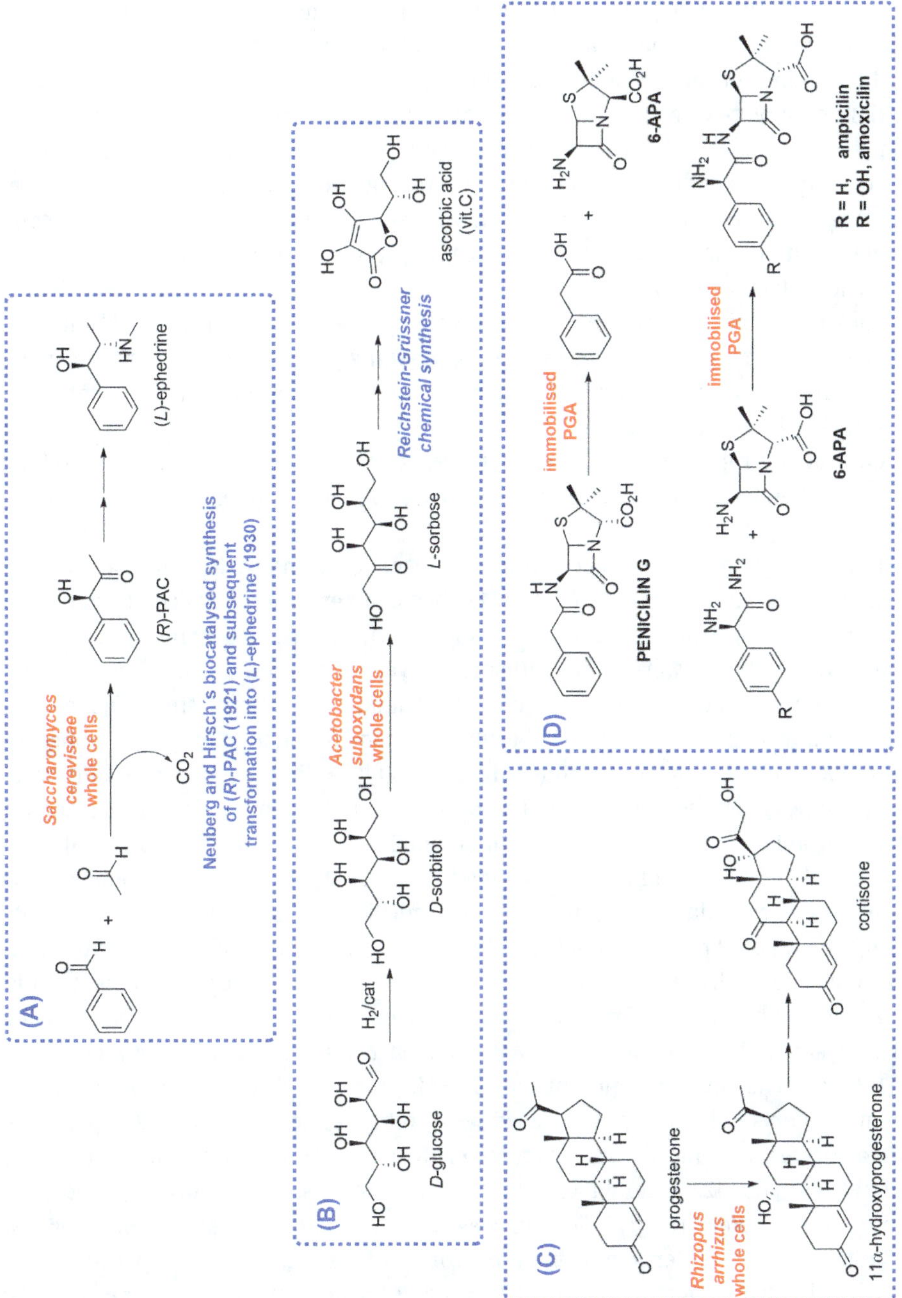


**Figure 5.7:** Some representative examples of the “first wave” of biocatalysis. (A) Ephedrine synthesis; (B) synthesis of vitamin C; (C) stereoslective hydroxylation of steroids; (D) preparation of high-fructose corn syrups (HFCS).

- Low stability under process conditions: Many enzymes were unstable or poorly active under industrial conditions such as high temperature, extreme pH, or in the presence of solvents and detergents.
- Narrow substrate scope: Enzymes typically catalyzed reactions involving their natural substrates, resulting in limited applicability for diverse or synthetic targets.
- Scale-up and production challenges: Culturing source organisms to extract enzymes was labor-intensive and costly; whole-cell biocatalysis suffered from complex fermentation and purification, leading to high costs and logistical barriers for industrial adoption.
- Lack of protein engineering capabilities: There was no available technology to modify enzymes for improved function, meaning process development depended on naturally occurring protein properties and "luck" in matching available enzymes to industrial needs [6].
- Control and reproducibility: Enzymatic reactions were difficult to control and optimize; batch-to-batch variability and low yields were common.

The first strategy to improve biocatalysts' performance in industrial processes was the use of immobilized enzymes [64, 68, 113, 114] (see Chapter 3 for a detailed description of this research area). From 1954 onwards, manufacturing routes to L-amino acids employing free aminoacylase were active industrially, yet they suffered from drawbacks including cumbersome product isolation, suboptimal conversions, steep enzyme expenses, and lack of recyclability. In the 1960s, Tanabe Seiyaku Co. in Japan pioneered its immobilized deployment, launching in 1969 the first commercial output of L-methionine via aminoacylase bound to DEAE-Sephadex within a fixed-bed reactor – the inaugural large-scale application of an immobilized enzyme (Figure 5.8A) [115].

Another industrial hallmark employing immobilized enzyme was undoubtedly the hydrolysis of penicillin G using immobilized penicillin acylases [116, 117] (Figure 5.8B), through the UNIFAR process, initiated in 1973, for the production of 6-aminopenicillanic acid (6-APA), a critical intermediate in the manufacture of semisynthetic penicillins. The enzymatic hydrolysis of penicillin was reported in the 1960s, selectively removing its side chain while preserving the core β-lactam structure essential for antibiotic activity [118–120]. The key innovation of the UNIFAR process lay in its use of immobilized enzyme systems, which improved enzyme stability, enabled continuous operation, and facilitated enzyme reuse – factors critical for industrial viability and cost-effectiveness [121, 122]. The enzyme was isolated from several sources (*Penicillium chrysogenum, Proteus rettgeri, Bacillus megaterium*, etc.) and covalently immobilized on Eupergit-C (Röhm, Germany), an epoxy-containing carrier. At first, operations ran in repeated batch fashion (80 g/L, pH 8.0, 30–35 °C), where the fixed enzyme was held back using a 400-mesh screen. Side-product phenylacetic acid underwent extraction for removal, while 6-APA crystallized out via metered nitric acid dosing to attain pH ~3.7. Yields exceeding 90% benefited from evaporation (under vacuum) or reverse osmosis of the mother liquor and/or fractionated streams. Meanwhile, Asahi Kasei

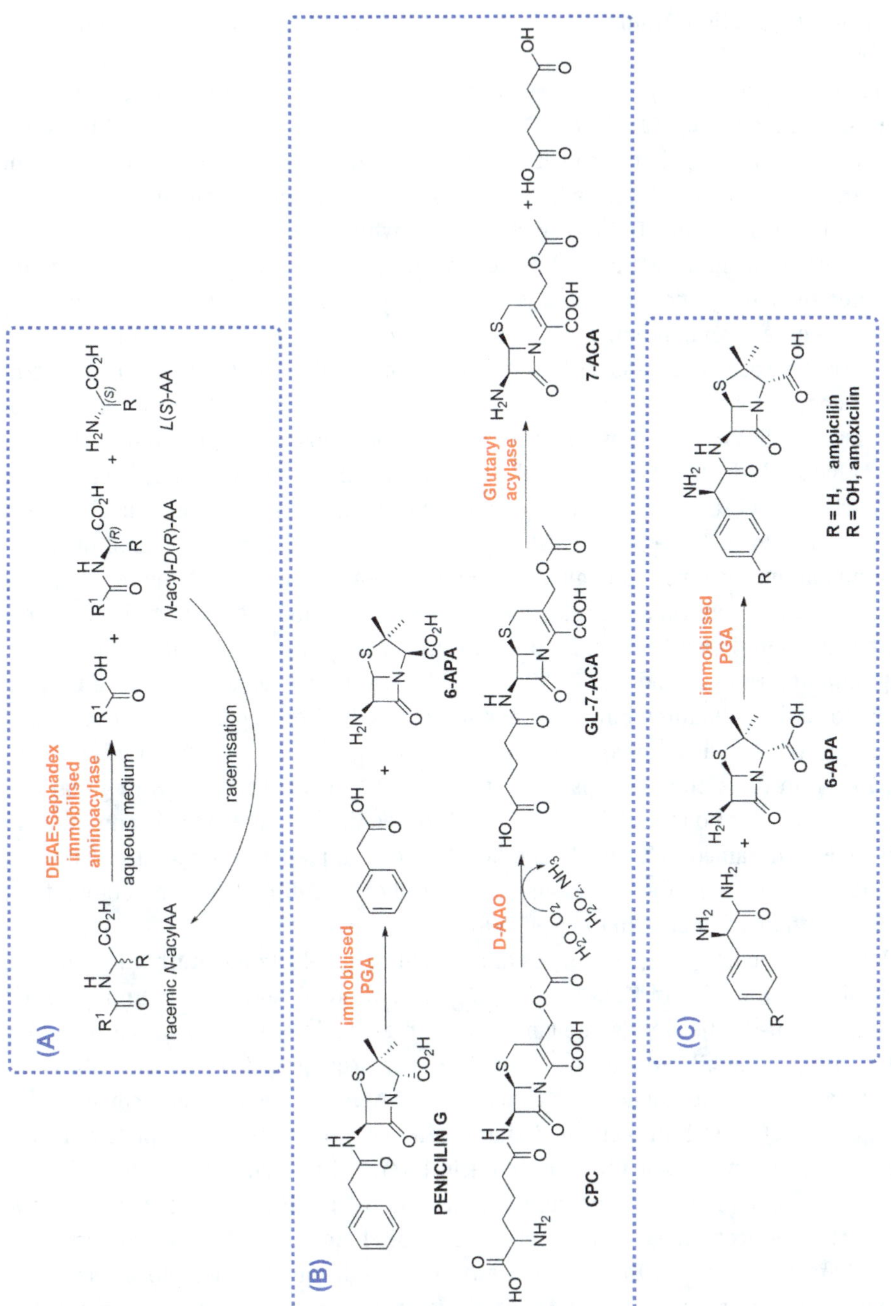


**Figure 5.8:** More representative examples of the "first wave" of biocatalysis. (A) Enzymatic synthesis of enantiopure amino acids; (B) enzymatic synthesis of 6-APA and 7-ACA; (C)) enzymatic synthesis of semisynthetic β.

Chemicals Corporation in Japan devised a parallel setup, deploying the enzyme – attached to aminated porous polyacrylonitrile fibers – across 18 linked columns in a recirculating system. With 30 L per column and reaction liquor (100 g/L, pH 8.4, 30–36 °C) pumped at 6,000 L/h, cycles lasted 3 h, affording 360 operational runs per unit. Downstream, 6-APA underwent isoelectric settling at pH 4.2, filtration, methanol rinse, and isolation [121, 122]. Tellingly, this biocatalytic route supplanted the 1960s chemical mainstay – the "Delft cleavage"[123] – which relied on perilous agents like phosphorus pentachloride and silyl protecting groups to furnish penicillin G's secondary amide selectively, yielding copious waste alongside cryogenic, energy-heavy steps.

On the other hand, although first implemented on a large scale in the 1990s (after the commonly recognized end of the first wave) to replace traditional chemical methods, the enzymatic synthesis of 7-aminocephalosporanic acid (7-ACA) from cephalosporin C (Figure 5.8B) can be considered another major industrial application of the first wave of biocatalysis [124], well before the modern era of directed evolution and enzyme engineering. This process typically involved two enzymes such as D-amino acid oxidase (DAAO) and glutaryl-7-ACA acylase (GLA) [125]. Thus, cephalosporin C, which is obtained by fermentation, is converted to glutaryl-7-ACA (GL-7-ACA) by DAAO, often immobilized on a support to allow reuse and facilitate product separation, in a stirred bioreactor with oxygen supply, and pH maintained around 7.5–8 with automatic titration (e.g., using ammonium hydroxide, at 25 °C). After the DAAO reaction, the mixture may be filtered to remove the immobilized DAAO; sometimes hydrogen peroxide (a byproduct) is added to convert any remaining side products (e.g., ketoadipyl-7-ACA) fully to glutaryl-7-ACA. In the second step, isolated glutaryl-7-ACA is then treated with glutaryl-7-ACA acylase (GLA, also immobilized), at a slightly alkaline pH (around 8.0), and the temperature is maintained between 25 and 30 °C. After completion, the mixture is filtered to remove enzymes, 7-ACA is crystallized, typically by acidification and cooling, then separated and dried (yields up to 85–90% for 7-ACA).

Likewise, the enzymatic synthesis of semisynthetic β-lactamic antibiotics, such as ampicillin and amoxicillin [126, 127] (Figure 5.8C), although industrially implemented in the late 1990s [128], can also be considered an example of the first wave, as it was applied using immobilized wild-type enzymes. The process typically starts with natural β-lactam antibiotic nuclei such as 6-APA or 7-ACA, obtained as described above, which are reacted with an activated side chain acyl donor (such as a phenylglycine methyl ester or amides), or other amino acid esters. The enzymatic synthesis is conducted in aqueous media under mild pH (around 6–8) and moderate temperatures, allowing for reduced energy consumption and less chemical waste compared to traditional chemical synthesis. Immobilized enzymes are frequently used to increase enzyme stability and allow reuse, and to improve yields, strategies like supersaturation of the nucleophile or one-pot reaction cascades (combining hydrolysis and synthesis in a single reactor) are employed. Product precipitation during synthesis helps shift equilibrium favorably and simplifies product recovery, to furnish the semisynthetic β-lactamic antibiotics in high yields (up to 85–90%).

### 5.3.4 Historical Perspective: The Second Wave

The second wave of biocatalysis represents a transformative period in the development of enzymatic technology, characterized by the integration of gene technology and early protein engineering methods that significantly expanded the scope and industrial applicability of biocatalysts [100, 102]. This wave commenced around the 1980s and continued into the early 2000s, setting the stage for the advanced techniques that define contemporary enzyme engineering. This wave was sparked by breakthroughs in molecular biology, particularly the advent of recombinant DNA technology. This allowed scientists to clone and express enzymes in microbial hosts, greatly facilitating their study and industrial production. Crucially, the ability to perform site-directed mutagenesis enabled researchers to modify enzymes at the genetic level to improve stability, activity, and substrate specificity. These technologies collectively provided a toolkit for rational and empirical enzyme optimization that was both faster and more effective than ever before, laying the groundwork for modern enzyme engineering. Please see Chapter 6 of this book for a detailed description of these techniques.

This period saw the rise of combinatorial protein engineering, also often referred to as directed evolution in its early stages. Techniques such as random mutagenesis and DNA shuffling were developed to create enzyme libraries with diverse mutations. These libraries could be screened for variants with enhanced or altered properties suitable for industrial applications. This marked a decisive shift from reliance on natural enzymes to the engineering of tailor-made biocatalysts addressing specific synthetic challenges. For instance, inside this second wave, we can include the large-scale industrial use of immobilized enzymes with improved operational stability and reusability, as for the enzymatic manufacture of semisynthetic beta-lactam antibiotics, mentioned in the previous paragraph [124, 126, 129–133].

Regarding hydrolases, many industrial processes belonging to this second wave were reported [20, 73–75, 100, 102, 134, 135]. Given their volume, we highlight the synthesis of (+)-(2*S*,3*R*)-*trans*-3-(4-methoxyphenyl)glycidic acid (MPGM), a vital chiral precursor for diltiazem – a benzothiazepine calcium-channel blocker prescribed for hypertension and angina relief. DSM-Andeno and Japan's Tanabe Seiyaku Co. Ltd. both leveraged lipases for resolving MPGM methyl ester: DSM-Andeno deployed *Candida antarctica* lipase [136], whereas Tanabe utilized one from *Serratia marcescens* Sr41 8000 [137, 138]. In each instance, the lipase selectively hydrolyzes racemic MPGM to "clean" the desired (2*S*,3*R*) enantiomer of the acid product from the enzymatic hydrolysis (2*R*,3*S*) enantiomer, which spontaneously decomposes to the corresponding aldehyde – a lipase inhibitor – so it must be removed by precipitation as a bisulfite combination (Figure 5.9).

The bioreactor used contains a membrane module made of polyacrylonitrile, with an asymmetric structure consisting of a dense layer on the lumen side and a spongy layer in the middle. Enzyme adsorption occurs through ultrafiltration of the

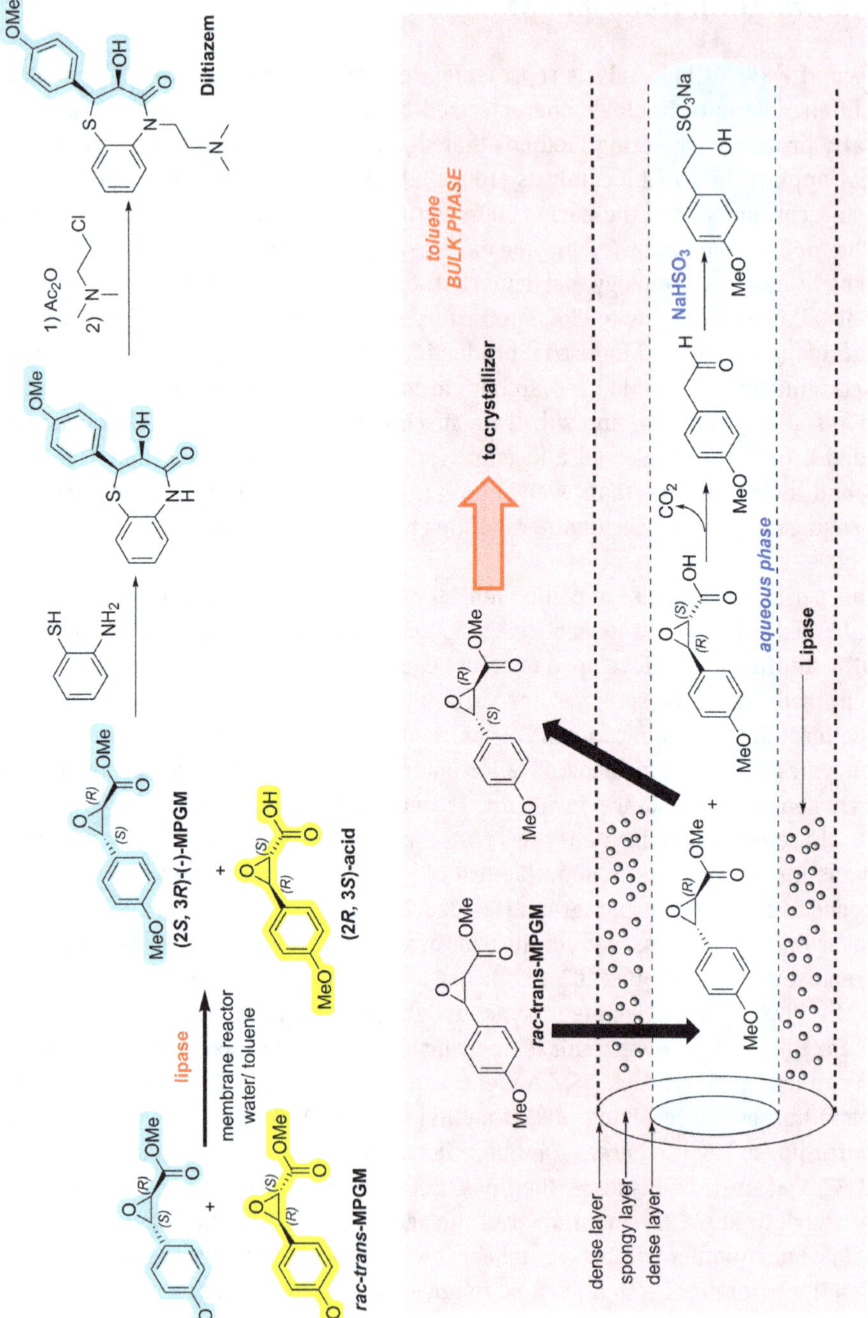


**Figure 5.9:** An industrial example of the use of lipases belonging to the "second wave": preparation of a key chiral synthon in the synthesis of diltiazem.

lipase solution, with the lipase being adsorbed within the inner spongy layer, at a loading of 3.0 g/m$^2$. In this setup, toluene serves to solubilize the racemate within the crystallizer and convey it to the membrane housing the immobilized lipase. The lipase enacts selective hydrolysis of the unwanted enantiomer, liberating the (2*R*,3*S*)-acid, which permeates into the aqueous compartment. This acid undergoes spontaneous decarboxylation to form the cognate aldehyde, which then binds bisulfite in the water layer. Without bisulfite, such aldehyde would inactivate the lipase. Meanwhile, the target (2*S*,3*R*) enantiomer persists in the toluene stream and recirculates to the crystallizer. This configuration delivers over 43% yield of crystalline (2*S*,3*R*)-(–)-MPGM at full enantiomeric purity. Though lipase potency wanes markedly after eight cycles, the membrane permits replenishment with fresh enzyme for continued service. Another remarkable industrial use of hydrolases, in this case dealing with the cosmetic industry, is the lipase-catalyzed synthesis of wax esters, long-chain esters formed from fatty acids and fatty alcohols, typically containing 12 or more carbon atoms (Figure 5.10A). These wax esters play a significant role in the cosmetics industry as emollients and texturizing agents. Their applications encompass skin creams, hair care, and lip products, driven by their desirable physicochemical properties – namely, lubricity, film-forming ability, and similarity to human sebum. Historically, the synthesis of wax esters was reliant on chemical catalysis, but growing emphasis on green chemistry and sustainability has shifted attention toward enzymatic routes, particularly employing lipases [139–142]; in fact, these enzymes catalyze the esterification of long-chain fatty acids with fatty alcohols under mild conditions and typically with high selectivity. The use of immobilized preparation allows for catalyst reuse and operational stability, addressing industrial scalability. In this way, useful compounds such as myristyl myristate [143] (an emollient used to soften and smooth skin and hair, an opacifier to improve product appearance, and a texture enhancer to provide body and spreadability to creams and lotions) or cetyl ricinoleate [144] (another preferred emollient due to its excellent spreading and moisturizing properties) could be produced. Another example is cetyl oleate, also a hydrating agent, stabilizer, lubricating agent, and shine imparting agent in creams and hair care products, which can be produced using an immobilized lipase [145] This eco-friendly and mild biocatalytic procedure, with a reduction in organic solvents and lower energy requirements, aligns with green chemistry standards, yielding high-purity products that meet the stringent purity specifications demanded by cosmetic formulators.

In parallel to enzyme engineering, advances in microbial strain optimization, fermentation technology, and bioprocess scale-up extended the practical reach of biocatalysis. This wave also witnessed the expansion of enzyme classes applied industrially beyond hydrolases to include other enzymes such as lyases, transferases, and oxidoreductases, enabling a broader range of chemical transformations, such as selective C–C bond formation and stereoselective oxidations and reductions. In redox biocatalysis, in this second wave, the development of effective cofactor regeneration systems [146–148], allowing the recycling of expensive cofactors like NAD(P)H

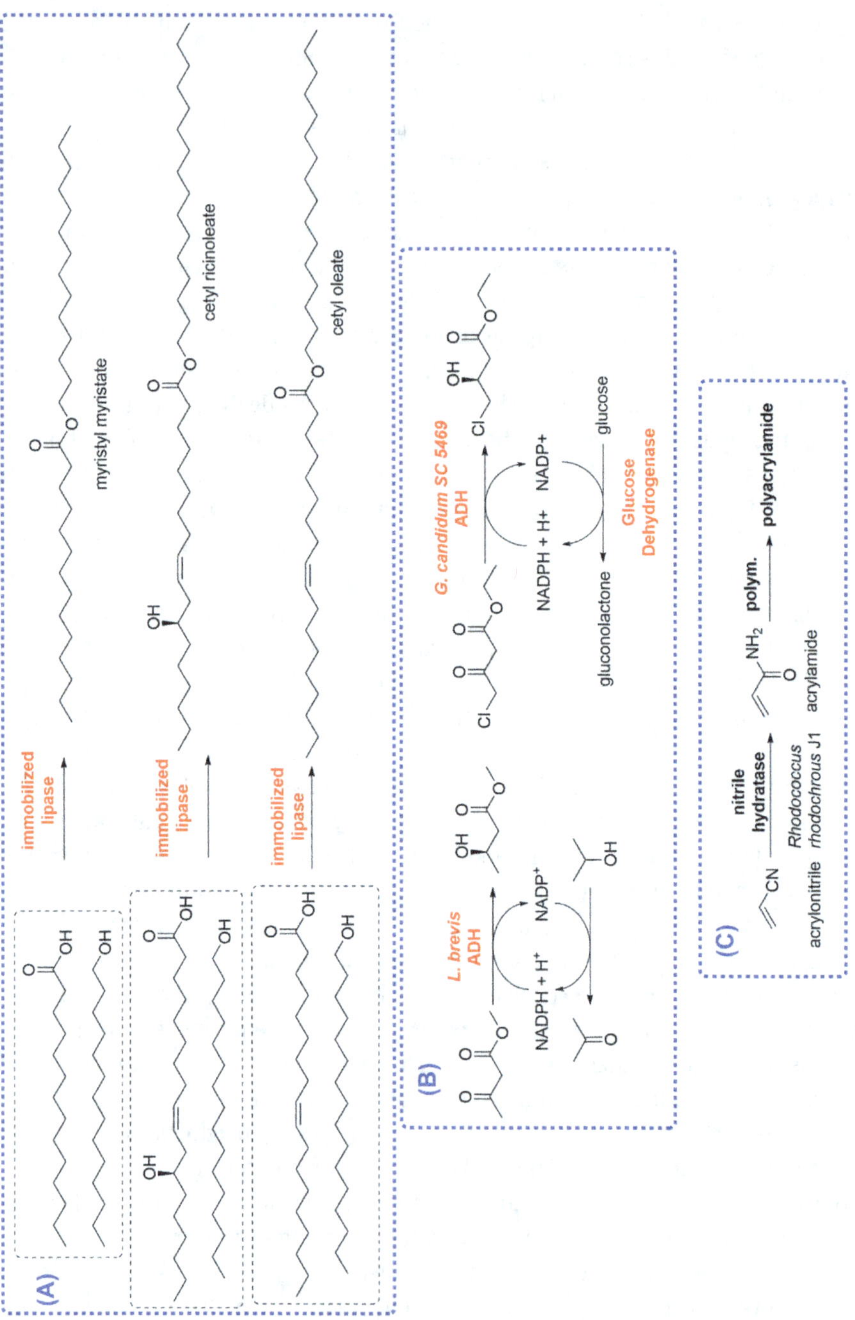


**Figure 5.10:** Industrial examples belonging to the "second wave": (A) preparation of wax esters; (B) bioreductions of β-oxoesters; and (C) enzymatic synthesis of acrylamide.

in situ, extended the practical use of oxidoreductases in synthesis, opening pathways to produce chiral alcohols, amines, and other valuable building blocks. As significant examples, we will mention the bioreduction of β-oxoesters using alcohol dehydrogenases (also known as ketoreductases), depicted in Figure 5.10B. In this way, by coupling the desired bioreduction with an *in situ* cofactor regeneration using an auxiliary substrate such as isopropanol (Figure 5.10B, left, a process developed by Julich using alcohol dehydrogenase from *Lactobacillus brevis* [149, 150]), or using a cofactor regeneration using an auxiliar substrate (glucose) and an auxiliar enzyme (glucose dehydrogenase (GDH)) in a bioreduction catalyzed by whole cells of *Geotrichum candidum SC 5469* (Figure 5.10B, right, by Bristol-Myers Squibb [151]).

We cannot exclude in this summary of the second wave of biocatalysis without mentioning the enzymatic synthesis of acrylamide, which exemplifies one of the most successful and large-scale applications of biocatalysis in the chemical industry (Figure 5.10C). This process replaces traditional chemical hydration of acrylonitrile with an environmentally benign approach that leverages microbial enzymes for efficient, selective production of acrylamide [152, 153]. At the core of the process is nitrile hydratase (NHase), a metalloenzyme first discovered in microorganisms such as *Rhodococcus rhodochrous* J1 and *Pseudomonas chlororaphis* B23 [154]. NHase catalyzes the hydrated addition of water to acrylonitrile, yielding acrylamide under mild conditions with high specificity and minimal by-product formation [155]. Among the advantages of this biotransformation, it is worthy to mention that the enzymatic method eliminates the need for corrosive acids and metal catalysts, drastically reducing environmental hazards and purification requirements; furthermore, NHase acts on acrylonitrile with near-stoichiometric conversion and high selectivity, eliminating the formation of unwanted side-products. Finally, the use of whole-cell catalysis protects the enzyme, enables repeated use, and operates effectively at low temperatures, further lowering energy demands. In fact, industrial production utilizes either free or immobilized whole cells containing active NHase, as the enzyme is often unstable in isolated form and is protected within the cellular environment. Batch and continuous flow bioreactors have been developed, with product concentrations exceeding 500 g/L and space-time yields above 0.1 kg/L · h reported for optimized processes using *R. rhodochrous* J1. These third-generation strains have replaced earlier, less robust catalyst systems and now dominate the global market for acrylamide production, which exceeds several hundred kilotons annually [154].

The second wave transformed enzyme engineering from a largely academic exercise into a practical industrial strategy. However, limitations remained in the size of enzyme libraries that could be feasibly screened and in the predictability of mutations' effects. These limitations were overcome with the "explosion" of the third wave of biocatalysis [100, 102].

### 5.3.5 Historical Perspective: The Third Wave

In 2012, Bornscheuer et al. published a seminal paper, which comprehensively reviewed the transformative advances in biocatalysis by that time, and coined the concept of "waves" in biocatalysis, thoroughly used in this chapter [100]. The onset of the third wave of biocatalysis was traced back to the 1990s, marked by the introduction of sophisticated protein engineering techniques such as DNA shuffling [156] and error-prone polymerase chain reaction [157]. These methods, combined with high-throughput screening approaches, paved the way to the concept of "directed evolution" [158], widely covered in Chapter 6. Thus, the field of biocatalysis greatly benefited from rapid advances in molecular biology, the significant reduction in genome sequencing costs, and especially the growth of bioinformatics and computational modeling tools. As a result, the design and optimization of wild-type enzymes – tasks that once required several years – could often be accomplished within a matter of months. In this "third wave," the (at that moment) unprecedented capability to engineer enzymes at the molecular level allowed the design of enzymatic reactions with remarkable specificity, efficiency, and under environmentally benign conditions, empowering scientists to tailor enzymes with customized functions, extending their applicability far beyond natural substrates to a wide range of synthetic targets. Literally, Bornscheuer et al. quoted: "In the past, an enzyme-based process was designed around the limitations of the enzyme; today, the enzyme is engineered to fit the process specification" [100].

A plethora of industrial examples belonging to this 'third wave' deserve mention [100, 102], but it would be completely impossible. Therefore, we will select some representative examples to illustrate the significant leap forward brought about by incorporating enzyme-engineering processes in cases of industrial interest. Particularly, as many examples of previous waves were dealing with hydrolases, we will focus on other types of enzymes. Thus, in the field of ketoreductases, a notable example showcasing the benefits of biocatalysis in the improved synthesis of a high-value active pharmaceutical ingredient is the production of montelukast sodium (Singulair, Figure 5.11A), a leukotriene receptor antagonist originally developed by Merck in 1991, patented in 1996, and prescribed for the management of asthma and seasonal allergic rhinitis. Annual output surpasses 66,000 kg, with global revenue reported to be valued at approximately USD 4.1–5.44 billion in 2023–2025, depending on the source [159]. Since montelukast is produced as a single enantiomer, a resolution stage was originally essential. However, implementing resolution toward the end of the synthesis was impractical due to the high cost and significant waste associated with recycling the unwanted enantiomer. Consequently, the stereogenic center was introduced early by enantioselective hydrogenation of the precursor ketone to furnish the desired (*S*)-alcohol, which constitutes the pivotal transformation in the synthetic pathway. The complexity of the ketone, containing several chemically reducible functionalities, necessitated not only high enantioselectivity but also extraordinary chemoselectivity in the reduction protocol. The traditional chemical approach employed asymmetric reduction

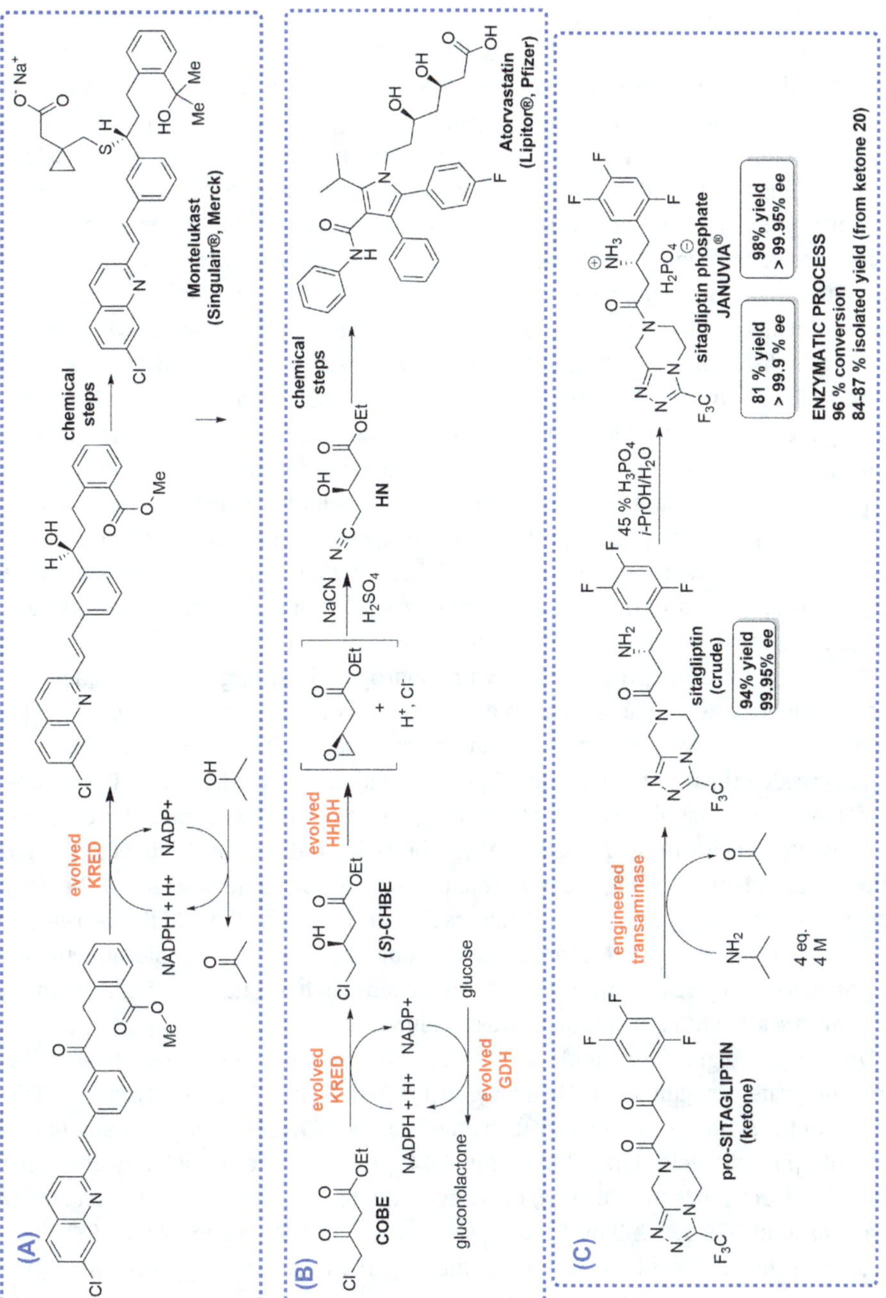


**Figure 5.11:** Three paradigmatic examples of biotransformations developed during the "third wave": (A) enzymatic synthesis of montelukast; (B) enzymatic synthesis of the lateral chain of atorvastatin; and (C) enzymatic synthesis of Sitagliptin.

with (–)-β-chlorodiisopropinolcampheylborane [(–)-DIPCl], affording the desired enantiopure alcohol in 90% yield after recrystallization [160]. The drawbacks of this method for large-scale operations, however, are considerable: substantial amounts (1.5–1.8 equivalents) of borane reagent are required, impacting atom economy and generating excessive waste. Furthermore, (–)-DIPCl is corrosive and sensitive to air and moisture, thus demanding strictly anhydrous conditions. To address these issues, catalytic alternatives were pursued, based on ruthenium-catalyzed asymmetric hydrogenations [161], but they necessitated operation under high hydrogen pressure, imposing substantial equipment and operational costs, and raising safety and toxicity concerns due to the nature of the metal catalysts. Driven by these limitations, significant efforts were undertaken to establish a biocatalytic solution. In 2010, Codexis Inc. reported the enzymatic reduction of a precursor ketone to yield the optically pure (*S*)-alcohol (Figure 5.11A), catalyzed by the ketoreductase KRED CDX-026 [162]. A set of related biocatalysts active on the substrate was subjected to rounds of directed evolution to enhance both activity and thermal stability, ultimately achieving a 1,000-fold improvement over the original enzyme. As emphasized earlier, both enzyme properties and reaction conditions are critical: due to the hydrophobic nature of substrate 15, organic cosolvents were required for adequate dissolution. Reactions were carried out at substrate concentrations of 100 g/L – comparable to the DIPCl-mediated process – in a toluene/isopropanol/triethanolamine-HCl buffer (pH 8.0, 1:5:3), with isopropanol serving as the regeneration agent for the enzyme's cofactor. This medium enabled ready crystallization of the (*S*)-alcohol, with full conversion attainable after precipitation. The yields (90–98%, 24 h) slightly exceeded those of the traditional process, with the added benefit that the target enantiomer was isolated directly, eliminating the need for postreaction resolution steps. Notably, process mass intensity (PMI) – the ratio of all material inputs to product obtained – highlighted the clear environmental and industrial advantages of the enzymatic method, showing a roughly 30-fold reduction in PMI and a 25-fold decrease in organic solvent use compared to the chemical route. Importantly, this biocatalytic process was successfully scaled up to 250 kg production batches, underscoring its viability for the large-scale synthesis of montelukast sodium.

Another paradigmatic example is the enzymatic preparation of ethyl (*R*)-4-cyano-3-hydroxybutyrate, also known as "hydroxynitrile" (HN, Figure 5.11B), a key chiral building block in the synthesis of the lateral chain of statins [163] such as atorvastatin (global market of approximately USD 4.6 billion in 2024, projected to grow steadily at a CAGR of about 3.8% from 2024 to 2031 [164]) or rosuvastatin (USD 8.5 billion in 2024, forecasted to reach around USD 12.2 billion by 2032, CAGR of about 5.0% during 2025–2032 [165]). The classical chemical methods for the synthesis of HN were not very sustainable [163], so Codexis addressed the challenge on a multiton scale by implementing a two-step process involving three enzymes [166], as illustrated in Figure 5.11B. The first phase utilizes a ketoreductase (KRED) for the biocatalytic reduction of ethyl 4-chloro-3-oxobutanoate (COBE), supported by glucose and NADP-reliant GDH for cofactor recycling. This generates ethyl (*S*)-4-chloro-3-hydroxybutanoate ((*S*)-CHBE) at 96% isolated yield and >99.5%

enantiomeric excess (e.e.). The following stage employs HHDH (halohydrin dehalogenase), which facilitates dehalogenation of vicinal haloalcohols to forge epoxide moieties [167], and mediates the nucleophilic substitution of chloride by cyanide, using hydrogen cyanide under neutral pH and room temperature conditions. The eco-benign character and superior performance of this approach – which secured Codexis the U.S. Environmental Protection Agency's Presidential Green Chemistry Challenge Award in 2006 [168] – arise from its stark departure from classical HN production pathways. Those legacy methods culminated in an $SN_2$ chloride-to-cyanide displacement under strongly basic conditions (pH 10) and high heat (~80 °C), problematic given the lability of both (S)-CHBE and HN under basic conditions, which spurred copious side reactions and waste [166]. Moreover, HN's high boiling point necessitated demanding high-vacuum fractional distillation for product recovery, leading to yield losses and substantial waste generation, thereby conflicting with the first and sixth principles of green chemistry [169]. By conducting cyanation at neutral pH through HHDH catalysis, the process markedly improves its sustainability. Initially, the wild-type forms of KRED, GDH, and HHDH presented low catalytic activities, requiring excessive enzyme loadings to achieve viable reaction rates. This challenge caused the formation of problematic emulsions that impeded downstream processing, coupled with severe product inhibition and insufficient enzyme stability. To realize a scalable and economically feasible process, these enzymes underwent in vitro evolution and gene shuffling, guided by specific criteria and process parameters. This optimization enhanced volumetric productivity approximately 2,500-fold in the cyanation step, including a 14-fold reduction in reaction time, a 7-fold increase in substrate concentration, a 25-fold reduction in enzyme usage, and a 50% increase in isolated yield [166].

The third wave's example to be commented on is the biocatalyzed synthesis of sitagliptin, illustrated in Figure 5.11C, catalyzed by an engineered ω-transaminase (also referred to as aminotransferases), which are by far the most widely used biocatalysts for obtaining optically pure amines [170, 171]. These enzymes employ pyridoxal phosphate as a cofactor and facilitate the transfer of an amino group from a suitable donor molecule to a ketone acceptor. Owing to the reversibility of this reaction, transaminases are applicable both to the enantioselective reduction of prochiral ketones and to the kinetic resolution of racemic amines [170]. Thus, a landmark study led by Savile et al., which represents a collaborative effort involving Solvias, Codexis, and Merck [172], reported the use of a transaminase-based route to synthesize the oral antidiabetic medication sitagliptin, which is sold as Januvia® by Merck & Co. In this work, a variant of an (*R*)-specific transaminase originating from *Arthrobacter* species (designated ATA-117) was tailored for the direct, enantioselective production of the chiral amine sitagliptin from its prochiral ketone precursor, prositagliptin (Figure 5.11C). Initially, the wild-type enzyme was unable to accept this substrate due to the bulky, fluorinated aromatic ring, which could not be accommodated in the binding pocket (wild-type enzymes only tolerated smaller methyl or ethyl substituents at this position [173]). As thoroughly commented in Chapter 6, there are essentially two main strategies for overcoming such substrate incompatibility: (i) a rational design approach, which involves using structural or sequence information to predict which amino acid residues should be altered, followed by targeted gene modification [174]; (ii) in contrast, directed evolution relies on generating random mutations, applying high-throughput screening to select superior variants over successive generations [158].

In optimizing a transaminase for prositagliptin, a hybrid methodology was employed. Computational modeling was first used to characterize the enzyme's active site, pinpointing two key regions that would need to be modified. Researchers initially designed and tested a truncated version of the substrate, for which limited enzyme activity was retained. Site-specific mutations then enabled catalysis of this intermediate, at which point the strategy moved to engineering a fit for the full substrate. This rational, stepwise process, referred to as "substrate-walking," involved rounds of mutation and rational alteration, eventually yielding an enzyme able to accept the complex structure of prositagliptin. Once substrate binding was achieved, multiple cycles of directed evolution were used to boost operational robustness [172]. As a result, an enzyme incorporating 27 separate mutations (after 11 cycles of evolution) was developed. This variant could operate in the presence of 50% DMSO at pH 10, facilitating excellent substrate dissolution while maintaining enzymatic stability. It was capable of converting concentrations of up to 200 g/L of the precursor at 45 °C, affording a yield of 92% with e.e. greater than 99.95%. This biocatalytic route afforded a yield improvement of 10–13% over the previously established rhodium-catalyzed chemical process, increased productivity by 53%, and reduced overall waste by 19%, eliminating the requirement for expensive precious metals or specialist hydrogenation equipment. The development and optimization of the engineered transaminase for sitagliptin took about 1 year. By the time the enzymatic process was ready for commercial use, demonstrated on a 45 kg scale [175], the product (Januvia®) had already been released (2006) and existing patents were in force [176–178], requiring additional regulatory approval for process changes [179]. As patent protection for sitagliptin ended in July 2022 for the free base, generic versions of sitagliptin are now available in several countries; so, in the United Kingdom and throughout the European Union, generic sitagliptin (such as Sitagliptin Accord) has been authorized and is recommended as a cost-effective treatment for type 2 diabetes since at least January 2024, following the loss of market exclusivity. In the United States, however, no therapeutically equivalent generic version of Januvia is currently available (patent ending in November 2026 for the phosphate salt [180]).

To conclude this section, going back to the pivotal paper of Bornscheuer et al. [100], these authors ended up highlighting ongoing challenges and future directions, such as improving enzyme robustness under industrial conditions, expanding enzyme function diversity through de novo design, and leveraging synthetic biology to create novel biosynthetic pathways. Thus, they argued that these developments will lead us to the fourth wave of biocatalysis, which will be commented on in the next section.

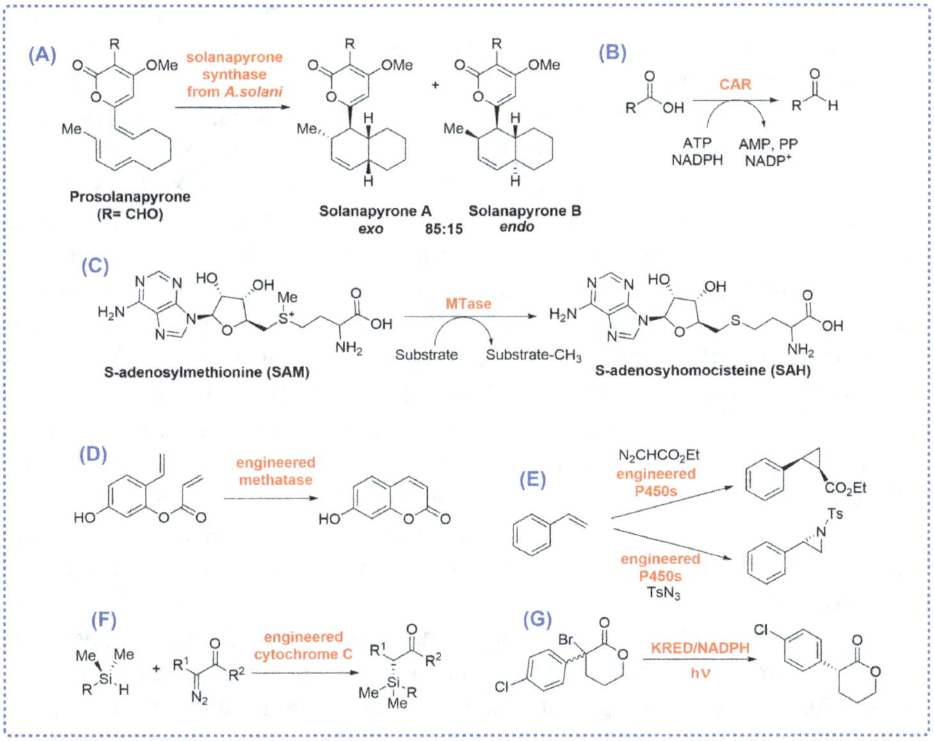

**Figure 5.12:** Novel enzyme classes were reported for the first time in the "fourth wave": (A) naturally evolved [4 + 2] cyclases (Diels-Alderases); (B) carboxylic acid reductases (CARs); (C), *S*-adenosyl methionine-dependent methyltransferases; (D) artificial metalloenzymes for catalyzing olefin metathesis; (E) engineered P450s for catalyzing "promiscuous" reactions such as cyclopropanations or aziridinations; (F) engineered cytochrome for Si–C bond formation; (G) radical-induced dehalogenation catalyzed by light-excited ketoreductases.

## 5.3.6 Current Scenario: The Fourth Wave

Similarly to what was referred to about the third wave, we owe the first reference to a fourth wave in an article from Bornscheuer in 2018 [101], in which this author argued that a fourth wave of biocatalysis was emerging due to the combination of advanced enzyme design with expanding reaction capabilities. This integrative approach enables the tailored design of enzymes rapidly for targeted reactions like asymmetric synthesis. Novel enzyme classes and nonnatural reactions were being incorporated into protein scaffolds, expanding the biocatalytic repertoire beyond what was previously possible.

In this sense, novel enzyme classes were reported for the first time, therefore expanding the reaction types. Amongst these, one interesting example is the use of naturally evolved [4 + 2] cyclases (Diels-Alderases) [181–183], illustrated in Figure 5.12A by the first case reported [184, 185], the cyclation of prosolanapyrone to furnish solanapyrones A and B by using a synthase from *Aternaria solani* (although this example would not belong to the fourth wave, if we attend to the publication dates!). Some other representative cases are carboxylic acid reductases (Figure 5.12B) [186] or *S*-adenosyl methionine-dependent methyltransferases (Figure 5.12C) [187].

On the other hand, engineering novel nonnatural reactions in enzymes allowed the implementation of innovative biotransformations, such as the use of artificial metalloenzymes for catalyzing olefin metathesis (Figure 5.12D) [188], engineered P450s for catalyzing "promiscuous" reactions such as cyclopropanations [189] or aziridinations (Figure 5.12E) [190], or novel Si–C bond formation by engineered cytochrome C (Figure 5.12F) [191]. Another very original process is the radical-induced dehalogenation enabled by light-excited ketoreductases [192]. In this innovative approach, a ketoreductase enzyme was used to tightly position the NAD(P)H cofactor near its active site. When blue light was used to excite the cofactor, this setup enabled the enzyme to catalyze a novel type of reaction: radical-driven, stereoselective dehalogenation (Figure 5.12G). Additionally, by choosing between (*R*)- and (*S*)-selective ketoreductases, the reaction consistently produced products with the corresponding absolute stereochemistry.

Conducting a brief survey of the numerous enzymatic processes that align with this fourth wave, the majority are predominantly related to the pharmaceutical industry [5, 6, 70, 193–198]. Once again, it would be impossible to comment on all of them, so we will mention some representative cases. For instance, researchers at Genentech and Roche developed a KRED process integral to the synthesis of ipatasertib [199], an investigational-targeted cancer therapy drug that works by inhibiting the AKT protein, a key component of the PI3K/AKT signaling pathway that is frequently overactive in many cancers and drives tumor cell growth and survival. (Figure 5.13A, left). A proprietary engineered biocatalyst from Codexis (which harnesses its CodeEvolver® directed evolution technology to iteratively refine native enzymes for enhanced performance, robustness, and tailored traits, having supplied recombinant variants for over two decades [200]) effected the diastereomerically precise reduction to the key alcohol, employing *iso*propanol as a sacrificial reductant to recycle NADPH. Early iterations relied on GDH/glucose cofactor loops, but the streamlined mono-enzyme format simplified execution. Substrate instability in preliminary runs prompted high-loading slurry operations, securing outstanding conversions. This biocatalytic tactic outshone a rival ruthenium-mediated asymmetric transfer hydrogenation owing to unmatched diastereocontrol and persistent metal residue challenges. Curiously, Genentech and Roche opted for nitrile kinetic resolution (Figure 5.13A, right), ramped to 200 g with dense substrate charging, using a c-LEcta-sourced enzyme to preferentially cleave the unwanted (*R*)-nitrile to its (*R*)-acid, preserving the target (*S*)-nitrile intact.

Amongst oxidative biocatalysts, monooxygenases stand out for their proficiency in inserting one oxygen atom from $O_2$ into diverse substrates, frequently exhibiting exceptional chemo-, regio-, and enantiocontrol. Baeyer-Villiger monooxygenases (BVMOs)

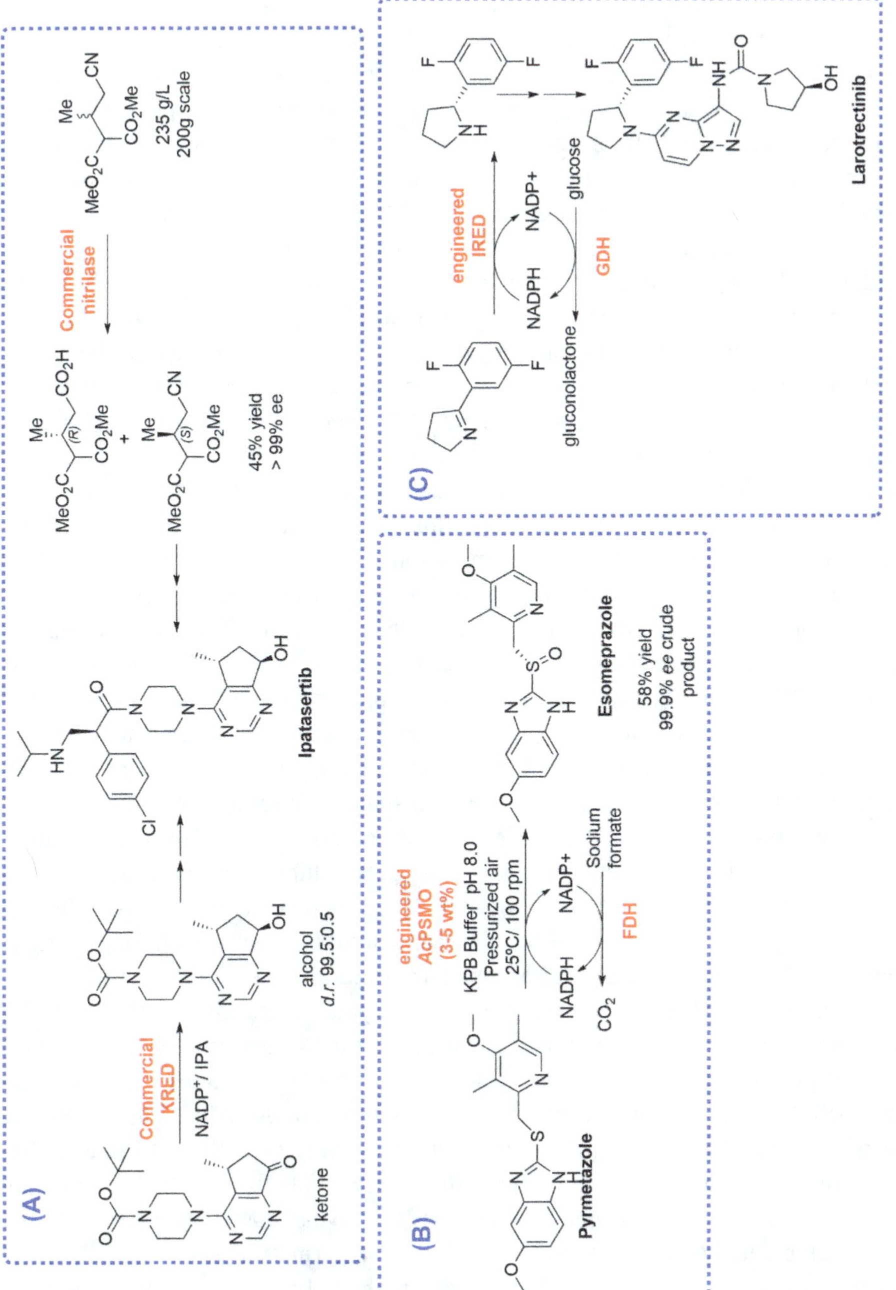


**Figure 5.13:** Illustrative cases of biocatalyzed synthesis of drugs: (A) enzymatic synthesis of Ipatasertib; (B) enzymatic synthesis of esomeprazole; and (C) enzymatic synthesis of larotrectinib.

constitute a specific category, enabling Baeyer-Villiger oxidations alongside heteroatom oxygenations, including sulfur [201, 202]. These flavoproteins accommodate flavin non-covalently and rely on NADPH as their cofactor, recycled through supporting enzyme cascades such as formate dehydrogenase (FDH), glucose or glucose-6-phosphate dehydrogenase (GDH or G6PDH), or ketoreductases (KREDs). Nevertheless, scaling up encounters hurdles, particularly the requisite for equimolar oxygen – rendering mass transfer of $O_2$ paramount – and the high cost of NADPH. A prime pharmaceutical instance of BVMO deployment is the synthesis of esomeprazole (Nexium®), the (*S*)-enantiomer of omeprazole [20], the pioneering member of proton pump inhibitors. Its manufacture entails preliminary chemical assembly of pyrmetazole (Figure 5.13B), a pivotal thioether precursor. Racemic omeprazole emerges from oxidation of the prochiral sulfide. Paralleling the montelukast biocatalytic strategy (Figure 5.11A), late-stage chiral resolution for esomeprazole proves impractical, demanding separation and reprocessing of the unwanted enantiomer alongside vast quantities of auxiliaries and solvents. Hence, asymmetric pyrmetazole oxidation emerged as the linchpin for securing the desired enantiomer. Conventional sulfoxidation – via a refined Kagan protocol with titanium catalysts and chiral modifiers – remained moisture-labile, susceptible to sulfone overoxidation, and reliant on aggressive, eco-harmful agents yielding copious residues [20]. To overcome these drawbacks, Codexis pioneered a biocatalytic route at scale in 2018, utilizing cyclohexane monooxygenase (CHMO) from *Acinetobacter calcoaceticus* NCIMB 9871 [203]. Recognized among the most versatile BVMOs for oxidizing both linear and cyclic ketones, as well as sulfoxides [204], CHMO's deployment at scale was initially curtailed by its vulnerability to heat and organic media, creating scope for biocatalyst refinement. Native CHMO delivered sluggish pyrmetazole turnover, spurring 19 iterative directed evolution cycles across myriad mutants to boost catalytic throughput, enantio- and chemocontrol, and suppress sulfone formation. The evolved variant achieved a staggering 140,000-fold activity surge relative to the parent. Optimization encompassed oxygen delivery, substrate loadings, catalase dosing to scavenge peroxide, and NADPH recycling via isopropanol with KRED. At optimized settings – 30 g scale in phosphate buffer (pH 9.0) plus 4% v/v *iso*propanol, featuring BVMO, KRED CDX-019, catalase, and $NADP^+$ – the process furnished extractive isolation (isobutyl methyl ketone) of product at 99% purity, 87% isolated yield, and 99.9% e.e. Subsequent tailoring recast CHMO as a streamlined pyrmetazole monooxygenase (AcPSMO) [205]. Pilot-scale biocatalysis (300 L) for omeprazole manufacture has employed the FDH/sodium formate regimen for cofactor turnover, providing irreversible formate-to-$CO_2$ oxidation sans the need for acetone stripping or elevated isopropanol doses typical of KRED setups [206]. Methanol (5% v/v) served as cosolvent for pyrmetazole solubilization; AcPSMO and FDH sustained near-complete activity, with methanol readily stripped at low temperature – vital given omeprazole's instability beyond 40 °C. Oxygen transfer intensified via elevated kLa and air pressurization over pure $O_2$. pH was tuned to 8.0, which preserved biocatalyst and NADPH integrity. From 600 g pyrmetazole input, biooxidation closed with NaOH addition yielding omeprazole sodium salt, harvested by membrane filtration to slash downstream ethyl acetate demand. Final drying delivered 380 g esomeprazole (58% yield) at 99.1% purity and >99% e.e. This debut large-scale AcPSMO deployment forged a green-by-design esomeprazole route, ripe for activity refinement.

Imine reductases (IREDs) are enzymes that catalyze the reduction of imines to amines, often utilized in the synthesis of chiral amines for pharmaceuticals and other applications [207]. They typically depend on the cofactor NADPH to carry out redox reactions involving carbon-nitrogen bonds. IREDs exhibit high enantioselectivity and have been used for reducing cyclic imines, as well as performing reductive amination of ketones with amines. Some IREDs can catalyze both imine formation and reduction and are termed reductive aminases (RedAms). In a very interesting process, an engineered IRED (ScIRED-R3-V4) obtained from the wild-type *Streptomyces clavuligerus* through three rounds of directed evolution showed a more than 100-fold increase in specific activity alongside a greater than 270-fold enhancement in stability. Using this refined enzyme, the cyclic imine substrate 2-(2,5-difluorophenyl)-pyrroline (Figure 5.13C) underwent complete reduction at 80 g/L concentration, furnishing kilogram quantities of a vital chiral amine for Larotrectinib (VITRAKVI®) synthesis (an oral, small-molecule, highly selective tropomyosin receptor kinase inhibitor [208]). This process delivered an 82.5% yield with e.e. surpassing 99.5%. Space-time yield reached 352 g/L per day. Moreover, crystallographic enzyme-substrate complex studies coupled with molecular dynamics simulations unveiled the molecular underpinnings of the boosted catalytic prowess.

Additionally, the implementation of cascades of evolved enzymes [209, 210] has enabled efficient syntheses of highly complex molecules, including pharmaceuticals, as well as bulk chemicals, through the integration of engineered enzyme pathways within microbial systems. Protein engineering and pathway optimization play a critical role in improving yields and selectivity. Some very illustrative examples can be found in the literature [196], e.g., the industrial synthesis of islatravir (the first example of a drug called nucleoside reverse transcriptase translocation inhibitor, used as an antiviral drug for HIV-1 infection and pre-exposure prophylaxis to prevent HIV infection); this biotransformation, developed by Merck & Co., Inc. and Codexis, Inc. [211], is an impressive example of enzymatic cascades. The procedure involves nine enzymes (five engineered and four auxiliary) working together in a streamlined, in vitro multistep pathway beginning from simple achiral building blocks. The main steps of the cascade (depicted in Figure 5.14A) are as follows:

1) *Oxidative desymmetrization of 2-ethynylglycerol to 2-ethynylglyceraldehyde catalyzed by an evolved galactose oxidase (GOase) variant.* This copper-dependent enzyme selectively oxidizes one primary alcohol to an aldehyde with reversed enantioselectivity (favoring the (*R*)-enantiomer) through multiple rounds of evolution. Auxiliary enzymes, catalase, and horseradish peroxidase remove hydrogen peroxide and maintain copper oxidation-state balance. This step establishes the first stereocenter with about 90:10 enantiomeric ratio, which is further improved by partial overoxidation during the reaction.

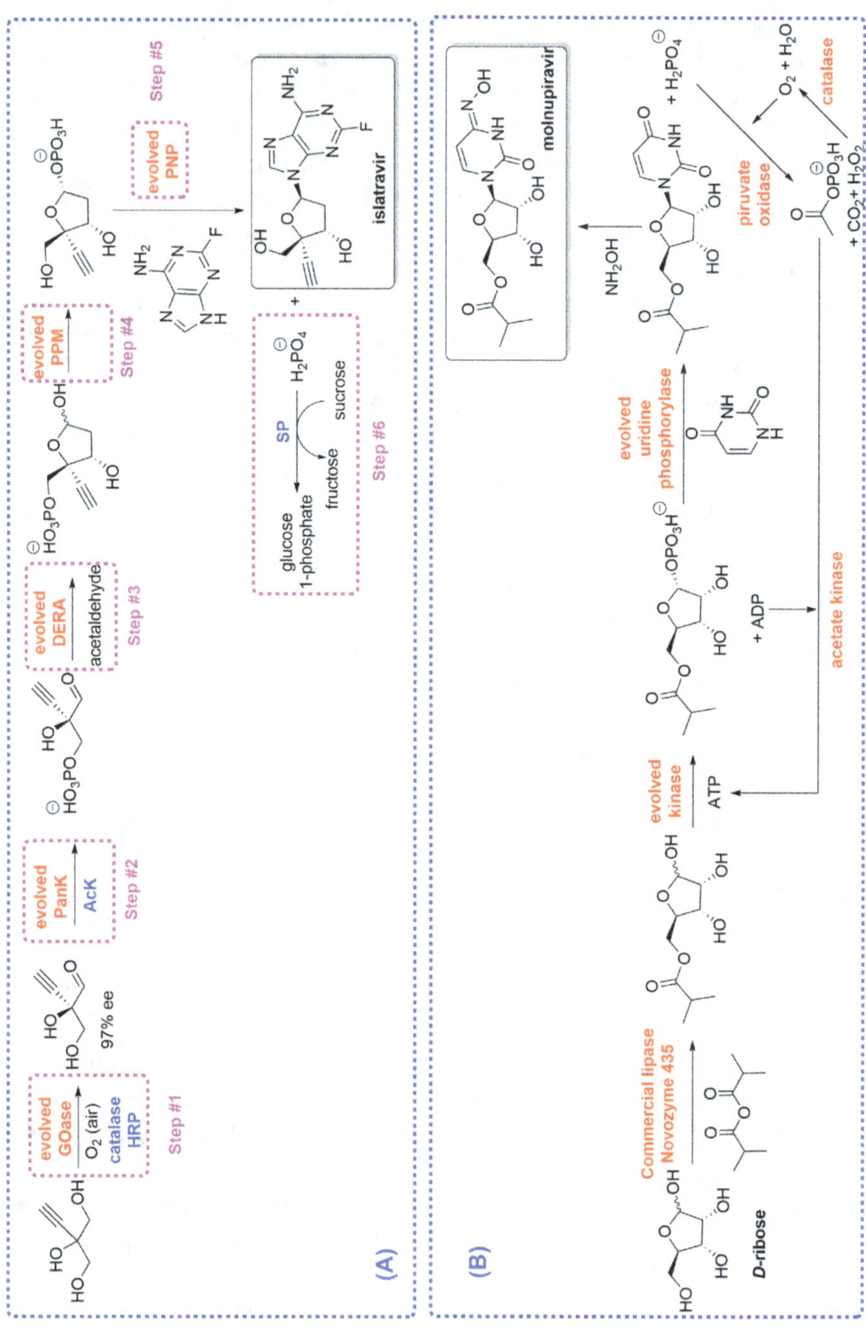


**Figure 5.14:** Enzymatic cascades to produce drugs: (A) fully enzymatic synthesis of islatravir and (B) enzymatic synthesis of molnupiravir.

2) *Phosphorylation of 2-ethynylglyceraldehyde to 2-ethynylglyceraldehyde 3-phosphate, catalyzed by an evolved pantothenate kinase (PanK) from Escherichia coli.* This enzyme was engineered to efficiently phosphorylate the aldehyde substrate and tolerate high substrate concentrations. ATP is regenerated *in situ* using acetyl phosphate as the phosphate donor with the help of acetate kinase (AcK) from *Thermotoga maritima.*
3) *Aldol addition by deoxyribose 5-phosphate aldolase (DERA).* The DERA enzyme from *Shewanella halifaxensis* catalyzed the stereoselective aldol addition of acetaldehyde to 2-ethynylglyceraldehyde 3-phosphate, generating a sugar phosphate intermediate with high diastereoselectivity and two stereocenters. The enzyme was evolved to tolerate high acetaldehyde concentrations and to maintain stereoselectivity.
4) *Phosphoryl transfer by phosphopentomutase (PPM).* PPM from *E. coli* converts 2-deoxyribose 1-phosphate to 2-deoxyribose 5-phosphate. This step is reversible and initially equilibrium-limited, so it benefits from coupling the subsequent glycosylation steps and phosphate removal to drive the reaction forward.
5) *Glycosylation by purine nucleoside phosphorylase.* Purine nucleoside phosphorylase catalyzes the formation of islatravir by coupling the sugar phosphate to the nucleobase. The enzyme was evolved for enhanced activity and stability under the reaction conditions.
6) *Phosphate removal and equilibrium shifting using sucrose phosphorylase.* Sucrose phosphorylase removes inhibitory inorganic phosphate formed during glycosylation by converting it to glucose 1-phosphate, helping to shift the equilibria of reversible reactions toward product formation and improving overall cascade efficiency.

Regarding cofactor regeneration and stability, the cascade operates with ATP regeneration via acetate kinase and acetyl phosphate, as well as enzymatic removal of harmful byproducts (such as hydrogen peroxide). Remarkably, the enzymes involved in the cascade were engineered for enhanced activity, stability, and selectivity, making the cascade compatible in a single reaction mixture under mild aqueous conditions without isolating intermediates. The overall cascade integrates these enzymatic steps into a three-step biocatalytic cascade yielding islatravir with over 50% overall yield, excellent stereochemical purity, and high atom economy, significantly reducing the number of synthesis steps compared to previous chemical routes.

Another recent cascade exemplifies the enzymatic manufacture of molnupiravir (Figure 5.14B), an antiviral prodrug targeting SARS-CoV-2's RNA-dependent RNA polymerase – a major leap forward [212]. Approved by the UK MHRA as Lagevrio in November 2021 and granted US FDA emergency use authorization on November 30, 2021, its initial chemical route spanned 10 steps with <10% overall yield. This was telescoped to a three-step pathway at 69% yield, isolating just one intermediate (Figure 5.14B), via innovative sustainable nucleotide assembly. Pivotal elements encompassed cheap feedstocks, engineered ribosyl-1-kinase and uridine phosphorylase, a phosphate loop with pyruvate oxidase and acetate kinase, plus scalable fine-tuning. Step one acylates D-ribose with *iso*butyric anhydride through lipase catalysis, affording high-yield 5-isobutyryl-D-ribose. Engineered ribosyl-1-kinase (from 5*S*-methyl-thioribose kinase) then phosphorylates it with ATP to 5-*iso*butyryl-D-ribose-1-phosphate, boasting 100-fold activity uplift and >90% conversion at <1 wt% loading. Uridine phosphorylase – boosted 80-fold over five evolution rounds – condenses this with uracil to 5-isobutyryluridine at full conversion and low enzyme use. The final chemical step deploys hydroxylamine, ammonium hydrogen sulfate, and imidazole in hexamethyldisilazane solvent to forge the oxime molnupiravir.

Finally, we must mention how the availability of massive sequence data from genomes and metagenomes has provided a rich source of new enzymes to be used in this fourth wave. However, the challenge lies in functionally annotating these sequences accurately and using bioinformatics tools to guide enzyme discovery and engineering (please see Chapter 6 for detailed information). A special mention must be done regarding the impressive advances in protein structure prediction driven by the progress in computational methods for predicting protein folding and dynamics, as recognized in 2024 with the Nobel Prize in Chemistry [213–216] to Demis Hassabis and John M. Jumper, who shared a quarter of the Nobel Prize in Chemistry for the creation of AlphaFold (which recently has been updated to AlphaFold3 [217]), while the other half was awarded to David Baker, creator of Rosetta [218]. These tools have boosted the field of rational enzyme design, overcoming the major bottlenecks caused by the large gap between known sequences and solved structures.

Regarding this last point, the tremendous advance in the rational design of biocatalysts, an archetypal example is the design of enzymes capable of efficiently catalysing the Kemp elimination, a chemical reaction in which a proton is removed by a base from carbon at position 3 of a benzisoxazole, resulting in the ring-opening to yield a phenolic product (Figure 5.15a). It is used as a classic model for studying proton abstraction and is particularly important in the field of enzyme design, as there are no natural [220]enzymes known to catalyze this reaction, making it a benchmark for testing artificial and computationally designed enzymes.

In fact, Baker and coworkers published in 2008 a seminal paper reporting the first example of the rational design of a "Kemp eliminase" [219]. The process started with the selection of catalytic mechanisms, guided by quantum mechanical transition state calculations to model ideal active sites for the reaction. Two mechanistic motifs were chosen to mimic the hypothetical transition state, either a carboxylate base (using the side chain of aspartate or glutamate) or a His-Asp dyad, by combining a histidine (as a catalytic base) polarized by an adjacent aspartate or glutamate. To fulfill with the canonical "three points rule," governing the enzymatic recognition of substrates, additional catalytic elements such as hydrogen bond donors and aromatic residues (for π-stacking) were incorporated to further stabilize the transition state. Subsequently, the RosettaMatch algorithm was deployed to identify suitable placements of these catalytic arrangements in a large set of protein scaffold crystal struc-

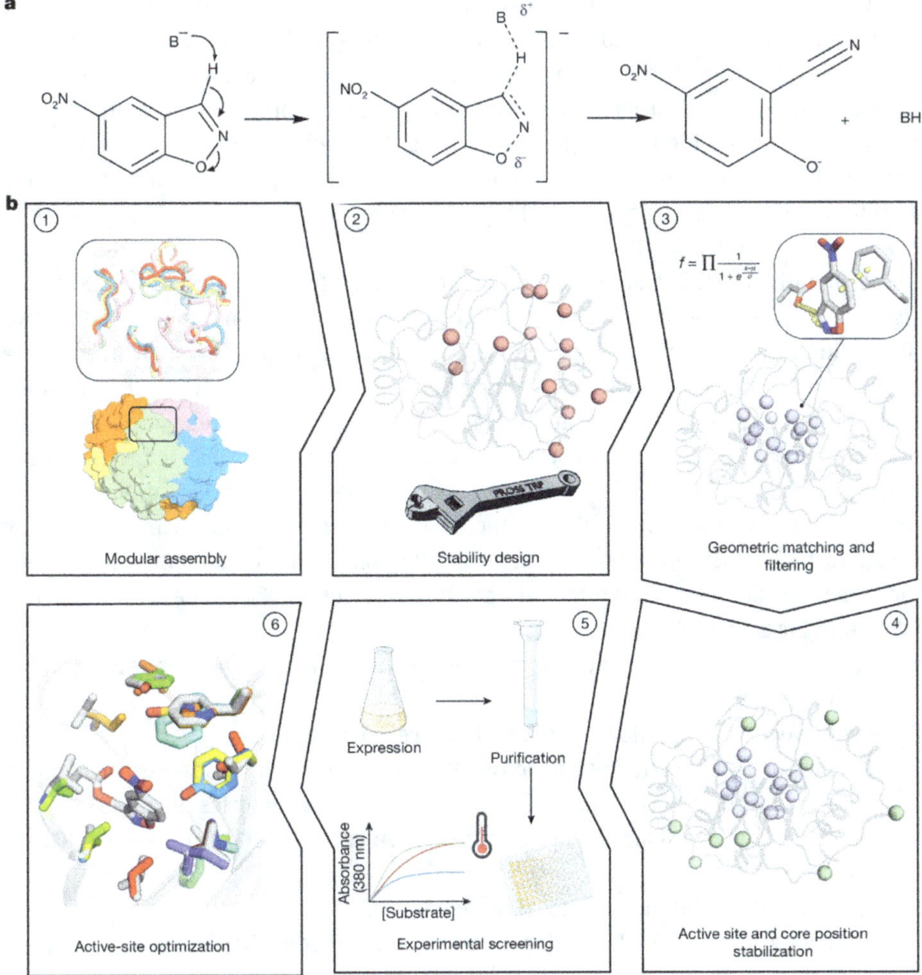


**Figure 5.15:** (A) Kemp elimination; (B) computational design of Kemp eliminases (KE). Reprinted from Listov et al. [222].

tures of various folds (e.g., TIM barrels, β-propellers, and Rossmann folds). In this way, surrounding residues were redesigned to maximize stability and transition state affinity using Rosetta's design tools, and models were filtered for catalytic geometry and computed transition state binding energy. For the experimental validation, 59 designs across 17 scaffolds were tested experimentally. Of these, eight showed measurable enzymatic activity toward the Kemp elimination, with both base mechanisms (carboxylate and His-Asp dyad) yielding successful enzymes. Kinetic characterization revealed rate enhancements up to $10^6$ fold over the uncatalyzed reaction and multiple catalytic turnovers. These results, although impressive, were leading to artificial

Kemp eliminases with relatively low catalytic efficiencies and turnover rates, typically in the range of $k_{cat}/K_M$ = 1–420 $M^{-1}$ $s^{-1}$ and $k_{cat}$ = 0.006–0.7 $s^{-1}$. Thus, to further improve the activity of the designed enzymes, the authors subjected one design (KE07) to seven rounds of directed evolution, combining random mutagenesis and DNA shuffling in *E. coli*, followed by high-throughput screening. Evolved variants carried 4–8 mutations and exhibited over 200-fold improvements in catalytic efficiency, exceeding a million-fold over background, reaching efficiencies similar to natural enzymes ($k_{cat}/K_M$ = $10^5$ $M^{-1}$ $s^{-1}$ and $k_{cat}$ = 10 $s^{-1}$).

As mentioned, this chapter set a new benchmark for computational enzyme design, but in general, as it was necessary to use directed evolution to have really active Kemp eliminases, it could be argued (as it was the case not only during the third wave [220] but also even in more recent times [221]) that rational design was still some steps behind directed evolution in the creation of fully functional enzymes. Nevertheless, the enormous increase in computational enzymatic design has led to balancing both strategies; in this sense, again a paper from Baker and coworkers described, also for the creation of Kemp eliminases, the first fully computational workflow for the design of highly efficient enzymes catalyzing the Kemp elimination without the need for experimental optimization by mutant-library screening [222]. The design pipeline (shown in Figure 5.13b) was:

1. Thousands of protein backbones arise via combinatorial backbone assembly (Figure 5.13.b1).
2. PROSS fortification follows (Figure 5.13.b2, red spheres).
3. Geometric docking and Rosetta active-site refinement (purple spheres) spawn millions of designs, filtered by stability and activity energy metrics (Figure 5.13.b3).
4. Dozens of top candidates undergo core (green spheres) and active-site polishing (Figure 5.13.b4).
5. Experimental validation ensues (Figure 5.13.b5).
6. Functional hits receive FuncLib active-site tuning (Figure 5.13.b6).

Thus, three engineered Kemp eliminases achieve catalytic efficiencies exceeding 2,000 $M^{-1}$ $s^{-1}$. The variant with the highest performance contains over 140 substitutions compared to any naturally occurring protein and featured an entirely new active site. This enzyme maintains exceptional thermal stability (above 85 °C) and demonstrates superior catalytic properties, with an efficiency of 12,700 $M^{-1}$ $s^{-1}$ and a turnover rate of 2.8 $s^{-1}$, providing an improvement of nearly 100-fold over previous computationally designed enzymes. Even more, introducing a residue previously regarded as indispensable in all Kemp eliminase designs further elevates efficiency above 105,000 $M^{-1}$ $s^{-1}$ and enhances the reaction rate to 30 $s^{-1}$, exceeding all prior de novo computational enzyme designs by up to two orders of magnitude. Then, we can be sure that the fourth wave of biocatalysis is underway, driving the field toward smarter, more adaptable, and industrially relevant biocatalytic solutions.

Another hallmark of the fourth wave is the full implementation of microbial cell factories – engineered microbes tailored for optimized production of target chemicals from renewables like nonedible biomass or $CO_2$ [223]. These microorganisms host biosynthetic pathways – natural, reconstructed, or fully synthetic – calibrated to channel low-cost, sustainable carbon feedstocks into target chemicals via fermentation or bioconversion. Relative to conventional chemical routes, microbial cell factories deliver benefits like mild operating regimes (reduced temperature/pressure), renewable inputs, and circumvention of noxious solvents or catalysts. They foster greener chemical production as a petrochemical alternative. Recent illustrative cases of microbial cell factories bearing engineered enzymes include:

1. *E. coli* and *Saccharomyces cerevisiae* capable to furnish amino acid and bio-based chemical production. Systems metabolic engineering in *E. coli* boosts L-valine, L-lysine, propan-1-ol, and mevalonic acid outputs through engineered enzymes and pathways, guided by genome-scale metabolic models for superior yields and fluxes [224].
2. Vitamin $B_{12}$ biosynthesis assembles an intricate de novo pathway in *E. coli*, merging endogenous genes with 28 heterologous ones from diverse species (*Sinorhizobium meliloti, Rhodobacter capsulatus, Salmonella typhimurium, Rhodopseudomonas palustris*, and *Brucella melitensis*). Modular partitioning facilitates microbial output of this vital cofactor (adenosylcobalamin) [225].
3. Carminic acid production. A biosynthetic pathway for carminic acid was constructed in *E. coli*, involving a combination of plant and bacterial genes. Key enzymes such as an engineered monooxygenase and glucosyltransferase were optimized by computational simulations, resulting in enhanced production of this natural red pigment [226]
4. Production of *cis,cis*-muconic acid (ccMA) – a prized platform chemical for sustainable polymers – employing an engineered *E. coli* strain [227]. Engineered *E. coli* strains drive ccMA production from vanillin (VA), a lignin-derived aromatic. This biological route outshines chemical synthesis by leveraging renewable biomass for greener outcomes. Cells express four recombinant enzymes: vanillin dehydrogenase (LigV), vanillic acid *O*-demethylase (VanAB), protocatechuate decarboxylase (AroY), and catechol 1,2-dioxygenase ($C_{12}O$) [228–230]. This pathway converts vanillin (VA) to ccMA at >95% yield in resting cell assays. Optimized growth and feeding conditions propelled growing cells to ~5.2 g/L ccMA from 40 mM VA, equating to ≈91.7% bioconvsersion yield. Notably, this growing cell bioprocess outpaced resting cells with 4-fold higher titer and 17-fold superior volumetric productivity (biomass-normalized), alongside greener metrics – markedly reduced E-factor and PMI for diminished waste and resource demands per product.

Then again, the growing implementation of flow biocatalysis can be considered part of the fourth wave of biocatalysis, which is characterized by the integration of biocatalysis with process intensification, automation, and advanced engineering techniques [231–236]. In fact, flow biocatalysis aligns well with this framework as it enables:

- Continuous operation with enhanced space-time yields
- Better control of reaction parameters
- Reduced enzyme inhibition through substrate feeding strategies
- Simplified product separation
- Increased catalyst stability and reuse through enzyme immobilization
- Process intensification facilitating scale-up and industrial relevance

## 5.4 Conclusion and Future Outlook

Industrial biocatalysis has reached a pivotal moment, demonstrating its transformative potential in enabling sustainable, high-efficiency manufacturing across pharmaceuticals, food, biofuel, textiles, and numerous other sectors. The integration of enzymes into industrial processes has drastically reduced energy use, hazardous reagents, and waste generation, while simultaneously improving product quality and economic viability. Through successive waves of technological advancement, from early immobilization techniques to the latest breakthroughs in computational design and multienzyme cascades, the field has overcome major obstacles related to scalability, stability, and process control. In fact, enzymatic processes have revolutionized industrial chemistry by providing precise, selective catalysis under mild reaction conditions. Notable successes include their use in producing high-value active pharmaceutical ingredients, environmentally benign detergents, and sustainable biofuels, catalyzed by robust and increasingly engineerable enzymes. Innovations like directed evolution and rational design have expanded the repertoire of functional enzymes, enabling tailored solutions for complex synthesis and even nonnatural transformations.

Despite the numerous benefits, several technical and economic challenges remain. The stability of enzymes under harsh industrial conditions is a longtime bottleneck; enzymes must often endure elevated temperatures, extremes of pH, and exposure to solvents without significant loss of activity. Cost-effective production and maintenance of enzymes, especially for high-volume applications, demands further optimization and improved process economics. Another critical challenge is the sustainability and scalability of enzyme supply chains, including life-cycle analysis, raw material sourcing, and consistent regulatory compliance. Finally, wider industrial adoption is often hampered by gaps in awareness, education, and investment, particularly among smaller enterprises and in regions outside North America and Europe.

The future of industrial biocatalysis is exceptionally bright, shaped by several forward-looking trends:

- **Multienzyme cascades:** Continued development of intelligent, modular enzyme cascades – supported by predictive modeling, coexpression in suitable hosts, and

flow biocatalysis – will enable one-pot syntheses of complex molecules and valorization of waste streams.
- **Sustainability metrics:** Life cycle assessment and routine use of "green chemistry metrics" (e.g., atom efficiency, *E*-factor, and solvent intensity) are guiding process optimization, ensuring industry targets both environmental and economic sustainability.
- **Computational and machine learning tools:** Computational modeling – epitomized by tools like AlphaFold and Rosetta – is unleashing rapid, precise enzyme design. AI-driven rational design promises the routine creation of enzymes for novel transformations, further boosting efficiency and selectivity.
- **Regulatory and market drivers:** Stricter environmental regulations and consumer demand for sustainable products are accelerating adoption, with projections indicating robust market growth in Asia-Pacific and other emerging economies.
- **Industrial resilience and longevity:** Advances in protein engineering are steadily improving enzyme stability, operational lifetimes, and adaptability to diverse industrial environments, unlocking new value propositions in both high- and low-margin markets.

Enzymes will continue to underpin the transition to greener, more circular manufacturing paradigms, combining technical sophistication with scalable, affordable, and eco-friendly solutions. The synergy of biotechnology, computational design, process intensification, and clear sustainability metrics promises to overcome current bottlenecks and foster transformative changes across industries worldwide. Cross-disciplinary collaboration, sustained investment, and policy support will be critical to unlocking the full potential of industrial biocatalysis and ensuring its future as a cornerstone of sustainable innovation.

## References

[1] Woodley JM. A perspective on process design and scale-up for biocatalysis. ChemCatChem 2025;n/a:e00794.
[2] Sharma N, Ahlawat YK, Stalin N, Mehmood S, Morya S, Malik A, et al. Microbial enzymes in industrial biotechnology: Sources, production, and significant applications of lipases. J Ind Microbiol Biotechnol 2025;52.
[3] Krueger J, Dieskau AP, Hassfeld J, Gries J, Block O, Weinmann H, et al. Chemical process development in the pharmaceutical industry in Europe – insights and perspectives from industry scientists. Angew Chem Int Ed 2025;64.
[4] Grigorakis K, Ferousi C, Topakas E. Protein engineering for industrial biocatalysis: Principles, approaches, and lessons from engineered PETases. Catalysts 2025;15.

[5] Falcioni F, Humphreys L, Lloyd RC, Wu H, Martinez I, Jones J, et al. The evolving landscape of industrial biocatalysis in perspective from the ACS green chemistry institute pharmaceutical roundtable. ACS Catal 2025;10780–10794.

[6] Bayer T, Wu S, Snajdrova R, Baldenius K, Bornscheuer U. An update: Enzymatic synthesis for industrial applications. Angew Chem Int Ed Engl 2025;e202505976.

[7] Lozano P, García-Verdugo E. From green to circular chemistry paved by biocatalysis. Green Chem 2023;25:7041–7057.

[8] Sheldon RA, Woodley JM. Role of biocatalysis in sustainable chemistry. Chem Rev 2018;118:801–838.

[9] Bilal M, Iqbal HMN. State-of-the-art strategies and applied perspectives of enzyme biocatalysis in food sector – Current status and future trends. Crit Rev Food Sci Nutr 2020;60:2052–2066.

[10] Londoño-Hernández L, Sampedro LJG, Restrepo LMS, Lopera LMS, Castellanos NAM, Madroñero J. Trends, technological developments, and challenges in food biocatalysts for industrial applications. In Enzymatic Processes for Food Valorization Elsevier; vol. 2024, p. 347–367.

[11] Borges S, Brassesco ME, Cunha SA, Coscueta ER, Pintado M. Recent trends in biocatalysis and its application in the food industry. In Enzymatic Processes for Food Valorization Elsevier; vol. 2024, p. 265–284.

[12] Kreuz A, Da Silva Db, Andreaus J. Enzymes for sustainable textile processing recent advances. In Advances in Renewable Natural Materials for Textile Sustainability 2024, CRC Press; p. 244–262.

[13] Zafar MG, Mumtaz A, Akbar A, Liaqat I, Hassan M, Mahmood M, et al. Distinctive role of enzymes in textile industry. In Arshad M, editors. Enzymes in Textile Processing: A Climate Changes Mitigation Approach: Textile Industry, Enzymes, and SDGs 2025, Singapore: Springer Nature Singapore; p. 1–17.

[14] Masood H, Khan S, SM K, Akbar A, Rehman AU, Rahman UU, et al. Enzyme applications in textile industry: A step toward sustainable development goals. In Arshad M, editors. Enzymes in Textile Processing: A Climate Changes Mitigation Approach: Textile Industry, Enzymes, and SDGs 2025, Singapore: Springer Nature Singapore; p. 19–33.

[15] Hameed U, Hayat MT, Fatima I, Makumbura MRC. Enzymes in textile – A step towards sustainability. In Arshad M, editors. Enzymes in Textile Processing: A Climate Changes Mitigation Approach: Textile Industry, Enzymes, and SDGs 2025, Singapore: Springer Nature Singapore; p. 35–85.

[16] Khan MF. Recent advances in microbial enzyme applications for sustainable textile processing and waste management. Sci 2025;7:46.

[17] Ishak SNH, Rnzra R, Kamarudin NHA, Leow ATC, Ali MSM. Expanding the horizon of biodiesel production via enzyme engineering. Int J Green Energy 2024;21:3367–3390.

[18] Ummalyma SB, Bhaskar T. Recent advances in the role of biocatalyst in biofuel cells and its application: An overview. Biotechnol Genet Eng Rev 2024;40:2051–2089.

[19] Simić S, Zukić E, Schmermund L, Faber K, Winkler CK, Kroutil W. Shortening synthetic routes to small molecule active pharmaceutical ingredients employing biocatalytic methods. Chem Rev 2022;122:1052–1126.

[20] de Gonzalo G, Alcántara AR, de María P D, Sánchez-Montero JM. Biocatalysis for the asymmetric synthesis of Active Pharmaceutical Ingredients (APIs): This time is for real. Expert Opin Drug Discov 2022;17:1159–1171.

[21] Fernandes GFS, Kim SH, Castagnolo D. Harnessing biocatalysis as a green tool in antibiotic synthesis and discovery. RSC Adv 2024;14:30396–30410.

[22] Aksenov AS, Sinelnikov IG, Shevchenko AR, Mayorova KA, Chukhchin DG, Osipov DO, et al. Enzymatic conversion of wood materials from the pulp and paper industry. Appl Biochem Microbiol 2024;60:448–456.

[23] Gupta GK, Dixit M, Pandey D, Kapoor RK, Kango N, Shukla P. Microbial enzyme bioprocesses in biobleaching of pulp and paper: Technological updates. In Microbial Bioprocesses: Applications and Perspectives Elsevier; vol. 2023, p. 319–337.

[24] Gupta GK, Dixit M, Kapoor RK, Shukla P. Xylanolytic enzymes in pulp and paper industry: New technologies and perspectives. Mol Biotechnol 2022;64:130–143.
[25] Podrepšek GH, Knez Ž, Leitgeb M. Industrial production of enzymes for use in animal-feed bioprocessing. Valorization of biomass to bioproducts: Biochemicals and biomaterials. Elsevier 2023:349–387.
[26] De Oliveira Sousa T, Da Silva N A, De Melo Oliveira V, Da Silva Ramos Av, Barbosa Filho JPM, Jmds B, et al. Use of proteases for animal feed supplementation: Scientific and technological updates. Prep Biochem Biotechnol 2025; in press.
[27] Ojha BK, Singh PK, Shrivastava N. Enzymes in the animal feed industry. In Enzymes in Food Biotechnology: Production, Applications, and Future Prospects 2018, Elsevier; p. 93–109.
[28] Plouhinec L, Neugnot V, Lafond M, Berrin JG. Carbohydrate-active enzymes in animal feed. Biotechnol Adv 2023;65.
[29] Sharma B, Dangi AK, Shukla P. Contemporary enzyme based technologies for bioremediation: A review. J Environ Manage 2018;210:10–22.
[30] Patel M, Kumar R, Kishor K, Mlsna T, Pittman CU, Mohan D. Pharmaceuticals of emerging concern in aquatic systems: Chemistry, occurrence, effects, and removal methods. Chem Rev 2019;119:3510–3673.
[31] Ghattavi S, Homaei A. Advanced Enzymatic Systems for Wastewater Treatment Future Prospects. Microbes and Enzymes for Water Treatment and 2025, Remediation: CRC Press; p. 99–117.
[32] Holtmann D, Hollmann F, Bouchaut B. Contribution of enzyme catalysis to the achievement of the United Nations' sustainable development goals. Molecules 2023;28:4125.
[33] Alcántara AR, Domínguez de María P, Littlechild J, Schürmann M, Sheldon RA, Wohlgemuth R. Biocatalysis as key to sustainable industrial chemistry. ChemSusChem 2022;e202102709.
[34] Vargas-Bernal R. Enzymes as emerging biocatalysts for biotransformation processes. In Zero Waste Management Technologies 2024, Springer Nature; p. 264–283.
[35] Shakilanishi S, Shanthi C. An overview on preparation of enzymes for industrial use. Biocatal Biotransform 2024;42:485–496.
[36] Sheldon RA. Green chemistry and biocatalysis: Engineering a sustainable future. Catal Today 2024;431.
[37] Bora D. Enzymes Market Size, Share Trends Analysis Report By Application (Industrial and Specialty Enzymes), By Source (Microorganisms, Plants, Animals), By Application (Food & Beverages, Household care, Bioenergy) and By Region (North America, Europe, APAC, Middle East and Africa, LATAM) Forecasts 2025–2033 2025, Straits Research.
[38] Enzymes Market Size, Share, Growth, Trends 2025 To 2034, 2025, Cervicorn Consulting.
[39] Enzymes Market Size, Share, Trends and Forecast by Type, Source, Reaction Type, Application, and Region 2025–2033, 2025, IMARC Group.
[40] Bidwai S, Shivarkar A. Enzymes Market Size, Share, and Trends 2025 to 2034, 2025, Precedence Research's.
[41] Sanders AD, Cheung LKY, Houfani AA, Grahame DAS, Bryksa BC, Dee DR, et al. Chapter 1 – A history of enzymes and their applications in the food industry. In Yada RY, Dee DR, editors. Improving and Tailoring Enzymes for Food Quality and Functionality 2024, Woodhead Publishing; Second Edition, p. 1–15.
[42] Kousoulis AA, Tsoucalas G, Armenis I, Marineli F, Karamanou M, Androutsos G. From the "hungry acid" to pepsinogen: A journey through time in quest for the stomach's secretion. Ann Gastroenterol 2012;25:119–122.
[43] Vandamme EJ, De Mol ML. Pioneer scientists at the origin of enzymology, biocatalysis and industrial enzymes 1833–1950. Biocatal Biotransform 2025.
[44] Payen A, Persoz JF. Mémoire sur la Diastase, les Principaux Produits de ses Réactions et leurs Applications Aux Arts Industrielles. Ann Chim 1833;53:73–92.

[45] Duclaux E. Traité de microbiologie 1899, Paris: Masson; vol. 2.
[46] Garg SK, Johri B. Rennet: Current trends and future research. Food Rev Int 1994;10:313–355.
[47] Kühne W. Über das Verhalten verschiedener organisirter und sog. ungeformter Fermente. In Verhandlungen des Naturhistorisch-medicinischen Vereins zu Heidelberg 1877;Neue Folge vol. 1: p. 190–193.
[48] Karplus PA, Pearson MA, Hausinger RP. 70 years of crystalline Urease: What have we learned?. Acc Chem Res 1997;30:330–337.
[49] Röhm O. Verfahren zur Reinigung von Textilien mittels tryptischer Fermente 1913, Germany: Röhm & Haas.
[50] Schweizerische Ferment AG. [Process for Producing Detergents Containing Bacterial Enzymes 1955, Netherlands: Schweizerische Ferment AG.
[51] Enzymes in Detergency 1997, New York: Marcel Dekker, Inc.; vol. 69.
[52] Greup DH, Hintzer HMR. The use of fungal enzymes for breadmaking purposes. In in Second International Congress for Fermentation Industries1952. Belgium: Central Institute for Nutrition Research T.N.O. Knocke; p. 232–338.
[53] Kornbrust BA, Forman T, Matveeva I. Applications of enzymes in breadmaking. In Breadmaking: Improving Quality Elsevier; vol. 2020, p. 415–440.
[54] Marshall RO, Kooi ER. Enzymatic conversion of D-glucose to D-fructose. Science 1957;125:648–649.
[55] Nam KH. Engineering Xylose Isomerase for industrial applications. Catalysts 2024;14.
[56] Barreto MQ, Garbelotti CV, Lopes DCB, Soares JDM, Ward RJ. Xylose isomerase: From fundamental research to applied enzyme technology. J Biotechnol 2025;404:39–54.
[57] Sternberg M. Microbial Rennets. Adv Appl Microbiol 1976;20:135–157.
[58] Abada EA. Application of microbial enzymes in the dairy industry. In Enzymes in Food Biotechnology: Production, Applications, and Future Prospects 2018, Elsevier; p. 61–72.
[59] Uniacke-Lowe T, Fox PF. Chapter 4 – Chymosin, Pepsins, and Other Aspartyl Proteinases: Structures, functions, catalytic mechanism and milk-clotting properties. In McSweeney PLH, et al. editors. Cheese (Fifth Edition) 2025San Diego: Academic Press; p. 73–131.
[60] Garg SK, Johri B. Rennet: Current trends and future research. Food Rev Int 1994;10:313–355.
[61] Akishev Z, Auyez M, Tursunbekova A, Khassenov B. Comparative biochemical properties of recombinant goat and calf chymosins and their implications in dairy processing. Sci Rep 2025;15.
[62] Maghraby YR, El-Shabasy RM, Ibrahim AH, Azzazy HME. Enzyme immobilization technologies and industrial applications. ACS Omega 2023;8:5184–5196.
[63] Motta JFG, De Freitas BCB, De Almeida AF, Martins GAS, Borges SV. Use of enzymes in the food industry: A review. Food Sci Technol 2023;43.
[64] Jothyswarupha KA, Venkataraman S, Rajendran DS, Shri SSS, Sivaprakasam S, Yamini T, et al. Immobilized enzymes: Exploring its potential in food industry applications. Food Sci Biotechnol 2025;34:1533–1555.
[65] Nelson JM, Griffin EG. Adsorption of invertase. J Am Chem Soc 1916;38:1109–1115.
[66] Tosa T, Mori T, Fuse N, Chibata I. Studies on continuous enzyme reactions. I. Screening of carriers for preparation of water-insoluble aminoacylase. Enzymologia 1966;31:214–224.
[67] Yushkova ED, Nazarova EA, Matyuhina AV, Noskova AO, Shavronskaya DO, Vinogradov VV, et al. Application of immobilized enzymes in food industry. J Agric Food Chem 2019;67:11553–11567.
[68] Robescu MS, Bavaro T. A comprehensive guide to enzyme immobilization: All you need to know. Molecules 2025;30:939.
[69] Buller R, Damborsky J, Hilvert D, Bornscheuer UT. Structure prediction and computational protein design for efficient biocatalysts and bioactive proteins. Angew Chem Int Ed 2025;64.
[70] Wohlgemuth R. Chapter 13 – Industrial asymmetric biocatalysis. In Gonzalo GD, Alcántara AR, editors. Biocatalysis in Asymmetric Synthesis 2024, Academic Press; p. 431–463.

[71] de Gonzalo G, Alcántara AR. Enzyme-catalyzed asymmetric synthesis. In Akiyama T, Ojima I, editors. Catalytic Asymmetric Synthesis 2022, John Wiley & Sons, Inc.; Fourth Edition.
[72] de María P D, de Gonzalo G, Alcántara AR. Biocatalysis as useful tool in asymmetric synthesis: An assessment of recently granted patents (2014–2019). Catalysts 2019;9:802.
[73] Alcántara AR. Biocatalysis and pharmaceuticals: A smart tool for sustainable development. Catalysts 2019;9:792.
[74] Hoyos P, Pace V, Alcántara AR. Chiral Building Blocks for Drugs Synthesis via Biotransformations. In Nag A, editors. Asymmetric Synthesis of Drugs and Natural Products 2018, Boca Raton, Florida: CRC Press; p. 346–448.
[75] Alcántara AR. Biotransformations in drug synthesis: A green and powerful tool for medicinal chemistry. J Med Chem Drug Des 2018;1:1–7.
[76] chemoselectivity. International Union of Pure and Applied Chemistry (IUPAC). 2025.
[77] Qiu J, Su E, Wang W, Wei D. Efficient asymmetric synthesis of D-*N*-formyl-phenylglycine via cross-linked nitrilase aggregates catalyzed dynamic kinetic resolution. Catal Commun 2014;51:19–23.
[78] Afagh NA, Yudin AK. Chemoselectivity and the curious reactivity preferences of functional groups. Angew Chem Int Ed 2010;49:262–310.
[79] Mondal D, Snodgrass HM, Gomez CA, Lewis JC. Non-native site-selective enzyme catalysis. Chem Rev 2023;123:10381–10431.
[80] Kinast G, Schedel M. A Four-Step Synthesis of 1-Deoxynojirimycin with a biotransformation as cardinal reaction step. Angew Chem Int Ed 1981;20:805–806.
[81] Sundaresan V, Abrol R. Biological chiral recognition: The substrate's perspective. In Proceedings of Chirality 2005, p. 15736174.
[82] Sundaresan V, Abro R. Towards a general model for protein-substrate stereoselectivity. Protein Sci 2002;11:1330–1339.
[83] Alcántara AR, de Gonzalo G. Chapter 1 – Introduction to asymmetric synthesis employing biocatalysts. In Gonzalo GD, Alcántara AR, editors. Biocatalysis in Asymmetric Synthesis 2024, Academic Press; p. 1–41.
[84] Pasteur L. Mémoire sur la relation qui peut exister entre la forme cristalline et la composition chimique, et sur la cause de la polarisation rotatoire. C R Acad Sci 1848;26.
[85] Flack HD. Louis Pasteurs discovery of molecular chirality and spontaneous resolution in 1848, together with a complete review of his crystallographic and chemical work. Acta Crystallogr A 2009;65:371–389.
[86] Vantomme G, Crassous J. Pasteur and chirality: A story of how serendipity favors the prepared minds. Chirality 2021;33:597–601.
[87] Sui J, Wang N, Wang J, Huang X, Wang T, Zhou L, et al. Strategies for chiral separation: From racemate to enantiomer. Chem Sci 2023;14:11955–12003.
[88] Piutti A. Una nuova specie di asparagine. L'Orosi-Giornale di Chimica, Farmacia e Scienze Affini 1886;9:198–202.
[89] Colli L, Guarna A. The dextrorotatory sweet asparagine of Arnaldo Piutti: the original product is conserved in Florence. Substantia. 2018;2(2):125–30. DOI: 0.13128/Substantia-66 CC BY
[90] Gal J. Pasteur and the art of chirality. Nat Chem 2017;9:604–605.
[91] Pasteur L. Mémoire sur la fermentation de l'acide tartrique. 1858.
[92] Pasteur L. Suite à une précédente communication sur les mycodermes: Nouveau procédé industriel de fabrication du vinaigre. 1862.
[93] Brown AJ. Enzyme action. J Chem Soc 1902;81:373–388.
[94] Fischer E. Einfluss der Configuration auf die Wirkung der Enzyme. Berichte der deutschen chemischen Gesellschaft 1894;27:2985–2993.
[95] Lichtenthaler FW. 100 Years "Schlüssel-Schloss-Prinzip": What Made Emil Fischer Use this Analogy?. Angew Chem Int Ed Engl 1995;33:2364–2374.

[96] Buchner E, Rapp R. Alkoholische Gährung ohne Hefezellen. Berichte der deutschen chemischen Gesellschaft 1899;32:127–137.
[97] Wöhler F, Liebig J. Ueber die Bildung des Bittermandelöls. Annalen der Pharmacie 1837;22:1–24.
[98] Rosenthaler L. Durch enzyme bewirkte asymmetrische synthesen. Biochem Z 1908;14:238–253.
[99] Pyser JB, Chakrabarty S, Romero EO, Narayan ARH. State-of-the-art biocatalysis. ACS Cent Sci 2021;7:1105–1116.
[100] Bornscheuer UT, Huisman GW, Kazlauskas RJ, Lutz S, Moore JC, Robins K. Engineering the third wave of biocatalysis. Nature 2012;485:185–194.
[101] Bornscheuer UT. The fourth wave of biocatalysis is approaching. Philos Trans R Soc A 2018;376:7.
[102] Hanefeld U, Hollmann F, Paul CE. Biocatalysis making waves in organic chemistry. Chem Soc Rev 2022.
[103] Warburg O, Christian W. Pyridin, der wasserstoffübertragende Bestandteil von Gärungsfermenten. Helv Chim Acta 1936;19:E79–E88.
[104] Sulek K. Nobel prize in 1931 for Otto Warburg for discovery of the respiratory enzyme. Wiad Lek 1968;21:329.
[105] Neuberg C, Hirsch J. Über ein kohlenstoffketten knüpfendes ferment (carboligase). Biochem Z 1921;115:282–310.
[106] Hilderbrandt G, Klavehn W. Verfahren zur Herstellung von 1-1-Phenyl-2-methylaminopropan-1-ol, 1930 Germany: Knoll AG.
[107] Kluyver AJ, De Leeuw FJG. *Acetobacter suboxydans*, een merkwaardige azijnbacterie. In Tijdschr. Verg. Geneesk. 10, 170 1924, Tijdschr Verg Geneesk; vol. 10, p. 170–281.
[108] Reichstein T, Grüssner A. Eine ergiebige Synthese der l-Ascorbinsäure (C-Vitamin). Helv Chim Acta 1934;17:311–328.
[109] Sarett LH. Partial synthesis of pregnene-4-triol-17(β),20(β),21-dione-3,11 and pregnene-4-diol-17 (β),21-trione-3,11,20 monoacetate. J Biol Chem 1946;162:601–631.
[110] Bhatti HN, Khera RA. Biological transformations of steroidal compounds: A review. Steroids 2012;77:1267–1290.
[111] Antrim RL, Colilla W, Schnyder BJ. Glucose isomerase production of High-Fructose Syrups. Appl Biochem Bioeng 1979;2:97–155.
[112] Vojcic L, Pitzler C, Körfer G, Jakob F, Maurer KH, Schwaneberg U. Advances in protease engineering for laundry detergents. New Biotechnol 2015;32:629–634.
[113] Bolivar JM, Woodley JM, Fernandez-Lafuente R. Is enzyme immobilization a mature discipline? Some critical considerations to capitalize on the benefits of immobilization. Chem Soc Rev 2022.
[114] Boudrant J, Woodley JM, Fernandez-Lafuente R. Parameters necessary to define an immobilized enzyme preparation. Process Biochem 2020;90:66–80.
[115] Chibata I, Tosa T, Sato T, Mori T, Yamamoto K. Applications of immobilized enzymes and immobilized microbial cells for L-Amino Acid production. In Weetall HH, Suzuki S, editors. Immobilized Enzyme Technology: Research and Applications 1975, Boston, MA: Springer US; p. 111–127.
[116] Illanes A, Valencia P. Industrial and therapeutic enzymes: Penicillin Acylase. In Current Developments in Biotechnology and Bioengineering: Production, Isolation and Purification of Industrial Products 2016, Elsevier Inc.; p. 267–305.
[117] Li K, Mohammed MAA, Zhou Y, Tu H, Zhang J, Liu C, et al. Recent progress in the development of immobilized penicillin G acylase for chemical and industrial applications: A mini-review. Polym Adv Technol 2020;31:368–388.
[118] Claridge CA, Gourevitch A, Lein J. Bacterial penicillin amidase. Nature 1960;187:237–238.
[119] Huang HT, English AR, Seto TA, Shull GM, Sobin BA. Enzymatic hydrolysis of the side chain of penicillins [8]. J Am Chem Soc 1960;82:3790–3791.

[120] Rolinson GN, Batchelor FR, Butterworth D, Cameron-wood J, Cole M, Eustace GC, et al. Formation of 6-aminopenicillanic acid from penicillin by enzymatic hydrolysis. Nature 1960;187:236–237.
[121] Verweij J, De Vroom E. Industrial transformations of penicillins and cephalosporins. Recueil des Travaux Chimiques des Pays-Bas 1993;112:66–81.
[122] Matsumoto K. Production of 6-APA, 7-ACA, and 7-ADCA by immobilized penicillin and cephalosporin amidases. Bioprocess Technol 1993;16:67–88.
[123] Weissenburger HWO, van der Hoeven MG. An efficient nonenzymatic conversion of benzylpenicillin to 6-aminopenicillanic acid. Rec Trav Chim Pays-Bas 1970;89:1081–1084.
[124] Rasyidah M, Sismindari S, Purwanto P. Cephalosporin C acylase: Important role, obstacles, and strategies to optimize expression in E. coli. J Appl Pharm Sci 2024;14:15–24.
[125] Parmar A, Kumar H, Marwaha SS, Kennedy JF. Recent trends in enzymatic conversion of cephalosporin C to 7-aminocephalosporanic acid (7-ACA). Crit Rev Biotechnol 1998;18:1–12.
[126] Pan X, Xu L, Li Y, Wu S, Wu Y, Wei W. Strategies to Improve the Biosynthesis of β-Lactam Antibiotics by Penicillin G Acylase: Progress and Prospects. Front Bioeng Biotechnol 2022;10.
[127] Rodriguez-Herrera R, Puc LEC, Sobrevilla JMV, Luque D, Cardona-Felix CS, Aguilar-González CN, et al. Enzymes in the pharmaceutical industry for β-lactam antibiotic production. In Enzymes in Food Biotechnology: Production, Applications, and Future Prospects 2018, Elsevier; p. 627–643.
[128] Bruggink A, Roos EC, de Vroom E. Penicillin acylase in the industrial production of beta-lactam antibiotics. Org Process Res Dev 1998;2:128–133.
[129] Li X, Wang J, Su W, Li C, Qu G, Yuan B, et al. Characterization and engineering of cephalosporin C acylases to produce 7-Aminocephalosporanic acid. Mol Catal 2023;550.
[130] Ibrahim AA, El-Housseiny GS, Aboshanab KM, Stratmann A, Yassien MA, NA H. Scaling up production of cephalosporin C by *Acremonium chrysogenum* W42-I in a fermenter using submerged fermentation. AMB Express 2024;14.
[131] Lin X, Kück U. Cephalosporins as key lead generation beta-lactam antibiotics. Appl Microbiol Biotechnol 2022;106:8007–8020.
[132] Lin X, Lambertz J, Dahlmann TA, Nowaczyk MM, König B, Kück U. A Straightforward Approach to Synthesize 7-Aminocephalosporanic Acid In Vivo in the Cephalosporin C Producer Acremonium chrysogenum. J Fungi 2022;8.
[133] Ma X, Deng S, Su E, Wei D. One-pot enzymatic production of deacetyl-7-aminocephalosporanic acid from cephalosporin C via immobilized cephalosporin C acylase and deacetylase. Biochem Eng J 2015;95:1–8.
[134] Hoyos P, Pace V, Hernáiz MJ, Alcántara AR. Biocatalysis in the pharmaceutical industry. A greener future. Curr Green Chem 2014;1:155–181.
[135] Muñoz Solano D, Hoyos P, Hernáiz MJ, Alcántara AR, Sánchez-Montero JM. Industrial biotransformations in the synthesis of building blocks leading to enantiopure drugs. Bioresour Technol 2012;115:196–207.
[136] Kierkels JGT, Peeters WPH. Process for the enzymatic preparation of optically active transglycidic acid esters 1994, DSM NV.
[137] Matsumae H, Shibatani T. Purification and characterization of the lipase from *Serratia marcescens* sr41-8000 responsible for asymmetric hydrolysis of 3-phenylglycidic acid-esters. J Ferment Bioeng 1994;77:152–158.
[138] Matsumae H, Furui M, Shibatani T, Tosa T. Production of optically-active 3-phenylglycidic acid ester by the lipase from *Serratia marcescens* on a hollow-fiber membrane reactor. J Ferment Bioeng 1994;78:59–63.
[139] Serrano-Arnaldos M, Máximo-Martín MF, Montiel-Morte MC, Ortega-Requena S, Gómez-Gómez E, Bastida-Rodríguez J. Solvent-free enzymatic production of high quality cetyl esters. Bioprocess Biosyst Eng 2016;39:641–649.

[140] Radzi SM, Basri M, Salleh AB, Ariff A, Mohammad R, Rahman MBA, et al. Large scale production of liquid wax ester by immobilized lipase. J Oleo Sci 2005;54:203–209.
[141] Tsujita T, Sumiyoshi M, Okuda H. Wax ester-synthesizing activity of lipases. Lipids 1999;34:1159–1166.
[142] Trani M, Ergan F, André G. Lipase-catalyzed production of wax esters. J Am Oil Chem Soc 1991;68:20–22.
[143] Garcia T, Martinez M, Aracil J. Enzymatic synthesis of myristyl myristate. Estimation of parameters and optimization of the process. Biocatal Biotransform 1996;14:67–85.
[144] Montiel MC, Serrano M, Máximo MF, Gómez M, Ortega-Requena S, Bastida J. Synthesis of cetyl ricinoleate catalyzed by immobilized Lipozyme® CalB lipase in a solvent-free system. Catal Today 2015;255:49–53.
[145] Khan NR, Rathod VK. Enzymatic synthesis of cetyl oleate in a solvent-free medium using microwave irradiation and physicochemical evaluation. Biocatal Biotransform 2020;38:114–122.
[146] Zhang J, Xu J-H, Biocatalysis FMC. Cofactor Regeneration. In Encyclopedia of Industrial Biotechnology 2009, John Wiley & Sons, Inc..
[147] Uppada V, Bhaduri S, Noronha SB. Cofactor regeneration – An important aspect of biocatalysis. Curr Sci 2014;106:946–957.
[148] Berenguer-Murcia A, Fernandez-Lafuente R. New trends in the recycling of NAD(P)H for the design of sustainable asymmetric reductions catalyzed by Dehydrogenases. Curr Org Chem 2010;14:1000–1021.
[149] Hummel W. New alcohol dehydrogenases for the synthesis of chiral compounds. Adv Biochem Eng Biotechnol 1997;58:145–184.
[150] Hummel W, Riebel B. Alcohol dehydrogenase and its use for the enzymatic production of chiral hydroxy compounds 1999, United States: Forschungszentrum Julich GmbH.
[151] Patel RN, McNamee CG, Banerjee A, Howell JM, Robison RS, Szarka LJ. Stereoselective reduction of β-keto esters by *Geotrichum candidum*. Enzyme Microb Technol 1992;14:731–738.
[152] Vithalani P, Mahla P, Padhiar J, Bhanushali U, Bhatt N. Microbial inventions and improvement in industrial bioprocess development. In Industrial Microbiology and Biotechnology: A New Horizon of the Microbial World 2024, Springer Nature; p. 551–572.
[153] May O. Industrial enzyme applications – Overview and historic perspective. In Industrial Enzyme Applications wiley; vol. 2019, p. 3–24.
[154] Cheng Z, Xia Y, Zhou Z. Recent advances and promises in Nitrile Hydratase: From mechanism to industrial applications. Front Bioeng Biotechnol 2020;8.
[155] Kobayashi M, Nagasawa T, Yamada H. Enzymatic synthesis of acrylamide: A success story not yet over. Trends Biotechnol 1992;10:402–408.
[156] Stemmer WPC. Rapid evolution of a protein in vitro by DNA shuffling. Nature 1994;370:389–391.
[157] Mullis KB. The Polymerase Chain Reaction (Nobel Lecture). Angew Chem Int Ed Engl 1994;33:1209–1213.
[158] Arnold FH. Innovation by evolution: Bringing new chemistry to life (Nobel Lecture). Angew Chem Int Ed 2019;58:14420–14426.
[159] Analytics MR. Montelukast Sodium market competitor insights: Trends and opportunities. 2025–2033, 2025.
[160] King AO, Corley EG, Anderson RK, Larsen RD, Verhoeven TR, Reider PJ, et al. An Efficient Synthesis of LTD4 Antagonist L-699,392. J Org Chem 1993;58:3731–3735.
[161] Bollikonda S, Mohanarangam S, Jinna RR, Kandirelli VKK, Makthala L, Sen S, et al. An Enantioselective formal synthesis of Montelukast Sodium. J Org Chem 2015;80:3891–3901.
[162] Liang J, Lalonde J, Borup B, Mitchell V, Mundorff E, Trinh N, et al. Development of a Biocatalytic process as an alternative to the (-)-DIP-Cl-mediated asymmetric reduction of a key intermediate of Montelukast. Org Process Res Dev 2010;14:193–198.

[163] Hoyos P, Pace V, Alcántara AR. Biocatalyzed synthesis of statins: A sustainable strategy for the preparation of valuable drugs. Catalysts 2019;9(3):260.
[164] Research CM Global Atorvastatin Market Report 2025, 2025.
[165] Insights BR. Rosuvastatin Calcium Market Size & Forecast [2025–2033, 2025].
[166] Ma SK, Gruber J, Davis C, Newman L, Gray D, Wang A, et al. A green-by-design biocatalytic process for atorvastatin intermediate. Green Chem 2010;12:81–86.
[167] Schallmey M, Floor RJ, Hauer B, Breuer M, Jekel PA, Wijma HJ, et al. Biocatalytic and structural properties of a highly engineered halohydrin dehalogenase. ChemBiochem 2013;14:870–881.
[168] Ritter SK. Going green keeps getting easier. Chem Eng News 2006;84:24–27.
[169] Anastas P, Eghbali N. Green chemistry: Principles and practice. Chem Soc Rev 2010;39:301–312.
[170] Kelly SA, Mix S, Moody TS, Gilmore BF. Transaminases for industrial biocatalysis: Novel enzyme discovery. Appl Microbiol Biotechnol 2020;104:4781–4794.
[171] Alcántara AR, de Gonzalo G. Green pathways: Enhancing amine synthesis using deep eutectic solvents. Catalysts 2025;15:586.
[172] Savile CK, Janey JM, Mundorff EC, Moore JC, Tam S, Jarvis WR, et al. Biocatalytic asymmetric synthesis of Chiral Amines from Ketones applied to Sitagliptin Manufacture. Science 2010;329:305–309.
[173] Park ES, Kim M, Shin JS. Molecular determinants for substrate selectivity of ω-transaminases. Appl Microbiol Biotechnol 2012;93:2425–2435.
[174] Song Z, Zhang Q, Wu W, Pu Z, Yu H. Rational design of enzyme activity and enantioselectivity. Front Bioeng Biotechnol 2023, p. 11.
[175] Savile C, Mundorff E, Moore JC, Devine PN, Janey JM. Construction of Arthrobacter KNK168 transaminase variants for biocatalytic manufacture of sitagliptin 2010 Codexis. Inc.
[176] Cypes SH, Chen AM, Ferlita RR, Hansen K, Lee I, Vydra VK, et al. Phosphoric acid salt of a dipeptidyl peptidase-IV inhibitor 2008, Rahway, NJ, US): United States: Merck & Co., Inc.
[177] Cypes SH, Chen AM, Ferlita RR, Hansen K, Lee I, Vydra VK, et al. Preparation of phosphoric acid salt of a β-amino acid amide dipeptidyl peptidase-IV inhibitor and its monohydrate, 2005 USA: Merck & Co., Inc.; p. 33.
[178] Edmondson SD, Fisher MH, Kim D, Maccoss M, Parmee ER, Weber AE, et al. Beta-amino heterocyclic dipeptidyl peptidase inhibitors for the treatment or prevention of diabetes 2004, Rahway, NJ): United States: Merck & Co., Inc.
[179] Truppo MD. Biocatalysis in the pharmaceutical industry: The need for speed. ACS Med Chem Lett 2017;8:476–480.
[180] Iqbal Z, Sadaf S. A patent-based consideration of latest platforms in the art of directed evolution: A decade long untold story. Biotechnol Genet Eng Rev 2022;1–114.
[181] Byrne MJ, Lees NR, Han LC, Van Der Kamp MW, Mulholland AJ, Stach JEM, et al. The Catalytic Mechanism of a Natural Diels-Alderase Revealed in Molecular Detail. J Am Chem Soc 2016;138:6095–6098.
[182] Zorn K, Back CR, Barringer R, Chadimová V, Manzo-Ruiz M, Mbatha SZ, et al. Interrogation of an enzyme library reveals the catalytic plasticity of naturally evolved [4+2] Cyclases. ChemBiochem 2023;24.
[183] Klas K, Tsukamoto S, Sherman DH, Williams RM. Natural Diels-Alderases: Elusive and Irresistable. J Org Chem 2015;80:11672–11685.
[184] Oikawa H, Suzuki Y, Naya A, Katayama K, Ichihara A. First Direct Evidence in Biological Diels-Alder Reaction of Incorporation of Diene-Dienophile Precursors in the Biosynthesis of Solanapyrones. J Am Chem Soc 1994;116:3605–3606.
[185] Oikawa H, Katayama K, Suzuki Y, Ichihara A. Enzymatic activity catalysing exo-selective Diels-Alder reaction in solanapyrone biosynthesis. J Chem Soc Chem Commun 1995;1321–1322.

[186] Basri RS, Rahman RNZRA, Kamarudin NHA, Ali MSM. Carboxylic acid reductases: Structure, catalytic requirements, and applications in biotechnology. Int J Biol Macromol 2023;240.
[187] Bennett MR, Shepherd SA, Cronin VA, Micklefield J. Recent advances in methyltransferase biocatalysis. Curr Opin Chem Biol 2017;37:97–106.
[188] Jeschek M, Reuter R, Heinisch T, Trindler C, Klehr J, Panke S, et al. Directed evolution of artificial metalloenzymes for in vivo metathesis. Nature 2016;537:661–665.
[189] Coelho PS, Brustad EM, Kannan A, Arnold FH. Olefin cyclopropanation via carbene transfer catalyzed by engineered cytochrome P450 enzymes. Science 2013;339:307–310.
[190] Farwell CC, Zhang RK, McIntosh JA, Hyster TK, Arnold FH. Enantioselective enzyme-catalyzed aziridination enabled by active-site evolution of a cytochrome P450. ACS Cent Sci 2015;1:89–93.
[191] Kan SBJ, Lewis RD, Chen K, Arnold FH. Directed evolution of cytochrome c for carbon–silicon bond formation: Bringing silicon to life. Science 2016;354:1048–1051.
[192] Hyster TK, Ward TR. Genetic optimization of Metalloenzymes: Enhancing enzymes for non-natural reactions. Angew Chem Int Ed 2016;55:7344–7357.
[193] Wu S, Snajdrova R, Moore JC, Baldenius K, Bornscheuer UT. Biocatalysis: Enzymatic synthesis for industrial applications. Angew Chem Int Ed 2021;60:88–119.
[194] Biocatalysis WR. Key enabling tools from biocatalytic one-step and multi-step reactions to biocatalytic total synthesis. New Biotechnol 2021;60:113–123.
[195] Alcántara AR, Domínguez de María P, Littlechild JA, Schürmann M, Sheldon RA, Wohlgemuth R. Biocatalysis as key to sustainable industrial chemistry. ChemSusChem 2022;COVER PROFILE: e202200709.
[196] France SP, Lewis RD, Martinez CA. The evolving nature of biocatalysis in pharmaceutical research and development. JACS Au 2023.
[197] Khanra M, Ravichandiran V, Swain SP. Lipase enzymes for sustainable synthesis of pharmaceuticals and chiral organic building blocks. Adv Sustain Syst 2025;9.
[198] Hecht K, Buller R. Industrializing biocatalysis. Chimia 2025;79:522–527.
[199] Han C, Savage S, Al-Sayah M, Yajima H, Remarchuk T, Reents R, et al. Asymmetric Synthesis of Akt Kinase Inhibitor Ipatasertib. Org Lett 2017;19:4806–4809.
[200] Shaw A. Codexis panels help clients find biocatalysts. Ind Bioprocess 2007;29:2.
[201] Fürst MJLJ, Gran-Scheuch A, Aalbers FS, Mw F. Baeyer-Villiger Monooxygenases: Tunable Oxidative Biocatalysts. ACS Catal 2019;9:11207–11241.
[202] De Gonzalo G, Alcántara AR. Multienzymatic Processes Involving Baeyer–Villiger Monooxygenases. Catalysts 2021;11:605.
[203] Bong YK, Song S, Nazor J, Vogel M, Widegren M, Smith D, et al. Baeyer-Villiger Monooxygenase-mediated synthesis of Esomeprazole As an alternative for Kagan Sulfoxidation. J Org Chem 2018;83:7453–7458.
[204] Maczka W, Wińska K, Grabarczyk M. Biotechnological methods of sulfoxidation: Yesterday, today, tomorrow. Catalysts 2018;8.
[205] Zhang Y, Wu YQ, Xu N, Zhao Q, Yu HL, Xu JH. Engineering of Cyclohexanone Monooxygenase for the Enantioselective Synthesis of (S)-Omeprazole. ACS Sustain Chem Eng 2019;7:7218–7226.
[206] Xu N, Zhu J, Wu YQ, Zhang Y, Xia JY, Zhao Q, et al. Enzymatic Preparation of the Chiral (S)-Sulfoxide Drug Esomeprazole at Pilot-Scale Levels. Org Process Res Dev 2020;24:1124–1130.
[207] Mangas-Sánchez J. Synthesis of chiral amines employing imine reductases and reductive aminases. In Biocatalysis in Asymmetric Synthesis: A volume in Foundations and Frontiers in Enzymology Elsevier; vol. 2024, p. 209–236.
[208] Lj S. Larotrectinib: First global approval. Drugs 2019;79:201–206.
[209] Rosenthal K, Bornscheuer UT, Lütz S. Cascades of evolved enzymes for the synthesis of complex molecules. Angew Chem Int Ed 2022;61.

[210] Reisenbauer JC, Sicinski KM, Arnold FH. Catalyzing the future: Recent advances in chemical synthesis using enzymes. Curr Opin Chem Biol 2024;83.
[211] Huffman MA, Fryszkowska A, Alvizo O, Borra-Garske M, Campos KR, Canada KA, et al. Design of an in vitro biocatalytic cascade for the manufacture of islatravir. Science 2019;366:1255–1259.
[212] McIntosh JA, Benkovics T, Silverman SM, Huffman MA, Kong J, Maligres PE, et al. Engineered Ribosyl-1-Kinase Enables Concise Synthesis of Molnupiravir, an Antiviral for COVID-19 2021, ACS Central Science.
[213] Baker HL. Hassabis, and Jumper win 2024 Nobel Prize in Chemistry. C&EN Glob Enterp 2024;102:4–4.
[214] Graham F. Daily briefing: AlphaFold developers share Nobel Prize in Chemistry. Nature 2024.
[215] Abriata LA. The Nobel Prize in Chemistry: Past, present, and future of AI in biology. Commun Biol 2024;7.
[216] Aguéro-Pizzolo S, Bettler E, Gouet P. Nobel Prize in chemistry 2024: The revolution of artificial intelligence in structural biology. Medecine/Sciences 2025;41:367–373.
[217] Abramson J, Adler J, Dunger J, Evans R, Green T, Pritzel A, et al. Accurate structure prediction of biomolecular interactions with AlphaFold 3. Nature 2024;630:493–500.
[218] Rohl CA, Strauss CEM, Misura KMS, Baker D. Protein Structure Prediction Using Rosetta. Methods Enzymol 2004;383:66–93.
[219] Rothlisberger D, Khersonsky O, Wollacott AM, Jiang L, DeChancie J, Betker J, et al. Kemp elimination catalysts by computational enzyme design. Nature 2008;453:190–194.
[220] Porter JL, Boon PLS, Murray TP, Huber T, Collyer CA, Ollis DL. Directed evolution of new and improved enzyme functions using an evolutionary intermediate and multidirectional search. ACS Chem Biol 2015;10:611–621.
[221] Hossack EJ, Hardy FJ, Green AP. Building enzymes through design and evolution. ACS Catal 2023;13:12436–12444.
[222] Listov D, Vos E, Hoffka G, Hoch SY, Berg A, Hamer-Rogotner S, et al. Complete computational design of high-efficiency Kemp elimination enzymes. Nature 2025;643:1421–1427.
[223] Cho JS, Kim GB, Eun H, Moon CW, Lee SY. Designing Microbial cell factories for the production of chemicals. JACS Au 2022;2:1781–1799.
[224] Gb K, Hr K, Lee SY. Comprehensive evaluation of the capacities of microbial cell factories. Nat Commun 2025;16.
[225] Fang H, Li D, Kang J, Jiang P, Sun J, Zhang D. Metabolic engineering of *Escherichia coli* for de novo biosynthesis of vitamin B12. Nat Commun 2018;9.
[226] Yang D, Jang WD, Lee SY. Production of Carminic Acid by metabolically engineered Escherichia coli. J Am Chem Soc 2021;143:5364–5377.
[227] Molinari F, Salini A, Vittore A, Santoro O, Izzo L, Fusco S, et al. Bio-based production of cis,cis-muconic acid as platform for a sustainable polymers production. Bioresour Technol 2024;408:131190.
[228] Rosini E, Molinari F, Miani D, Pollegioni L. Lignin valorization: Production of high value-added compounds by engineered microorganisms. Catalysts 2023;13.
[229] Molinari F, Pollegioni L, Rosini E. Whole-cell bioconversion of renewable biomasses-related aromatics to cis,cis-Muconic Acid. ACS Sustain Chem Eng 2023;11:2476–2485.
[230] Vignali E, Pollegioni L, Di Nardo G, Valetti F, Gazzola S, Gilardi G, et al. Multi-enzymatic cascade reactions for the Synthesis of cis,cis-Muconic Acid. Adv Synth Catal 2022;364:114–123.
[231] Bigliardi M, Contente ML. Flow biocatalysis: A pathway to sustainable innovation. Biochemist 2025;47:12–16.
[232] Benítez-Mateos AI, Contente ML, Roura Padrosa D, Paradisi F. Flow biocatalysis 101: Design, development and applications. React Chem Eng 2021;6:599–611.

[233] Tang Z, Oku Y, Matsuda T. Application of immobilized enzymes in flow biocatalysis for efficient synthesis. Org Process Res Dev 2024;28:1308–1326.
[234] Tamborini L, Molinari F, Pinto A. Chapter 12 – Development of asymmetric biotransformations: Flow biocatalysis, photobiocatalysis, and microwave biocatalysis. In Gonzalo GD, Alcántara AR, editors. Biocatalysis in Asymmetric Synthesis 2024, Academic Press; p. 403–429.
[235] Meyer L-E, Hobisch M, Kara S. Process intensification in continuous flow biocatalysis by up and downstream processing strategies. Curr Opin Biotechnol 2022;78:102835.
[236] Peñafiel I, Cosgrove SC. Biocatalysis in Flow for Drug Discovery 2021, Top Med Chem: Springer Science and Business Media Deutschland GmbH; p. 275–316.

# Chapter 6
# Biocatalyst Engineering

**Abstract:** This chapter provides a comprehensive overview of enzyme engineering, a multidisciplinary field that combines molecular biology, biochemistry, and computational biology to design and optimize enzymes with improved stability, specificity, activity, and efficiency for diverse industrial, medical, and research applications. Enzymes, nature's highly selective and efficient catalysts, often require enhancement to function effectively under rigorous industrial conditions such as extreme temperatures, pH, and the presence of organic solvents. This chapter delineates three principal enzyme-engineering approaches: rational design, semirational design, and directed evolution. On the one hand, rational design leverages detailed knowledge of enzyme sequences, structures, and catalytic mechanisms to introduce precise amino acid substitutions aimed at improving desired enzyme traits, such as thermostability, solvent tolerance, and altered substrate specificity. This approach benefits from advances in protein structure determination and computational modeling, including AI-driven tools like AlphaFold and Rosetta, which enable accurate prediction and manipulation of enzyme three-dimensional (3D) structures. On the other hand, semirational design, a hybrid strategy, combines structural and evolutionary data with computational algorithms to identify mutational "hotspots" for focused mutagenesis, generating manageable libraries enriched for beneficial variants. This method balances precision and exploration, increasing engineering efficiency. Finally, directed evolution mimics natural selection by generating mutant libraries through random mutagenesis, recombination, or focused saturation mutagenesis, followed by iterative screening to isolate improved enzyme variants. This approach requires minimal prior structural knowledge and has been instrumental in evolving enzymes with enhanced stability, activity, and stereoselectivity, often achieving improvements unattainable by rational methods alone.

This chapter presents case studies demonstrating these techniques to enhance enzyme thermostability, activity, stereoselectivity, and substrate scope, illustrating the synergistic use of computational design and laboratory evolution. Emerging advances include the incorporation of noncanonical amino acids and the design of artificial metalloenzymes, expanding the functional repertoire of biocatalysts beyond natural capabilities.

## 6.1 Introduction

Enzymes have evolved over millions of years, acquiring specific properties to perform highly specialized metabolic functions within defined cellular contexts [1, 2]. These biological catalysts exhibit remarkable catalytic power, demonstrating high chemo-, regio-, and stereo-selectivity and accelerating even the most complex reactions by

 | https://doi.org/10.1515/9783111122991-006

many orders of magnitude (up to $10^{19}$ times), far exceeding the rates of similar non-catalyzed reactions. They achieve this by reducing the activation energy required for reactions to proceed, stabilizing transition states, and reducing entropy, often through precise molecular interactions that align substrates optimally within their active sites [3], as schematically depicted in Figure 6.1. This alignment not only lowers activation energy but also enhances reaction specificity, enabling enzymes to function effectively under mild conditions where synthetic catalysts might require harsher environments. This acceleration is crucial for supporting the biochemical processes necessary for life, allowing reactions to occur at rates compatible with cellular function [4].

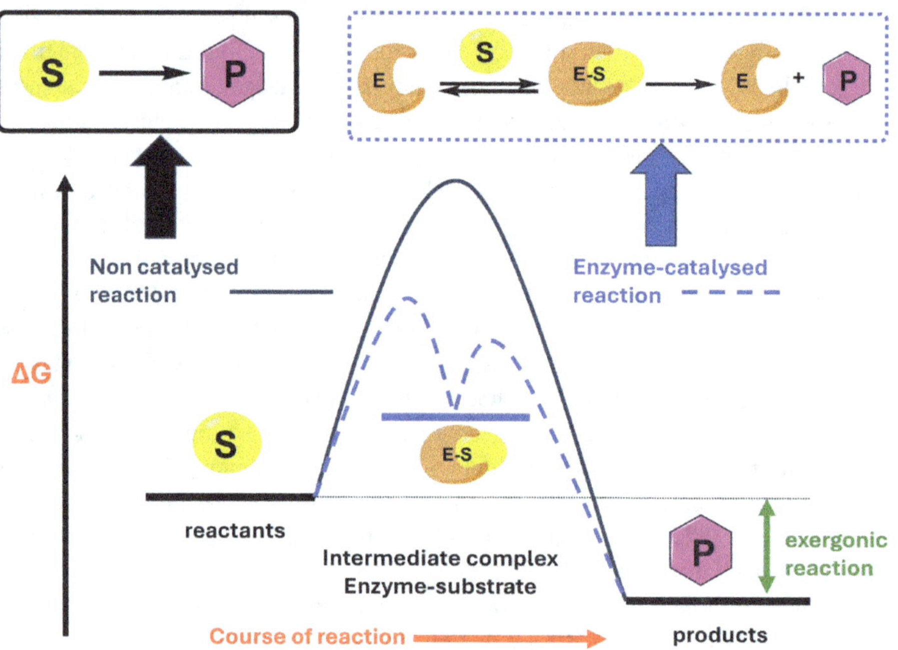


**Figure 6.1:** Schematic representation of enzyme-catalyzed versus noncatalyzed reactions.

Therefore, enzymatic catalysis represents an evolutionary optimization, allowing living organisms to perform complex chemical transformations quickly and efficiently, underscoring the unique role enzymes play in biocatalysis and metabolism.

In most organisms, enzymes are accustomed to functioning under mild reaction conditions (in aqueous solutions, at ambient temperature, and at neutral pH), also resulting in highly effective in *vitro* processes with a minimal environmental impact [5]. However, for the efficient biotechnological and industrial application of biocatalysts, enzymes must be active and stable under the rigorous conditions of these processes (presence of organic solvents or other nonaqueous media, high temperatures, extreme pH levels, etc.) [6–8]. Unfortunately, most enzymes, with the exception of a few from extreme environments [9, 10], tend to become inactive or denatured under demanding industrial conditions. In fact, the limitations of enzymes in catalysis have been widely acknowledged [11], including several critical challenges:

- Insufficient robustness: Enzymes often lack the stability needed to withstand harsh operating conditions such as high temperatures, extreme pH levels, or the presence of organic solvents.
- Limited activity: Many enzymes exhibit low catalytic efficiency under nonoptimal conditions, reducing their industrial applicability.
- Narrow substrate scope: Enzymes are highly specific, which can limit their ability to process a broad range of substrates, restricting their versatility.
- Suboptimal stereoselectivity: In some cases, enzymes fail to provide the desired stereochemical outcome, leading to insufficient or incorrect stereoselectivity.
- Suboptimal regioselectivity: Enzymes may also show insufficient or incorrect regioselectivity, making them unsuitable for certain complex transformations.
- Product inhibition: Some enzymes are inhibited by the very products they generate, hindering reaction progress and reducing overall yields.

Additionally, a significant barrier has been the lack of familiarity among organic chemists, largely using homogeneous and/or heterogeneous synthetic catalysts, with enzymatic processes [12]. This negative bias is further reinforced by the rather common idea (although not fully accurate) that enzymes are unable to catalyze many of the key reaction types that are pivotal to contemporary synthetic chemistry; this impression is being reconsidered, as the creation of à la carte artificial enzymes for catalyzing nonnatural enzymatic reactions is accepted nowadays [13–16]. In any case, these factors have historically constrained a broader implementation of enzymatic catalysis in industrial and synthetic applications [17, 18].

Consequently, over the past decades, research in protein engineering has been motivated by the need to generate more active, stable, and selective biocatalysts, capable of working under experimental conditions far from their usual ones. This progress has led to the creation of different methods for developing or enhancing new enzymatic activities and properties of biotechnological importance, while also shedding light on the structure-function relationship of many proteins. Enzyme engineering primarily relies on three approaches: rational design, directed evolution, and semirational design, a hybrid method that integrates elements of both. The choice of strategy is influenced by factors such as the feasibility of high-throughput screening (HTS) technology, the availability of structural and functional data on the protein, and a clear understanding of the desired trait (Figure 6.2). Selecting the most appropriate method for enzyme modification requires careful consideration of various scientific and technical aspects related to engineered enzymes. These different and complementary methods will be presented in the next sections.

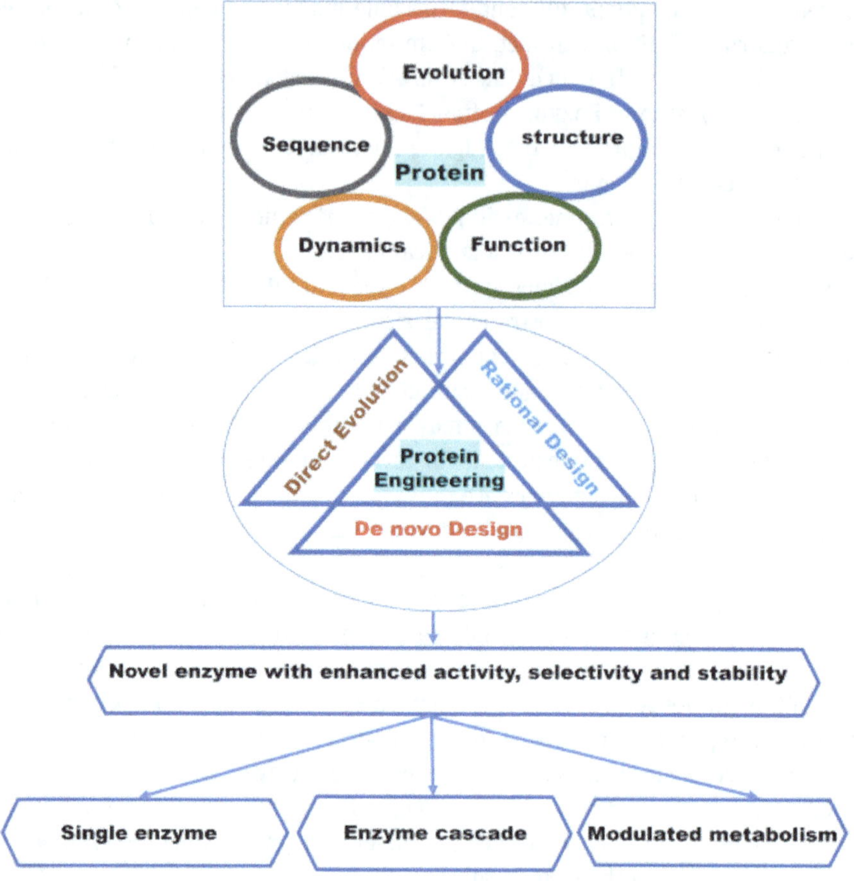


**Figure 6.2:** Different strategies for enzyme engineering. Reproduced from Nndochinwa et al. [18]. OPEN ACCESS.

## 6.2 Rational and Semirational Design of Enzymes

### 6.2.1 Rational Design of Enzymes

In the rational design of enzymes, by using information of the sequence, mechanism, and/or 3D structure of an enzyme, the aim is to perform a substitution of some amino acids at a given position of the enzyme for some others rationally selected, in order to obtain a new catalytic protein with improved properties, specifically leading to more robust and efficient biocatalysts for applications in organic chemistry and biotechnology [19, 20].

This approach was first envisioned in the 1970s [18]; However, a crucial step for the full implementation of this technique was the introduction of site-directed mutagenesis (SDM), as pioneered by Michael Smith in the early 1980s [21, 22], which earned him a Nobel Prize in 1993 [23], shared with Kary B. Mullis for his invention of the polymerase chain reaction (PCR) [24]. Using this technique (Figure 6.3), it became possible to make real what had previously been designed, allowing the substitution of any amino acid in a protein with one of the 19 other canonical amino acids.

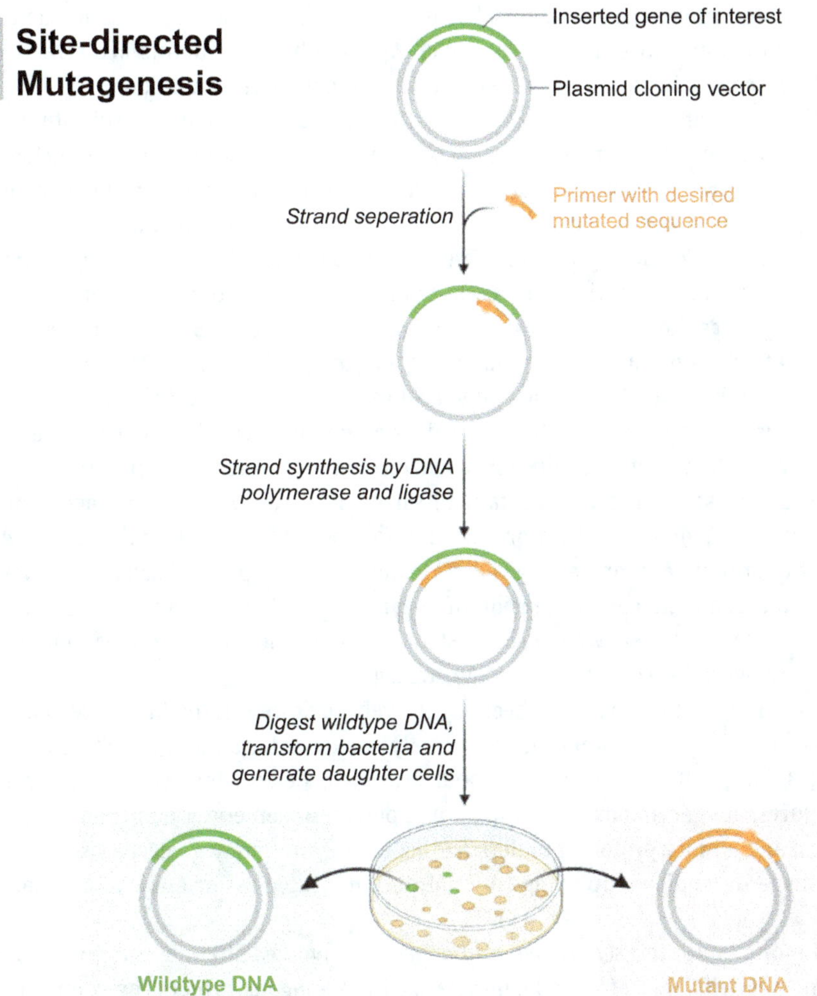


**Figure 6.3:** Site-directed mutagenesis (https://app.biorender.com/biorender-templates/figures/all/t-641db8b980ee2c368dafc89a-site-directed-mutagenesis). Reprinted from "Site-directed mutagenesis," by BioRender.com (2025). Retrieved from https://app.biorender.com/biorender-templates.

The overall process involves designing primers with the desired mutation(s) and using PCR to amplify and introduce changes into the enzyme's gene. In fact, the process begins with the creation of a short DNA primer, specifically designed to incorporate the desired mutation. This primer is synthesized to match the template DNA surrounding the mutation site, enabling it to hybridize effectively with the target gene. The mutation can take various forms, including a single nucleotide alteration (point mutation), multiple base modifications, deletions, or insertions. Once hybridized, the primer is extended by a DNA polymerase, which replicates the remainder of the gene sequence. The resulting DNA strand contains the intended mutation and is subsequently inserted into a vector for introduction into a host cell and cloning. Finally, the mutants are screened through DNA sequencing to confirm the presence of the intended mutation.

Identifying specific residues that serve as potential "hot spots" – where substitutions are most likely to enhance catalytic efficiency – remains a significant challenge. Achieving this requires precise knowledge of the enzyme's structure and mechanism, enabling informed predictions about which amino acids are essential for substrate recognition and transformation. Ideally, optimal efficiency would be attained through a single, well-informed prediction, eliminating the necessity of testing all 19 possible mutations at a given site – a process known as site-saturation mutagenesis (SSM) [25, 26], a semirational approach that will be discussed later. Enhancing enzyme performance is further complicated by the fact that altering a single amino acid residue is often insufficient.

In any case, rational design offers a distinct advantage over the other main engineering approach, directed evolution, as it circumvents the laborious and time-intensive task of screening vast mutant libraries. Rational design approaches are highly versatile, efficient, and hold promise for advancement into algorithmic models capable of quantitatively forecasting the performance of engineered sequences. Nevertheless, an incomplete comprehension of enzyme catalytic mechanisms, along with the intricate relationship between protein structure and function, can impact both the precision and overall success of this design strategy.

Rational design of proteins has been extensively reviewed in the literature [19, 27, 28]. There are *two general approaches*, namely *sequence-based* and *structure-based methods* [18, 19]. For the development of both strategies, the growing availability of protein structures, advancements in computational power, the emergence of sophisticated algorithms, and a more profound understanding of enzyme catalytic mechanisms have led to a surge in studies employing rational design strategies for enzyme engineering [19].

On the one hand, the *sequence-based approach* involves aligning the target protein's sequence with that of a well-characterized homologous protein; by comparing these sequences, key amino acids that could be modified to enhance enzymatic properties can be identified [29–31]. This approach relies on the fact that enzymes showing a high degree of sequence identity and structural resemblance will predictably exhibit similar functions. For this purpose, multiple sequence alignment (MSA) has been extensively employed in enzyme engineering to enhance functional properties [32]; thus, mutations are envisioned in the target enzyme to align with conserved sequences found in homologous proteins with desirable characteristics. One of the key objectives of MSAs is to pinpoint positions that remain conserved across a set of sequences yet differ in the target protein. These positions have been termed "conserved but different" (CbD) sites [33]; once these CbD sites are identified, mutating the original residues at these positions to the equivalent consensus residues can enhance desired functional properties. On the other hand, as those amino acids occurring most frequently at a specific position within a family of homologous proteins do contribute significantly to structural stability, an approach of "back to consensus" mutations has been widely employed to enhance enzyme properties [34]. Finally, another sequence-based approach is the ancestral sequence reconstruction technique, which deduces the sequences of putative ancestral enzymes based on amino acid sequences of modern or known proteins; thus, candidates obtained from this approach are generally considered to be more stable and can be subsequently used as model scaffolds for further studies [35]. As a representative example, Bornscheuer and coworkers reported the use of sequence alignment to shift the pH optima of (*R*)-selective transaminases toward neutral pH, thereby improving their compatibility in enzyme cascades and industrial applications [36]. The focus was on modifying amino acid residues that influence the pH-dependent activity; thus, three highly homologous (*R*)-selective transaminases were selected for the study (ATA-Afu (from *Aspergillus fumigatus*) – pH optimum: 8.5; ATA-Gze (from *Gibberella zeae*) – pH optimum: 7.5; and ATA-Ate (from *Aspergillus terreus*) – pH optimum: 7.5). Despite their structural similarity, they exhibited different pH optima. The authors performed protein structure alignment and found that three residues in the substrate tunnel differed among these enzymes: E49 (ATA-Afu) versus Q49 (ATA-Gze), T123 (ATA-Afu) versus K123 (ATA-Gze), and G127 (ATA-Afu) versus E127 (ATA-Gze), as shown in Figure 6.4 (marked with triangles). These residues were hypothesized to influence the enzyme's electrostatic environment and, consequently, its pH preference.

Therefore, to test the impact of these residues, SDM was performed, creating six single-point mutants and three double mutants. Each variant was expressed in *Escherichia coli*, purified, and subjected to pH activity profiling, as well as computational calculations such as molecular docking and p$K_a$ prediction. Finally, these authors identified key residues that affect the pH profiles of (*R*)-selective transaminases and demonstrated a shift of pH optima resulting in improved catalytic activity. The possible mechanism behind the different pH optima found for the ATA variants was explored and confirmed by activity tests of different variants, together with molecular modeling analysis. This study successfully engineered an (*R*)-selective transaminase (ATA-Afu-E49Q) with a shifted pH optimum from 8.5 to 7.5, making it more suitable for enzyme cascade reactions.

On the other hand, the *structure-based approach* employs a pre-existent enzyme's 3D model to redesign its structure in order to generate an improved catalyst. In either case, modifications may involve substitutions, deletions, insertions, or the replacement of residues with amino acids of varying physicochemical properties and sizes [37, 38]. To obtain the initial 3D templates, the Research Collaboratory for Structural Bioinformatics-Protein Data Bank (PDB, https://www.rcsb.org/), initially established in

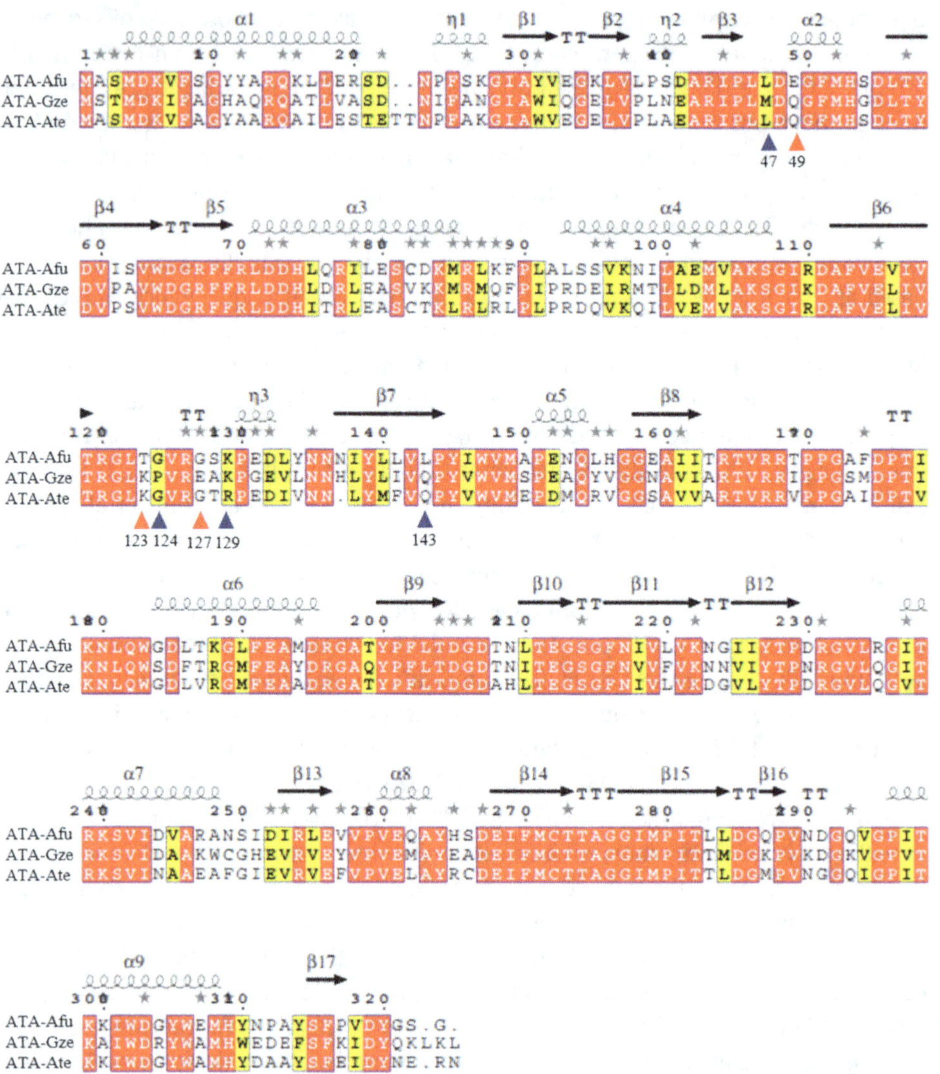

**Figure 6.4:** Amino acid sequence alignments of ATA-Afu, ATA-Gze, and ATA-Ate, highlighting residues in the substrates tunnel with triangles. Reprinted from Xiang et al. [36]. OPEN ACCESS.

1971 at Brookhaven National Laboratory as the pioneer international repository for 3D structure data of biological macromolecules [39], has documented more than 230,000 protein structures (elucidated through techniques such as X-ray crystallography, cryo-electron microscopy, and nuclear magnetic resonance) [40, 41]. Recent advancements in artificial intelligence, particularly AlphaFold2 [42] or RoseTTAFold [43–45], have demonstrated remarkable accuracy in predicting protein structures, offering a potential solution to forecast the structure of billions of known protein sequences yet to have their structures determined. In 2024, Demis Hassabis and John M. Jumper shared one-half of the Nobel Prize in Chemistry for the creation of AlphaFold2 (which recently has been updated to AlphaFold3 [46]), while the other half was awarded to David Baker, creator of Rosetta [45] (please go to Section 6.2.6 for a detailed discussion on this topic). In this sense, the PDB also allows access through computed structure models to more than 1 million known protein sequences predicted with the previously mentioned methods. Currently, there is no universal framework governing structure-based enzyme engineering, as functionally relevant residues and structural motifs vary among different enzymes. However, an accepted consensus suggests that targeting amino acid residues in proximity to the active site is generally more effective in altering properties such as catalytic activity and stereoselectivity [19]. Conversely, modifications in distant loop regions may influence attributes such as thermostability and substrate specificity [37], while mutations within access tunnels, interfacial regions, and remote gating sites that regulate entry to the active center play a crucial role in modulating enzyme catalytic properties [33].

### 6.2.2 Semirational Design of Enzymes

Therefore, with the significant advances in computer science applied to enzyme design reported in recent years, the strategy known as **semirational design** has experienced a remarkable rise. In fact, semirational protein design represents a sophisticated hybrid approach that merges computational insights with experimental library screening to engineer proteins with enhanced or novel functions. By leveraging structural, evolutionary, and mechanistic data, this method bridges the gap between purely rational design and random directed evolution (which will be further commented on in Section 6.3), enabling efficient optimization of biocatalysts for industrial, medical, and research applications [47–49]. In fact, semirational protein design combines existing knowledge (such as structure, sequence, and function) with computational and evolutionary tools to guide targeted mutagenesis, as shown in Figure 6.5.

This approach focuses on mutating selected "hotspot" residues that are likely to influence the desired property, as identified by analysis and algorithms. The result is a small, high-quality mutant library – usually fewer than 1,000 variants – which is larger than what is typically generated in rational design but much smaller and more manageable than random mutagenesis libraries. While some experimental screening is still required, it is moderate compared to random approaches. Semirational design does not require as complete structural information as rational design, making it more flexible. It uses computational predictions, evolutionary analysis, and sometimes machine learning to select which sites to mutate. Controlled randomness is introduced by targeting only certain regions for mutagenesis, allowing for efficient exploration of sequence space. The main goal is to balance precision with diversity, increasing the chances of finding improved or altered protein functions. A direct com-

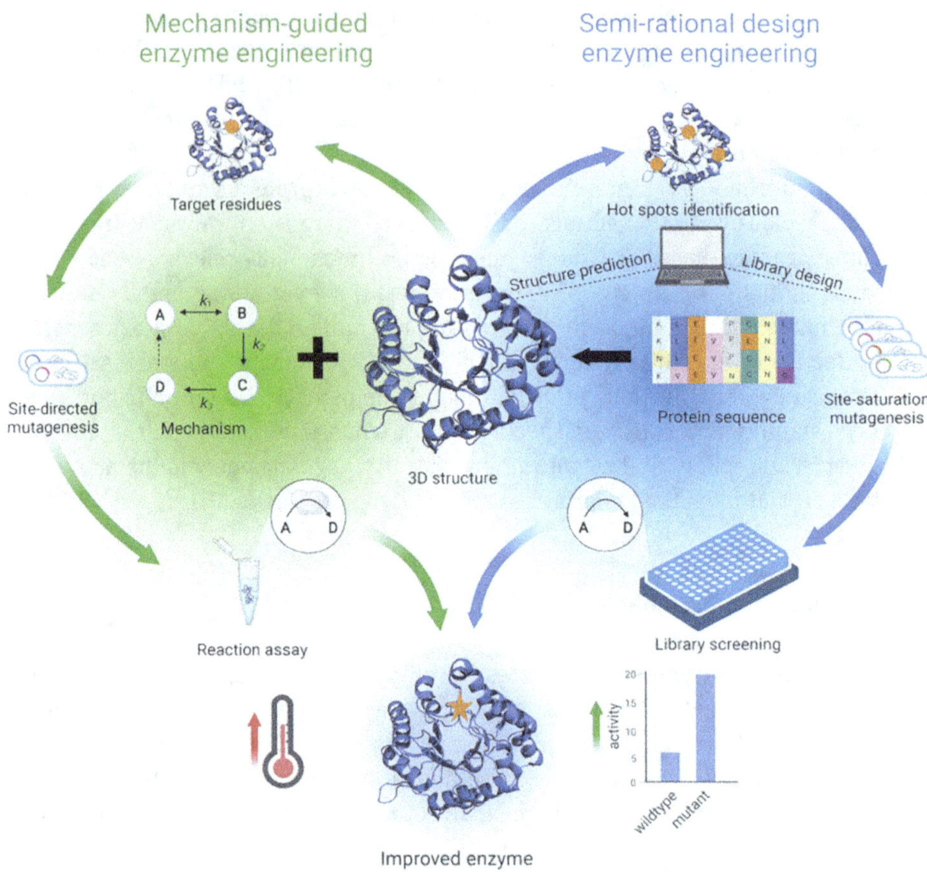


**Figure 6.5:** Rational versus semirational design of proteins. Reproduced from Phintha and Chaiyen [27] with permissions.

parison of rational and semirational design is presented in Table 6.1, based on literature data [19, 20, 47–49], and graphically depicted in Figure 6.5, taken from the review of Phintha and Chaiyen [27].

In summary, rational design is a precise, hypothesis-driven approach that makes deliberate changes based on deep knowledge of protein structure and function, while semirational design is a guided yet exploratory approach that uses available knowledge to focus randomization on promising regions, creating manageable libraries that are more likely to yield improved proteins. Rational design is best when you know exactly what to change; semirational design is best when you want to efficiently explore options without testing every possibility.

As mentioned, semirational protein engineering integrates evolutionary sequence conservation data and 3D structural insights to systematically identify amino acid positions with high potential for functional optimization, known as mutational hotspots.

**Table 6.1:** Comparison between rational and semirational design.

| Aspect | Rational design | Semirational design |
|---|---|---|
| **Basis** | Relies on structural, functional, and computational models of known proteins | Combines structural/evolutionary data with predictive algorithms to guide mutagenesis |
| **Key features** | – Site-directed mutagenesis<br>– De novo or protein redesign<br>– Fixed backbone assumptions | – Focused mutagenesis at predicted "hotspot" residues<br>– Smaller, functionally enriched libraries |
| **Advantages** | – High precision in targeting specific residues<br>– Reduces experimental screening load | – Balances exploration and efficiency<br>– Incorporates evolutionary variability data |
| **Limitations** | – Requires high-resolution structural data<br>– Limited ability to predict long-range effects | – Still requires partial structural/ functional knowledge<br>– May miss unpredictable beneficial mutations |
| **Computational tools** | – Molecular dynamics simulations<br>– Rotamer libraries<br>– Energy minimization algorithms | – Machine learning models<br>– Phylogenetic analysis<br>– Docking simulations |
| **Typical use cases** | – Creating novel folds (e.g., Top7)<br>– Redesigning binding sites<br>– Stabilizing proteins | – Optimizing enzyme activity/specificity<br>– Engineering pH/temperature stability<br>– Directed evolution prescreening |
| **Library size** | Minimal (often single mutations) | Moderate (100–1,000 variants) |

This approach combines nature's evolutionary blueprint with structural mechanistic understanding to focus mutagenesis efforts on residues most likely to improve target properties while maintaining protein stability. Evolutionary conservation patterns serve as primary guides for hotspot identification [33]. The 3DM database exemplifies this by tracking evolutionary variability across superfamilies, enabling identification of:

- CbD sites [33] – positions conserved within subfamilies but divergent between them, indicating potential functional switches.
- Correlated mutation networks – evolutionarily coupled residues maintaining structural or functional complementarity [50].
- Subfamily-specific positions – those amino acid residues that distinguish functional classes within protein families [51]. They are also known as specificity-determining positions.

On the other hand, 3D structural data provides spatial context to evolutionary patterns, revealing:

- Active site architecture – catalytic residues and substrate-binding pockets amenable to stereoselectivity engineering [52].
- Access tunnels – gating regions controlling substrate entry/product release [48].
- Allosteric networks – distal residues dynamically coupled to active sites through conformational motions [52].
- Interface regions – surfaces mediating protein-protein or domain-domain interactions [53].

The integration of these datasets enables predictive hotspot selection through tools like ConCavity (combines conservation scores with structural pocket detection to map functional sites [52]), HotSpot Wizard (identifies tunnels and channels through structural analysis [54]), or COSMIS (quantifies 3D mutational constraint using evolutionary and structural features [55]). By this methodology, it is possible to design reduced library sizes (<1,000 variants [56]), which enable detailed characterization using low-throughput assays, such as spectroscopy or chromatography [57]. An exhaustive description of all available software would be nearly impossible; therefore, some reviews should be consulted to get an extended vision on this research field [58–63].

On the other hand, for carrying out library building, once the target residues to be mutated have been identified, focusing on predefined positions (e.g., substrate-binding tunnels), the use of SSM is a classical approach [25, 26]. In SSM, a specific codon (or set of codons) within a gene is systematically replaced with all possible amino acid codons, generating a comprehensive library of protein variants at that position. This approach allows researchers to explore the functional impact of every possible amino acid substitution at a chosen site, making it a powerful tool for mapping sequence-function relationships and optimizing protein properties [64, 65]. The implementation of the Combinatorial Active-Site Saturation Test (CAST) and iterative saturation mutagenesis (ISM), developed by Manfred T. Reetz and coworkers, implied a deep step forward in the semirational directed evolution strategy, and for the engineering of enzyme active sites for improved or novel catalytic properties, such as stereoselectivity, substrate scope, or activity [20]. For active-site targeting, CAST focuses on multiresidue sites (labeled A, B, C, etc.) around the enzyme's binding pocket, identified via X-ray crystallography, homology models, or docking simulations; each site comprises between 1 and 4 amino acid residues critical for substrate binding or catalysis. Once identified, SSM is applied to the residues within these sites, which are randomized using reduced amino acid alphabets (e.g., 5–12 amino acids) instead of all 20 canonical options, using codon degeneracy (e.g., NDT encoding 12 amino acids) for minimizing library size while retaining functional diversity [66]. Then, ISM applies CAST in cycles, as the best variant from one site serves as the template for mutating the next site, creating an evolutionary tree with multiple pathways. This stepwise approach captures synergistic (epistatic) effects between mutations, enhancing catalytic efficiency or stability [67, 68].

Another semirational method to be mentioned is OmniChange, a sequence-independent, multisite saturation mutagenesis method designed to simultaneously and efficiently randomize up to five independent codons within a gene, regardless of their sequence position or distance from each other [69]. Unlike traditional approaches that often require sequential mutagenesis or are limited by gene sequence constraints, OmniChange enables the generation of highly diverse mutant libraries in a single day and without the need for restriction enzymes, ligases, or additional PCR amplification steps. OmniChange involves four main steps:

1. Fragment generation: PCR is used to create DNA fragments, each carrying a targeted codon replaced with an NNK-degenerate codon (where N = A, T, G, or C; K = G or T), allowing for maximal amino acid diversity at each site.
2. Chemical cleavage: The PCR products are chemically cleaved to produce complementary single-stranded 5′-overhangs using phosphorothiolated primers and iodine/ethanol treatment. This enables precise fragment assembly.
3. Assembly: The fragments are hybridized together in a one-pot reaction to form a full-length gene with up to 10 DNA nicks (two per fragment junction) but without requiring enzymatic ligation.
4. Transformation and nick repair: The assembled, nicked plasmid is transformed into *E. coli*, where the host's endogenous machinery repairs the nicks, resulting in a functional, circular plasmid containing all the targeted mutations.

There are many different methods available for creating enzyme mutant libraries, which will be also commented on in Section 6.3. The comprehensive and comparative guide published by Alejaldre et al. [70], structured to serve both novices and experts, presents those methods with a focus on introducing genetic diversity for enzyme engineering, foundational overviews, and highlighting cutting-edge developments in the field.

Thus, in the following section, we will present some examples showing how rational and semirational design has been used to increase the catalytic properties of enzymes, namely stability, activity, and selectivity.

### 6.2.3 Rational and Semirational Design for Improving Protein Stability

Pure rational design was initially utilized to improve enzyme thermostability under extreme conditions [20, 71]. For instance, Matthews et al. reported in 1987 how the mutations of only two amino acidic residues (A82P and G77A) stabilized lysozyme toward reversible and irreversible thermal denaturation at physiological pH, with no apparent effect on enzymatic activity. For this same enzyme, Nicholson et al. described in 1988 how mutants S38D and N144D, rationally designed to interact by electrostatic interaction with α-helix dipoles, increased the thermal stability of the protein, without establishing hydrogen bonding between the substituted amino acid and the end of the α-helix [72].

After some other pioneering works [73, 74], a systematic methodology (structure-based strategy) was proposed by Eijsink et al. in 2004 for enhancing enzyme thermostability, based on the rational introduction of hydrogen bonds and salt bridges on the protein's surface through SDM [75], aiming to promote a rigidification of the flexible regions of the protein [76]. In fact, naturally evolved thermophilic proteins have acquired additional charged residues on their surfaces [77], enabling the formation of a greater number of salt bridges compared to their mesophilic counterparts [78]. This fact has also been confirmed using sequence-based engineering protocols [79], based on the axioma that some conserved regions in different protein families become heir to thermal resistance from ancestral proteins, thus upholding structural stability [80]. From a thermodynamic perspective, the improved thermal stability conferred by salt bridges must be assigned to a reduction in the heat capacity associated with protein unfolding [81]. Based on this, the rational generation of salt-bridge networks can effectively counteract the increased molecular disorder associated with thermal motion, thereby reinforcing not only thermostability [81, 82] but also the enzymatic tolerance to organic solvents [83]. Many examples can be found in the literature, for instance, the notorious increase reported in the melting temperature of a mesophilic β-glucosidase due to the deliberate incorporation of mutants adding six salt bridges [82] or the described increased thermostability of rationally designed mutants of 1,4-α-glucan branching enzyme (EC 2.4.1.18, GBE) from *Geobacillus thermoglucosidans* STB02 [84].

This fact leads to a general agreement that protein stability can be effectively improved by a rational modification of its net charge, which plays a crucial role in regulating electrostatic interactions both within the protein itself and between the protein and its surrounding environment. Increasing the number of charged residues, particularly those that contribute to salt bridge or ion pair formation, can significantly enhance structural stability. These electrostatic interactions are vital for preserving protein integrity under extreme conditions, such as elevated temperatures or fluctuating pH levels, as reported for improving the thermostability of mutants of IsPETase from *Ideonella sakaiensis* 201-F6 displaying a stronger salt bridging network [85], also leading to a 3.6-fold increase in its efficiency in degrading polyethylene terephthalate (PET). A very recent article reviews rational modifications of PETases for improving their catalytic behavior [86].

Beyond charge modification, targeting intrinsically disordered regions – commonly found in protein termini and loops – can further reinforce protein stability. These flexible regions are prone to conformational changes that can lead to denaturation under challenging environmental conditions. Introducing charged or hydrophobic residues, such as proline, phenylalanine, valine, isoleucine, leucine, and tyrosine, can stabilize both local structures and the overall electrostatic balance of the protein, thereby optimizing net charge distribution [87].

Another structure-based strategy for increasing proteins' stability is based on the analysis of the B-factor, also referred to as the temperature factor or Debye-Waller factor, which represents the reduction in electron density caused by thermal motion or positional disorder in protein crystallography; this analysis has been extensively utilized in enzyme engineering to identify flexible regions within proteins [88], since its initial description by Reetz and Carballeira [68, 89] for enhancing the thermostability of Lipase A (LipA) from *Bacillus subtilis*. Thus, the authors employed B-FITTER software to pinpoint 10 residues exhibiting the highest averaged B-factors in the enzyme: G13, R33, D34, K35, K69, K112, M134, Y139, I157, and Q164; these residues were organized into eight clusters labeled A (G13), B (R33, D34, K35), C (K69), D (K112), E (M134), F (Y139), G (I157), and H (Q164). For each cluster, SSM-based libraries were generated using NNK degenerate codons (N: A/T/C/G; K: G/T) to encode all 20 amino acids. An HTS approach utilizing 96-well plates was implemented, with Lip A activity assessed *via* absorbance measurements at 405 nm during hydrolysis of *p*-nitrophenyl caprylate. On the other hand, thermostability was evaluated by assessing the temperature leading to a reduction in initial activity by 50% after 15 and 60 min of heat exposure (namely, $T_{50}{}^{15}$ and $T_{50}{}^{60}$, 50 and 48 °C, respectively, for the wild-type (WT) enzyme). Initial screening revealed enhanced mutants in all clusters except C and H. To further optimize stability, the M134D variant (group E's most thermostable mutant, designated variant IV) served as the template for ISM, as depicted in Figure 6.6 (taken from the excellent review of Phinthia and Chaiyen [27]).

**Figure 6.6:** Target conformational flexibility hotspots using B-FITTER-identified residues (high B-factor regions) for iterative saturation mutagenesis (ISM). Reproduced from Phintha and Chaiyen [27] with permissions.

This process yielded variants X and XI, which demonstrated remarkable $T_{50}{}^{60}$ values of 89 and 93 °C, far exceeding the WT's 48 °C. At 55 °C, variants X and XI displayed half-lives ($t_{1/2}$) of 905 and 980 min, respectively, compared to the WT's $t_{1/2}$ of under 2 min. Kinetic analyses measuring $K_M$ and $k_{cat}$ values at 25 °C using *p*-nitrophenyl acetate as the substrate showed no significant differences between the WT and engineered variants, confirming that thermostability improvements did not compromise catalytic efficiency. Subsequent biophysical characterization revealed that, despite undergoing unfolding at elevated temperatures, the variants exhibited markedly reduced irreversible aggregation of denatured enzymes compared to the WT. Similarly, structural analysis revealed that the enhanced stability of LipA mutants was once again associated with an increased number of salt bridges and hydrogen bonds on the enzyme's surface.

Some other analyses derived from the use of the B-factor are focused on the modifications of flexible loops (loop-grafting strategy) [90], aided by MORPHING (Mutagenic Organized Recombination Process) method [91].

Disulfide bridges arise from chemical interactions between sulfur-containing groups (–SH) on cysteine amino acids, creating covalent links that stabilize the 3D architecture of proteins. These bonds are critical for enhancing structural rigidity in many biomolecules. Computational approaches, including tools like Disulfide by Design 2 (DbD2), MODIP, SSbondPre, and Yosshi, enable researchers to predict optimal sites for disulfide bond insertion, thereby improving heat resistance and the functional versatility of enzymes [27]. To briefly comment on some of them, DbD2 evaluates geometric parameters such as the Cβ–Sγ–Sγ–Cβ (c3) torsion angle – a measure of atomic rotation – and assesses energy profiles and B-factors (atomic displacement metrics) to rank stabilizing mutations [92]. MODIP, meanwhile, identifies viable disulfide pairs by analyzing spatial alignment of sulfur atoms while accounting for steric clashes [93]. Some studies have integrated both tools to design disulfide bonds in enzymes like the (*R*)-selective amine transaminase from *A. terreus* [94], as shown in Figure 6.7. Using the enzyme's crystal structure (PDB: 4CE5), MODIP and DbD2 predicted 153 and 40 candidate residue pairs, respectively, with seven overlapping suggestions. Experimental validation revealed that three variants (N25C-A28C, R131C-D134C, and M150C-M280C) exhibited enhanced thermal stability. Subsequent molecular dynamics (MD) simulations at 313 and 400 K revealed that stabilized variants like M150C-M280C maintained lower root-mean-square deviation and reduced flexibility near the engineered disulfide bonds. The M150C-M280C mutation, located in a long loop (130 residues), and R131C-D134C, positioned in a flexible surface region, both restricted structural fluctuations, enhancing stability. This study demonstrates the value of combining computational predictions to guide disulfide engineering. However, while the introduction of disulfide bonds generally stabilizes proteins, they may inadvertently disrupt folding pathways or promote misfolded states. To mitigate risks, integrating multiple prediction algorithms and mechanistic insights is recommended when selecting mutation sites [27].

Replacing specific amino acid residues with proline is a widely used rational approach for enhancing enzyme thermostability; in fact, proline contributes to increased conformational rigidity in the protein backbone [95], being particularly effective in β-turns and at the *N*-terminal cap of α-helices, with loops being especially responsive to such modifications. A refinement of this approach, known as the "H-bonding criteria," has been employed to selectively filter mutation sites. This principle suggests that proline substitution at residues already forming hydrogen bonds may lead to protein destabilization [37].

Needless to say, computational tools have significantly advanced the (semi)rational design and targeted mutagenesis of proteins, offering an unparalleled level of precision in predicting and modeling mutations to enhance protein stability. These tools, either usable as local software or online servers, facilitate the identification of optimal

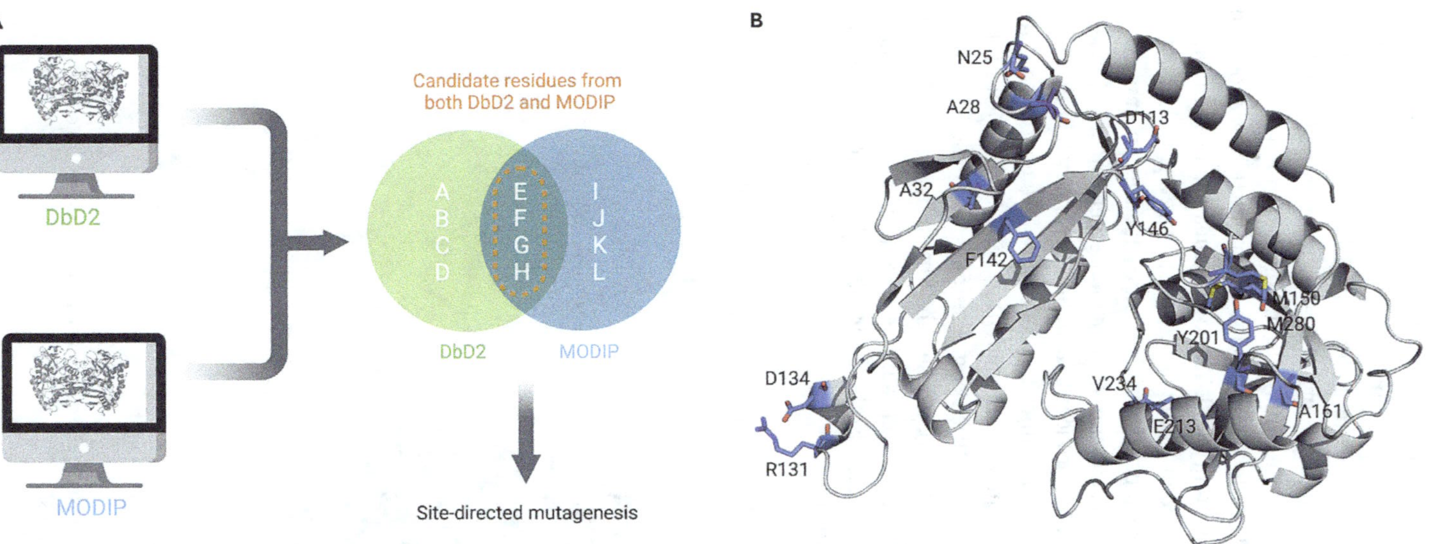


**Figure 6.7:** Protein engineering through disulfide bond introduction using computational tools. (A) Potential disulfide bonds identified using DbD2 and MODIP software. Residue pairs predicted by both tools were selected for subsequent site-directed mutagenesis. (B) Target residues for mutation in the (*R*)-selective amine transaminase from *Aspergillus terreus* (PDB ID: 4CE5), as determined by the analysis with DbD2 and MODIP. Reproduced from Phintha and Chaiyen [27] with permissions.

mutation sites, enabling more efficient and effective protein engineering strategies. It would be almost impossible to mention all of them (see, for instance, review by Phintha and Chaiyen [27]), although maybe Rosseta, FoldX, and Discovery Studio (local software) and PoPMuSiC or FireProt (online platforms) are the most powerful [87].

As an elegant example, the rational design and structure-based engineering of alkaline pectate lyase (PelN) from *Paenibacillus* sp. 0602 was reported by Zhou and Wang aiming to improve its thermostability [96]. These enzymes are used in ramie degumming, a process that removes pectin from plant fibers; however, its industrial application is hindered by limited thermostability. This study focuses on improving PelN's thermostability using rational design and structure-based engineering, by applying computational and experimental approaches, including Discovery Studio 4.1 for mutation energy calculation, PoPMuSiC algorithm to predict the free energy changes due to protein folding mutations, and molecular modeling using SWISS-MODEL for structural analysis. Hence, the authors identified some residues (Figure 6.8) in an α-helix (Gly90, Gly241, Gly316, Pro84, and Pro 219), whose substitution by Ala or Val could be beneficial to stabilizing this α-helix, as well some others (Glu137, Ser92, Lys93, Asp178, or Gly246), whose substitution could also be desirable to create a more robust enzyme suitable for industrial conditions.

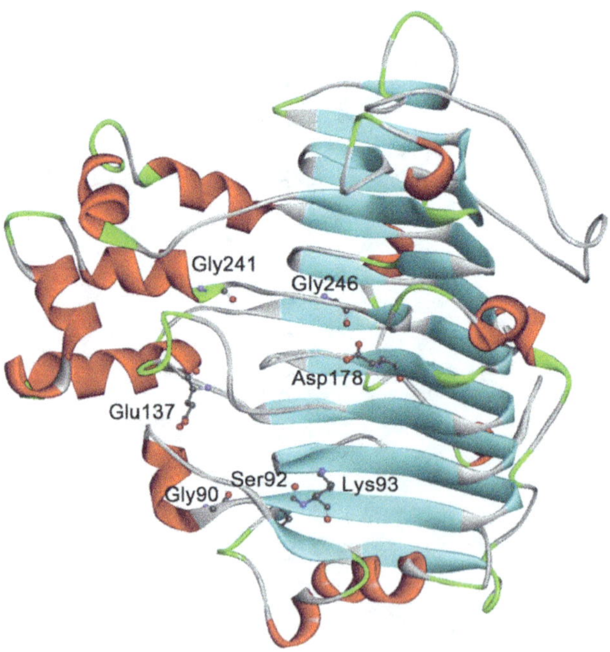


**Figure 6.8:** Selected residues in wild-type PelN susceptible to be mutated to get a more stable protein. Reprinted from Zhou and Wang [96]. OPEN ACCESS.

Glycine (G241) and lysine (K93) were identified as key residues affecting enzyme stability, and a total of 26 PelN mutants were generated, among which two (G241A and G241V) exhibited significant improvements in thermostability. Additionally, a combination mutation (K93I/G241A) showed enhanced thermostability and specific activity. In fact, while the half-life of WT PelN at 60 °C was 33.5 min, the G241A and G241V mutants showed increased half-life by 28.4 and 13.6 min, respectively, while the double mutant K93I/G241A exhibited a 15.9 min increase in half-life and a 1.6 °C increase in melting temperature ($T_m$). On the other hand, while the WT enzyme had an optimal temperature of 67.5 °C, the K93I mutation lowered the optimal temperature to 60 °C, making it more suitable for industrial applications; conversely, the G241A and G241V mutants retained their optimal temperature but exhibited improved thermal resistance. From a structural point of view, G241A mutation increased the number of salt bridges and hydrophobic interactions, stabilizing the α-helix structure. The G241V mutation promoted strengthened intramolecular interactions, reducing protein flexibility and enhancing stability. The K93I mutation improved enzymatic efficiency by modifying substrate binding interactions, while K93I/G241A double mutation combined the benefits of K93I's improved activity with G241A's enhanced stability. Finally, the mutant enzymes were tested in ramie degumming at 50 °C for 12 h, proving that the double mutant K93I/G241A exhibited superior degumming efficiency compared to the WT PelN.

Needless to say, both sequence-based and structure-based methods can be used simultaneously. For instance, the rational design to improve the catalytic efficiency and stability of arginine deiminase (ADI) reported by Zhang et al. [97] combined both strategies to modulate ADI from *Enterobacter faecalis*, an enzyme that catalyzes the hydrolysis of arginine to citrulline and ammonium and is especially useful in targeting arginine-deficient tumors such as hepatocellular carcinoma and melanoma. Additionally, L-citrulline, the hydrolysate of arginine, has multiple physiological benefits, including its role in nitric oxide synthesis, cardiovascular health, and immune enhancement. However, for industrial and pharmaceutical applications, ADI requires enhanced enzymatic efficiency and stability. These authors employed several computational tools (SWISS-MODEL for protein structure simulation, MSA for identifying mutation sites, DeepDDG and FireProt to predict protein stability, mCSM-PPI2 for analyzing protein-protein interactions and affinity changes, molecular docking using AutoDock Vina to analyze substrate binding, and MD simulation for stability analysis) for rational design and site-specific mutations, resulting in 24 ADI mutants. Some of the rational mutations of ADI are shown in Figure 6.9. Thus, while the WT ADI had an optimal temperature of 60 °C, differential scanning calorimetry revealed that the thermal melting points ($T_m$ values) of T340I, F44W, and E220I increased, indicating enhanced stability, while mutants F44W and T340I retained high thermal stability, maintaining 80% activity at 70 °C. Structural studies revealed that the stability-enhancing mutations of F44W and T340I was caused by a strengthened hydrophobic interactions at the dimer interface, reducing structural unfolding, while E220I caused a reinforcement of hydrophobic interactions between β5 and β8 folds of the enzyme, leading to a higher stability.

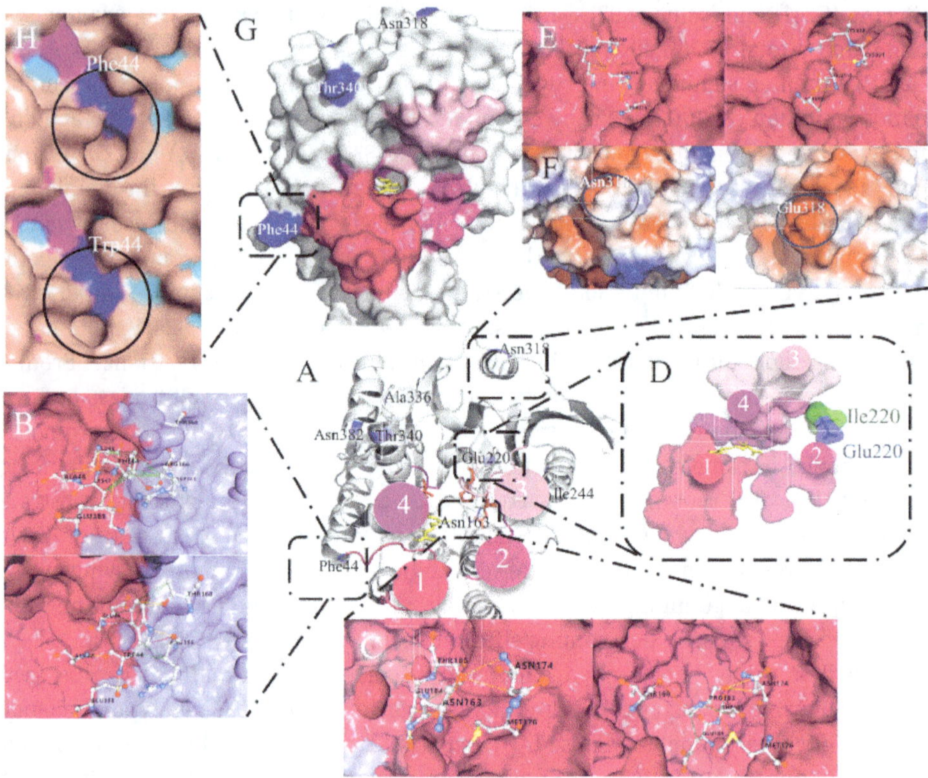


**Figure 6.9:** Analysis of the structure of ADI mutants and mutation site structure. (A) Structural representation of wild-type ADI, showing some areas with the correspondent mutations: F44W (B); N163P (C), E220I (D), and N138E (E) and its change in net surface charge of N138E (F). On the other hand, the surface structure of ADI is shown in (G), while changes in the surface of crucial F44W is shown in (H). Reprinted from Zhang et al. [97] with permissions.

Additionally, it was observed how these three mutants (F44W, E220I, and T340I), as well as some other ones (N163P, E220L, N318E, A336G, and N382F), also displayed activity increases ranging from 1.33 to 2.53 times compared to WT ADI. This enzyme activity improvement was linked to mutations near the four catalytic loop regions, facilitating substrate binding and product release. Furthermore, an analysis of the modification of the optimal pH of the mutants was also carried out; thus, while WT ADI functioned optimally at pH 7.5, with rapid loss of activity beyond this range, mutant F44W exhibited an optimal pH of 8.0, retaining high activity up to pH 9.0. The mutants E220I, N382F, A336G, and I244V also showed improved activity across broader pH ranges. As a conclusion, F44W, E220I, and T340I exhibited the best overall performance, with improved activity, stability, and optimal pH range.

To finalize this section, we recommend a recent review from Shi et al. [98] for an updated vision of the methodology.

### 6.2.4 Rational and Semirational Design for Improving Enzyme Activity

This last example illustrates that not only stability but also enzyme activity can be enhanced by rational design. Needless to say, there is a plethora of examples reported in the literature using rational design (sequence-based, structure-based, or a mix of both strategies) to increase enzymatic activity, so that it would be impossible to summarize all of them. Some recent papers have reviewed this item [19, 37].

Considering sequence-based approach using MSA, this technique has proven to be valuable for increasing the activity of several enzymes, for instance, bacillus-like esterase (EstA), which exhibits minimal catalytic activity toward tertiary alcohol esters. Thus, the MSA alignment of 1,343 sequences of homologous enzymes revealed a highly conserved GGG motif in the oxyanion hole, while EstA contains a serine at the third position, resulting in a GGS motif. By substituting this serine with glycine, the engineered mutant EstA-GGG exhibited a 26-fold increase in conversion efficiency toward tertiary alcohol esters compared to the WT enzyme [99]. A similar approach was employed to enhance the activity of glutamate dehydrogenase from *Pseudomonas putida* (*Pp*GluDH) [100], fructosyltransferase from *Aspergillus niger* [101], and many other enzymes [19].

In another example, the same strategy was applied to engineer an amidase (AmdA) from *Agrobacterium tumefaciens* d3 to enhance its ability to degrade ethyl carbamate (EC), which is a probable human carcinogen (IARC Group 2A) commonly found in fermented foods and alcoholic beverages [102]. Traditional methods to reduce EC focus on precursor management and fermentation process optimization, but none can directly degrade already-formed EC in beverages. Urethanase enzymes, which hydrolyze EC to ethanol, $CO_2$, and $NH_3$ offer a promising biocatalytic solution. However, AmdA, while resistant to ethanol, has limited catalytic activity toward EC, restricting its industrial application. Thus, to increase AmdA activity, its amino acid sequence was aligned with three known urethanases capable of EC degradation, derived from *Rhodococcus equi* strain TB-60, *Lysinibacillus fusiformis* SC02, and *Candida parapsilosis* (Figure 6.10A).

Sequence analysis revealed that the catalytic triad of AmdA (Lys98-Ser173-Ser197) was conserved across all aligned sequences. Using MSA, CbD sites adjacent to the catalytic triad were identified, leading to the design of six mutations (R94P, P163A, A172G, N175G, G195A, and L200C), using EVcouplings server to analyze a large pool of 80,105 homologous sequences to validate the consensus positions. AlphaFold 2 was used to predict the high-quality 3D structure of AmdA, outperforming previous SWISS-MODEL predictions in accuracy, and 21 amino acids within 5 Å of the catalytic triad were identified as mutagenesis targets (Figure 6.10B). Finally, SSM was performed at these positions using NNK codon degeneracy, followed by high-throughput colorimetric screening for improved activity. The ability of WT and mutant enzymes (both whole-cell and purified) to degrade EC in model rice wine (15% ethanol, pH 4.0, 50 mg $L^{-1}$ EC) was evaluated. Of the six consensus-guided mutations, only G195A

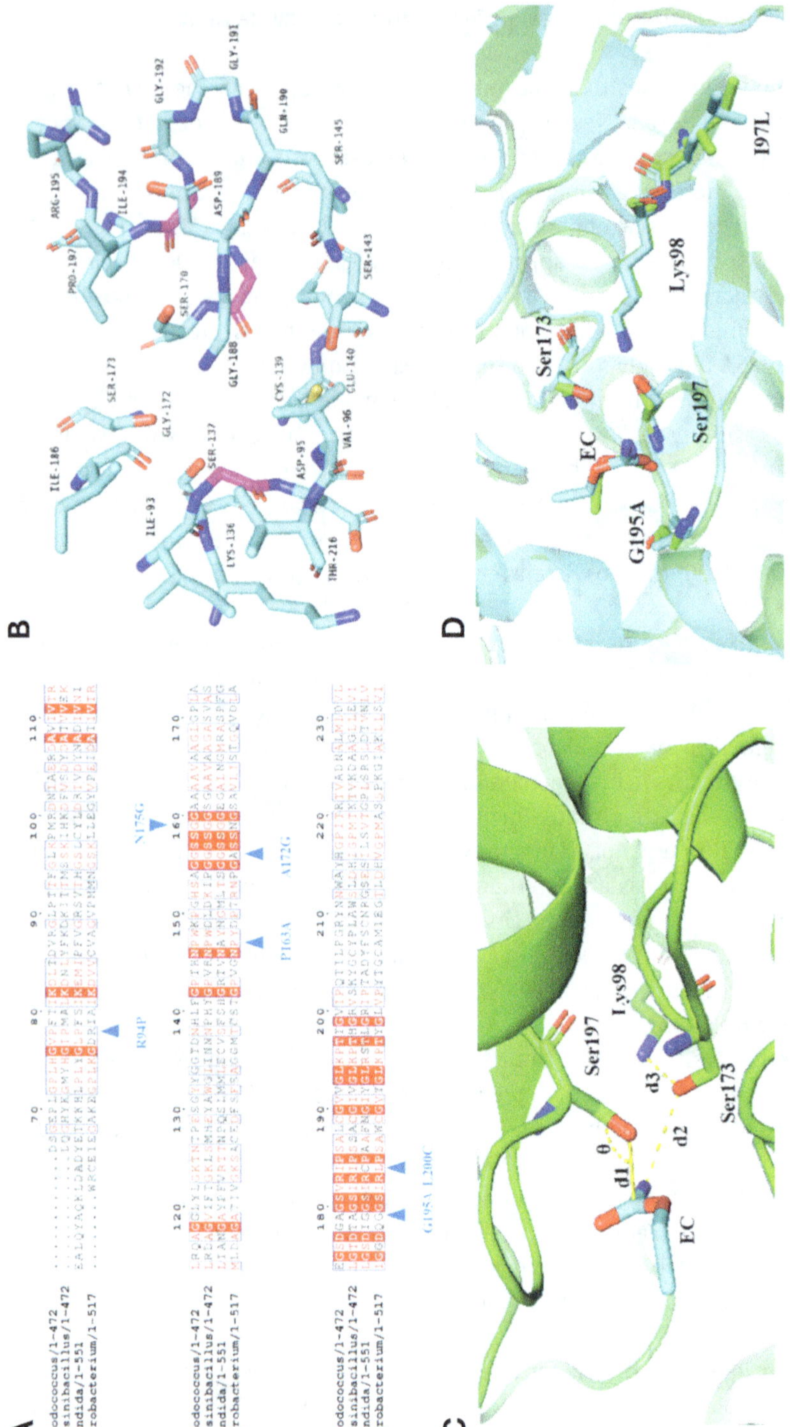


**Figure 6.10:** (A) Sequence comparison of AmdA with other urethanases. (B) Mutation sites close to the catalytic triad. (C) Structure alignment between WT and I97L/G195A. (D) Definition of attack angles and distances. Reprinted from Yao et al. [102], with permissions.

(glycine to alanine at position 195) significantly improved activity (4.9-fold in crude extracts). Additionally, SSM identified three additional beneficial mutations (I97L, S177C, and R94P), with G195A remaining the most effective. Combining I97L and G195A yielded a double mutant (I97L/G195A) with a 3.1-fold increase in specific activity and a 1.5-fold increase in ethanol tolerance compared to the WT. Regarding stability and tolerance trade-offs, G195A and I97L/G195A mutants had lower thermostability (half-life at 60 °C: 39.5 and 43 min vs. 98.6 min for the WT); on the other hand, I97L and I97L/G195A showed improved ethanol tolerance, retaining 58% activity at 40% ethanol (vs. 40% for the WT). The optimal pH shifted slightly (from 7.5 to 7.0 for G195A and I97L/G195A), and the mutants had reduced activity at extreme pH values. To explain the activity, MD simulations (100 ns, triplicate) were conducted on WT and the best mutant complexes with EC, analyzing near-attack conformations, hydrogen bond formation, and flexibility (RMSF) to elucidate mechanisms behind improved activity and altered stability. Thus, comparing the WT (Figure 6.10C) and I97L/G195A mutant (Figure 6.10D), it was observed that the beneficial mutations shortened the distance between catalytic residues and the substrate, enhancing the occurrence of a critical hydrogen bond (Ser173-Lys98) in the active site. Also, despite a lower proportion of near-attack conformations in the mutant (1.2% vs. 20.2% in WT), the increased proximity and hydrogen bonding facilitated higher turnover numbers. Finally, increased flexibility in specific protein regions (Trp339-Glu345, Gly351-Gly356, and Ala376-Ala386) in the mutants explained the observed decrease in thermostability, with local unfolding detected near the mutation site. As a final test, authors tested the enzymatic performance in model rice wine, observing that whole-cell catalysts expressing I97L/G195A or G195A removed >96% of EC within 10 h, compared to 29% for the WT. Conversely, purified mutant enzymes also outperformed the WT but did not achieve complete EC removal, likely due to lower enzyme concentrations and acid sensitivity.

### 6.2.5 Rational Design for Improving Enzyme Stereoselectivity

For this rational approach, different methodologies can be employed, based on sequence alignment, reshaping steric hindrance, remodeling interaction networks, or by means of dynamics modification.

#### 6.2.5.1 Rational Design Based on Sequence Alignment

This technique serves as a valuable tool for pinpointing the key amino acids responsible for the enantioselectivity of enzymes. In fact, aligning sequences with highly enantioselective homologues can inform strategies for engineering the selectivity of target enzymes. For instance, Liu et al. [103] observed that the analysis of enzyme sequences showed that (*R*)-selective styrene monooxygenases (SMOs) form a distinct cluster separate from their (*S*)-selective counterparts. By comparing the sequences of 9 (*R*)-SMOs with 12 (*S*)-SMOs, researchers identified 13 amino acid positions that are fully conserved in (*R*)-SMOs but differ in the (*S*)-selective enzymes. Using a highly active (*R*)-SMO from *Streptomyces exfoliatus* as a model, these 13 residues were substituted with the most common amino acids found at the corresponding positions in (*S*)-SMOs. Several of the resulting mutants exhibited enhanced enantioselectivity compared to the WT enzyme, with the W86I variant achieving 98% enantiomeric excess (e.e.). Interestingly, the A219V mutation reversed the enzyme's stereoselectivity, producing (*S*)-styrene oxide with 43% e.e. Further improvements in enantioselectivity were achieved through saturation mutagenesis at these positions, and similar effects were observed when these mutations were introduced into other (*R*)-SMOs [103].

Some examples can be found for SDM of epoxide hydrolases, for instance, the enzyme from *Sphingomonas* sp. HXN-200 (SpEH). In this case, guided by the alignment of 64 related enzymes, six nonconserved residues near the active site were targeted. Mutating these positions to hydrophobic amino acids produced a triple mutant (V196A/N226A/M332A) with a 2.8-fold increase in enantioselectivity for *rac*-phenyl glycidyl ether (*rac*-PGE), favoring the (*S*)-enantiomer [104]. Similarly, using sequence alignment, the enantioselectivity of *Phaseolus vulgaris* epoxide hydrolase (PvEH2) toward *rac*-1,2-epoxyhexane was increased by 11.5-fold [105].

For other hydrolases, such as lipases, similar examples can be found, So, *Candida antarctica* lipase B (CAL-B) typically displays low enantioselectivity toward secondary alcohols with small substituents, whereas its homolog from *Pseudozyma brasiliensis* (PBL) shows strong selectivity for (±)-but-3-yn-2-ol. Sequence comparison between CAL-B and PBL, which share 72% sequence identity, highlighted differences at residues 42, 43, 45, and 47 within the substrate-binding pocket. Replacing CAL-B's residues 42–47 with those from PBL resulted in a mutant (CAL-B-42–47) with a 40- to 50-fold increase in selectivity for (±)-but-3-yn-2-ol and a 3- to 4-fold increase for (±)-butan-2-ol, relative to the WT [106]. Also, for lipases, Bustos-Baena et al. [107] recently reported engineering of the LipTioCCR11 lipase from *Geobacillus thermoleovorans* CCR11 to enhance its enantioselectivity. The authors employed rational design guided by sequence alignment and structural bioinformatics to modify key residues in the catalytic cavity, aiming to improve substrate recognition and selectivity. Hence, six amino acids in the catalytic cavity were substituted based on alignment with enantioselective homologs (*Burkholderia cepacia* BCL and *Pseudomonas aeruginosa* PAL): F156M (oxyanion hole), L310F, F320L, I459V (hydrophobic cavity), V311A, and M313F (acyl-binding pocket). With these mutations, catalytic cavity volume increased from 358 (WT) to 406 Å$^3$ (mutant); additionally, flexibility was increased (MSF values for catalytic triad residues Ser253, Asp457, and His498), enhancing substrate accommodation, while negative $\Delta\Delta G$ values (e.g., −1.06 kcal/mol for Phe320Leu) indicated improved thermodynamic stability. Also, the optimal temperature was increased from 40 (WT) to 60 °C, as well as organic solvent tolerance (from 84% activity in hexane for the WT up to 118–182% activity in heptane/octanol for the mutant). Regarding enantioselectivity, upon lipase-catalyzed esterification of (*R*) and (*S*)-malic acid with two equivalents of methanol, the mutant achieved 94.5% e.e. for (*R*)-dimethyl malate in methanol, surpassing performance of both WT and CALB; this fact was explained by molecular docking, as expanded hydrophobic cavity accommodated bulky *R*-enantiomers, while aromatic residues (Phe313 and Leu310) stabilized substrate orientation.

#### 6.2.5.2 Rational Design Based on Steric Hindrance

This last example also illustrates rational design based on steric hindrance. In fact, a critical factor influencing enzyme function is steric hindrance within the active site. The concept of steric hindrance (when molecular groups occupy positions closer than their combined van der Waals radii) is particularly useful when designing enzymes to accept bulky or nonnatural substrates. In many cases, reducing steric hindrance through targeted mutations allows for increased reaction efficiency. Conversely, introducing larger residues to restrict access to certain parts of the active site can enhance enantioselectivity. By altering the size, shape, and flexibility of binding pockets, steric constraints can be optimized to improve enzyme activity and selectivity. Furthermore, molecular docking studies play an essential role in predicting steric hindrance effects. By simulating enzyme-substrate interactions, researchers can identify steric clashes and design mutations that improve substrate accessibility and binding affinity. High-resolution structural studies, such as X-ray crystallography and cryo-electron microscopy, further aid in validating these predictions and refining enzyme modifications. In enzymatic catalysis involving artificial substrates, mismatches between the substrate's dimensions and the enzyme's entry channel or active site can create physical obstructions. These clashes destabilize substrate-enzyme binding, often reducing catalytic efficiency or altering stereochemical precision. Consequently, enzyme engineering strategies that adjust the architecture of the binding cavity or molecular channels – enhancing substrate compatibility or product exit – can optimize catalytic performance.

Many examples can be found in the literature illustrating this approach, for instance, by modifying the binding pocket's size. A seminal example was the modification of a bacterial phosphotriesterase (PTE, also known as parathion hydrolase) to alter its selectivity for organophosphotriesters [108]. In a more recent paper dealing with the same enzyme, Jaito et al. [109] described a focused cohort of mutants of PTE from *Brevundimonas diminuta* (formerly known as *Pseudomonas diminuta*) rationally designed to assess stereochemical preference for hydrolyzing (*Rp*)-configured diastereomers of precursors of phosphoramidate nucleoside prodrugs (ProTide) such as remdesivir and sofosbuvir (denoted (*Rp*)-rem and (*Rp*)-sof). As already known [108], PTE shows a tripartite substrate-binding architecture comprising three distinct pockets (small, large, and leaving group) to facilitate organophosphate hydrolysis.

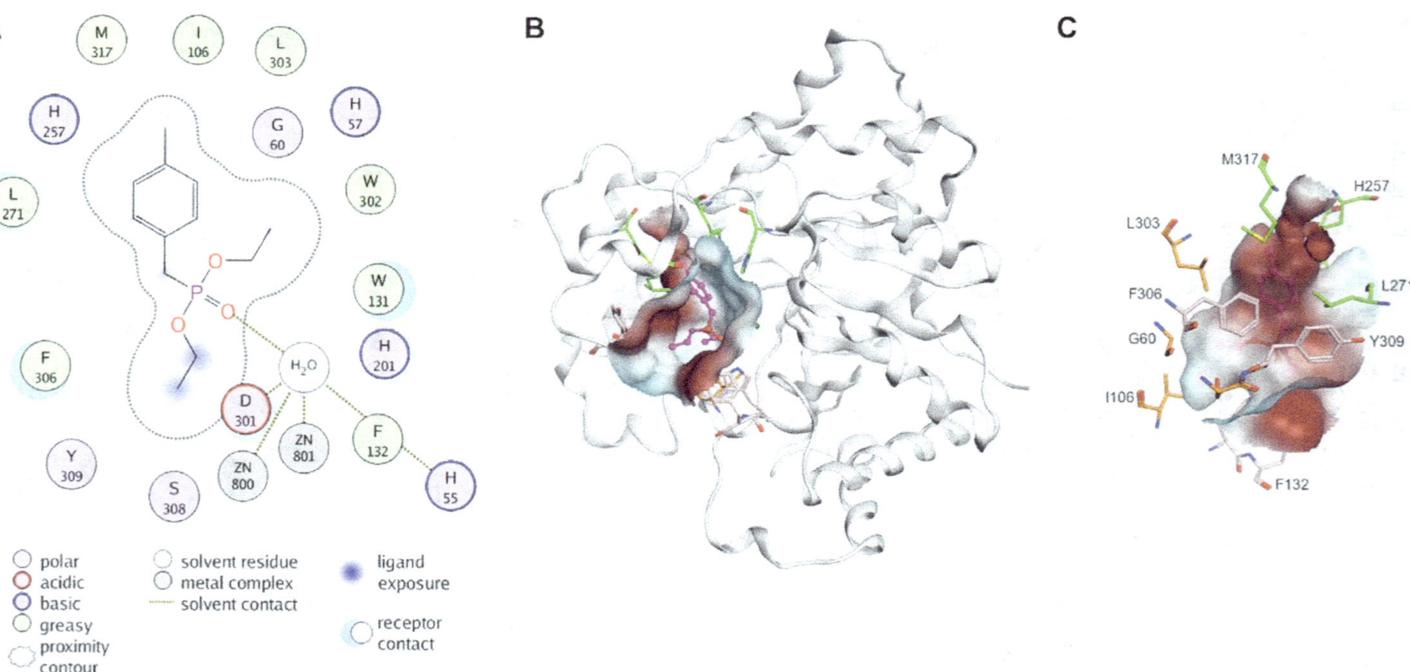


**Figure 6.11:** Structural architecture of phosphotriesterase (PTE). (A) Ligand-binding interface analysis between PTE and the paraoxon-mimetic compound diethyl 4-methylbenzylphosphonate (PDB: 1DPM); residue physicochemical properties and exposure dynamics are annotated using a chromatic code for clarity. (B) Hydrophobic gradients in the catalytic site: red, hydrophobic regions (e.g., aromatic/aliphatic residues); blue, hydrophilic zones (e.g., polar/charged residues), ligand is depicted as a purple ball-and-stick model. (C) Binding cleft partitioned into three functionally distinct regions: small pocket (orange), large pocket (green), and leaving group pocket (pink). From Jaito et al. [109]. OPEN ACCESS.

As shown in Figure 6.11A, the small pocket (residues G60, I106, L303, and S308) governs stereochemical selectivity through steric and hydrophobic interactions; on the other hand, the large pocket (formed by H254, H257, L271, and M317) is solvent-exposed and modulates substrate orientation. Finally, the leaving group pocket (also exposed to solvent, formed by a hydrophobic cavity lined by W131, F132, F306, and Y309) accommodates departing substituents, These authors analyzed PTE-substrate interactions using X-ray structure of a complex (PDB: 1DPM [110]) of PTE and paraoxon (diethyl 4-nitrophenyl phosphate, a parasympathomimetic drug which acts as a cholinesterase inhibitor), and performed molecular docking and MD simulations to identify key residues in active site pockets, and generated 22 variants *via* SDM targeting small pocket (I106A/V, L303A/F, S308A/Y), the leaving group pocket (W131M) and the large pocket (L271E/N/F, H254Y). The W131M mutation replaces tryptophan with methionine at position 131, expanding the leaving pocket to better accommodate bulky prodrug moieties (e.g., remdesivir and sofosbuvir precursors). This enhances hydrolysis of (*Rp*)-diastereomers, while the small pocket hydrophobicity adjustment, by the I106A mutation (isoleucine to alanine), altered hydrophobicity in the small subsite, favoring (*Rp*)-diastereomer orientation. Regarding the I106A/W131M double mutant, it synergically combined small- and leaving group pocket modifications, achieving 333.79-fold and 995.58-fold improvements in catalytic efficiency ($k_{cat}/K_M$) for (*Rp*)-remdesivir and (*Rp*)-sofosbuvir, respectively, and high selectivity ratios (*Rp*/*Sp*) of 840 for sofosbuvir and >1,000 for remdesivir, a remarkable increase compared to WT (*Rp*/*Sp* = 3.5).

Enhancing enantioselectivity can also be achieved by introducing steric barriers that favor the desired substrate while excluding others from the active site. For instance, a lipase from *Pseudomonas alcaligenes* (PaL), known for its ability to hydrolyze L-menthyl propionate ((1*R*,2*S*,5*R*)-2-*iso*propyl-5-methylcyclohexyl propionate) to produce the valuable (1*R*,2*S*,5*R*)-menthol (L-menthol), was engineered for higher diastereoselectivity using a rational design based on steric hindrance [111, 112]. Menthyl propionate has three chiral centers, resulting in eight possible isomers, but only the hydrolysis of L-menthyl propionate yields the target L-menthol ((1*R*,2*S*,5*R*)-2-*iso*propyl-5-methylcyclohexan-1-ol), a compound widely used in oral care products. Although PaL is highly selective for resolving the racemic mixture of D- and L-menthyl propionate and barely acts on D-menthyl propionate (1*S*,2*R*,5*S*), its diastereomeric discrimination is suboptimal. Structural studies revealed that cavities near the substrates' stereocenters contributed to this limitation. By substituting residues V180 and A272 with bulkier amino acids, these cavities were filled, restricting substrate orientation and increasing steric exclusion. This led to reduced binding of nontarget isomers and improved diastereoselectivity, so that he double mutant V180L/A272F enhanced selectivity toward L-menthyl propionate by 4.7 times. For enhancing the stereoselectivity in the enzymatic resolution of a similar substrate (D,L-menthyl acetate), a *para*-nitrobenzyl esterase from *B. subtilis* 168 (pnbA-BS) was successfully cloned and subsequently subjected to protein engineering to rise its L-enantioselectivity [113], *via* modification of steric hindrance and improve structural flexibility in regions proximal to the substrate-binding site. Therefore, the replacement of alanine at position 400 with proline (A400P) resulted in a dramatic increase in the enantioselectivity (E value), from 1.0 to 466.6. The A400P variant was isolated and rigorously validated for its strict L-enantioselectivity in the selective hydrolysis of racemic acetate. Nevertheless, the improvement in selectivity was accompanied by a reduction in overall enzymatic activity. To establish a more efficient, user-friendly, and environmentally sustainable catalytic system, authors eliminated organic solvents and implemented a continuous substrate-feeding strategy in a whole-cell biocatalysis setup. Under these conditions, the selective hydrolysis of 1.0 M D,L-menthyl acetate over 14 h achieved a conversion rate of 48.9%, an e.e. of product greater than 99%, and a space-time yield of 160.52 g/L per day.

For enzymes other than hydrolases, it is also possible to use this strategy. For instance, for the old yellow enzyme OYE3 (catalyzing C=C reduction) from *Saccharomyces cerevisiae* S288C, which reduces (*E*/*Z*)-citral to (*R*)-citronellal but with low stereoselectivity (Figure 6.12A). Through semirational design, the substrate binding region was modified to favor the (*R*)-enantiomer [114]. Mutating W116 to smaller residues expanded the binding pocket, promoting a flipped binding mode for (*Z*)-citral and achieving over 99% (*R*)-selectivity (Figure 6.12B). The S296F/W116G double mutant showed strict (*R*)-selectivity for (*E*)-citral and (*E*/*Z*)-citral, with no activity toward (*Z*)-citral (Figure 6.12C).

Many other examples can be found in the excellent revisions of Song et al. [19] and Xu et al. [37].

#### 6.2.5.3 Rational Design Based on Remodeling Interaction Networks

The enzyme's active site serves as the central hub for the entire catalytic process, encompassing substrate attachment, transition state formation, and eventual product release. Various interactions – including hydrogen bonding, hydrophobic forces, and salt bridges – occur between the substrate and the active site, playing a vital role in positioning the substrate correctly for catalysis. Therefore, deliberately redesigning these interaction networks between the substrate and the enzyme's active site can be an effective approach to alter substrate binding strength, ultimately improving the enzyme's catalytic efficiency and selectivity for specific enantiomers, or even reversing stereoselectivity. As lipases-binding cavities are well known, it is not surprising that they had been frequently modified using rational design, as reviewed by Maldonado et al. [115]. In fact, the catalytic cleft of lipases contains four functionally conserved regions, as shown in Figure 6.13:

- Catalytic triad: Composed of Ser-Asp/Glu-His residues, responsible for nucleophilic attack and hydrolysis [116]. Catalytic Ser (Figure 6.13, blue) is located inside a highly conserved pentapeptide GlyXSerXGly, where X denotes any aminoacid residue [117].
- Oxyanion hole (Figure 6.13, red): Stabilizes the tetrahedral intermediate *via* hydrogen bonding [118, 119] (e.g., Gly in GX or GGGX classes, Tyr/Asp in Y-class li-

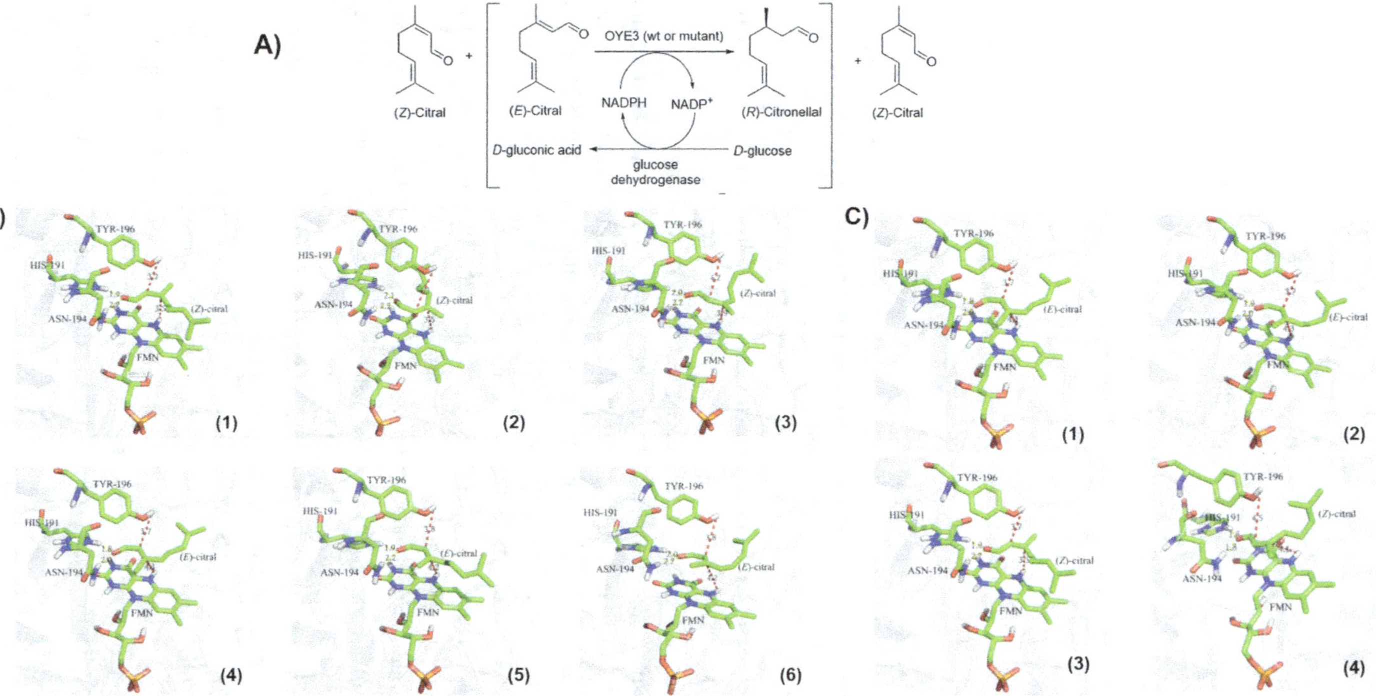


**Figure 6.12:** (A) The OYE3 variant-mediated reduction of (*E*/*Z*)-citral to (*R*)-citronellal and the retained (*Z*)-citral with the assistance of glucose dehydrogenase-catalyzed NADPH regeneration. (B) Binding modes of citral isomers in OYE3 and its variants: (1) OYE3 and (*Z*)-citral; (2) W116A and (*Z*)-citral, flipped binding orientation; (3) S296F and (*Z*)-citral, flipped binding orientation; (4) OYE3 and (*E*)-citral; (5) W116A and (*E*)-citral, preserved binding orientation; and (6) S296F and (*E*)-citral, preserved binding orientation. (C) Docking analysis of the difference of OYE3 and its double mutant S296F/W116G: (1) OYE3 and (*E*)-citral; (2) S296F/W116G and (*E*)-citral; (3) OYE3 and (*Z*)-citral; and (4) S296F/W116G and (*Z*)-citral. Modified from Wang et al. [114] OPEN ACCESS.

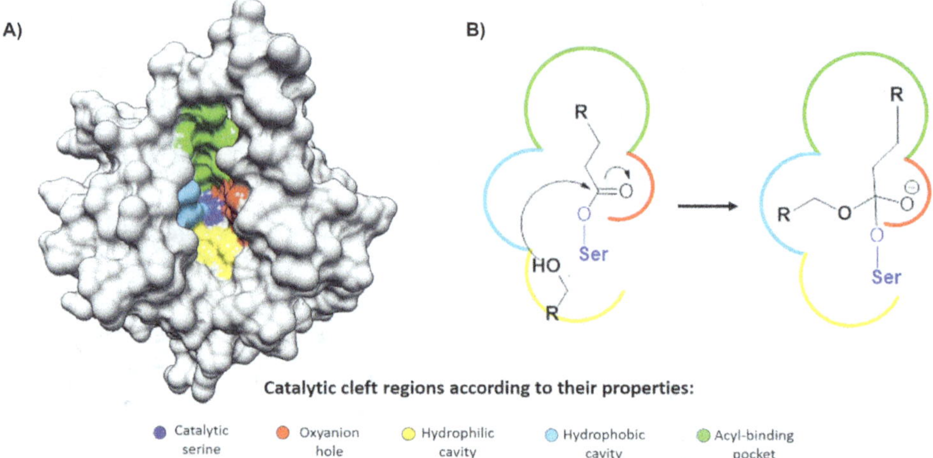


**Figure 6.13:** (A) Different areas in the catalytic cleft of lipases, exemplified for the metagenomic lipase LipC12 with the lid in open conformation. (B) Schematic representation of the areas, in different color codes. Reprinted from Maldonado et al. [115], with permissions.

pases [120]). Mutating this area is very delicate, as the residues that build this cavity are essential for catalysis, so that it is very possible that any change could imply negative effects. Anyhow, some cases are reported; for instance, for lipase Lip from *Thermomyces lanuginosus*, Li et al. [121] introduced mutations at amino acid residues A99 and V116, situated within the acyl-binding pocket, as well as S88, which is part of the oxyanion hole. The combined mutation S88T/A99N/V116D led to an enhanced ee of the *S*-enantiomer during the kinetic resolution of racemic 2-carboxyethyl-3-cyano-5-methylhexanoic acid esters.

- Acyl-binding pocket (Figure 6.13, green): A nonpolar region that accommodates the acyl moiety of substrates (fatty acids). Residues here (e.g., W104, L144, V190 in CALB) dictate steric complementarity. Thus, by simply replacing W104 in CALB with smaller polar residues (Cys/Ser/Thr), it was possible to improve *S*-enantioselectivity (up to 99% e.e.) in the resolution of racemic esters bearing bulky di(hetero)aryl methanol moieties, by strengthening electronic interactions between the substrate's pyridyl group and the redesigned binding pocket [122].
- Hydrophilic cavity (Figure 6.13, yellow): Interacts with nucleophiles (e.g., OH group of alcohols in (trans)esterifications), directing them toward catalytic serine. In CALB, S47 is key for the chiral recognition, for instance, mutant S47A displayed reversed stereoselectivity upon alcohol esterifications [123].
- Hydrophobic cavity (Figure 6.13, cyan): A large hydrophobic surface that interacts with large substituents on the substrate's chiral center. For carboxylic acids, the hydrophobic cavity (e.g., F344/F345 in CRL [124]) often determines *R*/*S* preference.

As a general conclusion derived from the excellent review of Maldonado et al. [115], when dealing with a racemic carboxylic acid substrate, the most effective mutations tend to occur within the acyl-binding pocket, the hydrophobic cavity, or the oxyanion hole. Conversely, in the case of racemic alcohol substrates, the beneficial mutations are typically found within both hydrophobic and hydrophilic regions of the binding site, and occasionally also involve residues in the oxyanion hole. Furthermore, in resolution processes involving chiral carboxylic acid groups, the reaction pathway – whether synthetic or hydrolytic – plays a key role in determining suitable mutation sites for enhancing enantioselectivity. During synthesis, the acyl donor initially occupies the acyl-binding pocket, leading to the formation of an acyl-enzyme intermediate. Therefore, alterations within the acyl-binding pocket are likely to have the most pronounced effect on the enzyme's capacity to differentiate between enantiomers of the chiral carboxylic acid. In contrast, during the hydrolysis of chiral esters, the substrate engages both the acyl-binding pocket and the hydrophobic cavity simultaneously, irrespective of whether the stereocenter resides in the acid or alcohol portion. As a result, when aiming to improve the enantioselectivity of lipases in hydrolytic reactions, introducing mutations in both of these regions may prove beneficial. For the WT LipC12 lipase depicted in Figure 6.13, molecular docking of transition-state complexes highlighted residue V261 in the WT LipC12 as a key determinant of enantioselectivity in both the transesterification of racemic-1-phenylethanol and the hydrolysis of *p*-nitrophenyl octanoate [125]. SDM at this position yielded three LipC12 variants with enhanced hydrolytic activity, the most active being LipC12V261F, which exhibited a 5.8-fold increase in activity compared to the WT. In terms of enantioselectivity, the most notable improvement was observed in the variant LipC12V261Q, displaying an enantiomeric ratio (E-value) five times greater than that of LipC12wt. However, an inverse relationship between catalytic activity and enantioselectivity was apparent: the variants with heightened activity showed reduced enantiomeric discrimination, while the enantioselective variant LipC12V261Q demonstrated hydrolytic activity that was four times lower than the WT.

#### 6.2.5.4 Rational Design Based on Remodeling Enzyme Dynamics

The full catalytic cycle of an enzyme includes substrate binding, formation of the enzyme-substrate complex for catalysis, and subsequent release of the product – processes that depend on conformational shifts involving opening and closing movements of the enzyme. Enzymes exhibit dynamic behavior across a broad spectrum of timescales, and substantial structural rearrangements have been documented during catalysis. These conformational dynamics are now recognized as playing a critical role in regulating every stage of the catalytic process. Effective enzymatic function relies not only on the catalytic site itself but also on the surrounding dynamic network, including substrate and product channels and structural domains that facilitate conformational changes. As such, rational or semirational design approaches that target enzyme dynamics offer promising strategies for fine-tuning catalytic stereoselectivity.

Enzymes operate across diverse timescales, with large-scale structural rearrangements observed during catalysis. These dynamic processes are now recognized as essential regulators of every catalytic phase, influencing substrate access and positioning, transition-state stabilization, and finally, product expulsion. In fact, effective catalysis relies on a dynamic architecture surrounding the active site, including substrate/product channels (gateways for molecular traffic), flexible domains (enabling structural transitions), and allosteric networks (long-range communication pathways). Thus, targeting these dynamic features through rational design strategies offers precise control over enzymatic behavior, either by modifying gating mechanisms in transport channels, stabilizing transition-state conformations, or by optimizing domain flexibility for faster substrate turnover [19, 37].

We had already mentioned that replacing specific amino acid residues by proline is a widely used rational approach for enhancing enzyme thermostability, as this cyclic aminoacid increases conformational rigidity in the protein backbone [95]. Regarding stereoselectivity, prolines are also key residues, as described by Qu et al. for the zinc-dependent alcohol dehydrogenase from *Thermoanaerobacter brockii* (TbSADH) [126]. These authors reported the PiLoT (proline-induced loop engineering) approach to enhance the dynamic fluctuations of an active-site loop of WT TbSADH. This enzyme features a flexible loop spanning residues 84–92, situated near the substrate binding pocket and adjacent to the β6 strand. Notably, a proline at position 84 imparts significant rigidity to this loop. To modulate loop dynamics, researchers performed saturation mutagenesis and site deletion at this position (ΔP84). MD simulations showed that five variants – P84G, P84S, P84V, P84Y, and ΔP84 – exhibited increased flexibility, as evidenced by higher RMSF values for residue 84 and the entire loop compared to the WT. Experimental assays revealed that P84S and P84Y mutants acquired (*S*)-selectivity with moderate conversion rates, while the ΔP84 mutant displayed (*R*)-selectivity. Further combinatorial saturation mutagenesis at positions 85 and 86, using P84S and ΔP84 as templates for (*S*)- and (*R*)-selectivity, respectively, yielded two double mutants (P84S/I86L and P84S/I86A) with outstanding (*S*)-selectivity (>99% e.e. at >99% conversion) and one double mutant (ΔP84/A85G) with high (*R*)-selectivity (97% e.e. at 99% conversion). Additional MD simulations confirmed that the top variants, P84S/I86L and ΔP84/A85G, demonstrated greater loop mobility at residues 84 and 85, resulting in an expanded and more flexible substrate-binding pocket.

MD simulations have proven essential for probing protein flexibility and screening variants with altered dynamics. For example, conformational dynamics engineering was applied to lipase CALB to enhance *R*-enantioselectivity in the esterification of 3-*tert* butyl-dimethyl-silyloxy glutaric anhydride to (*R*)-3-*tert*butyl-dimethyl-silyloxy glutaric acid methyl monoester guided by MD simulations [127]. The previously developed CALB variant EF5 showed high *R*-enantioselectivity (98.5% e.e.) at 5 °C, but this dropped to just 8% at 30 °C, even as catalytic yield rose from 13.2% to 80% with increasing temperature. MD simulations indicated that higher temperatures increased the conformational dynamics of the active pocket and tunnel, suggesting that controlling these dynamics could improve *R*-enantioselectivity at elevated temperatures. A total of 13 residues in the substrate pocket and channel were targeted for in silico alanine scanning or serine substitution, identifying D223 and A281 as key positions based on notable RMSF changes in D223A and A281S mutants. Saturation mutagenesis at these sites, performed computationally, led to the selection of D223V and A281S variants with reduced flexibility. When tested experimentally, these single mutants and their combination (D223V/A281S) dramatically improved *R*-enantioselectivity at 30 °C – from 8% e.e. in the parent EF5 to 93.5% e.e. (A281S), 95.8% e.e. (D223V), and over 99% e.e. (D223V/A281S). This demonstrates that reducing the conformational dynamics of the binding pocket and channel is an effective strategy for enhancing the *R*-enantioselectivity of CALB [127].

### 6.2.6 Computational Protein Design

Proteins, the workhorses of biology, derive their functions from precise 3D structures. Two intertwined challenges have driven decades of research: predicting a protein's structure from its sequence and designing novel proteins with tailored functions. Creating new enzymes with significantly improved activity or stereoselectivity through purely computational means seemed to be an extremely challenging task; anyhow, recent advances in computational methods, particularly deep learning, have transformed this research field, enabling unprecedented accuracy and innovation, although the growing integration of many computing tools has led to an increasing number of validated successes in recent years.

From a historical perspective, the first protein structures, such as myoglobin [128] and hemoglobin [129], were solved in the late 1950s and early 1960s using X-ray crystallography. These structures revealed the complexity of protein folding and the importance of 3D conformation for biological function. In 1972, Christian Anfinsen demonstrated that a protein's sequence determines its structure, suggesting that it should be possible, in principle, to predict structure from sequence [130]. However, the sheer number of possible conformations (known as Levinthal's paradox [131]) made this a formidable computational problem. Over the following decades, experimental techniques like X-ray crystallography, NMR spectroscopy, and cryo-electron microscopy provided detailed structures for thousands of proteins, but predicting structure from sequence remained elusive.

In 1974, Chou and Fasman analyzed a set of 15 proteins with established structures to estimate the likelihood of each of the 20 standard amino acids appearing in α-helices or β-sheets [132]. By averaging these α- and β-propensity values across segments of a protein sequence, they developed a method to predict the secondary structure of polypeptide chains. However, the predictive accuracy of this approach was limited – largely because it only considered the linear amino acid sequence, neglecting the critical influence of 3D (tertiary) interactions on secondary structure formation. At that time, the Protein Data Bank contained only a few hundred experimentally determined protein structures, restricting the statistical power of such analyses [40, 41]. It was not until the 1990s, with advancements in protein crystallography, that the database grew significantly. Despite the scarcity of structural data in those early years, some general physicochemical principles emerged. For example, hydrophobic amino acid side chains were typically found buried within the protein core, shielded from water, while polar and charged side chains were more often exposed on the protein surface, interacting with the solvent. Occasionally, polar residues were also observed in the protein interior, but they tended to form hydrogen bond networks to offset the loss of solvation energy during folding. These observations led to initial efforts in protein design, where researchers aimed to create polypeptide sequences that followed the basic rule: nonpolar residues inside, polar residues outside [133]. With this understanding, amphiphilic helical structures became an ideal target for protein design – specifically, α-helices with one hydrophilic, solvent-exposed face and one hydrophobic face that could pack against other hy-

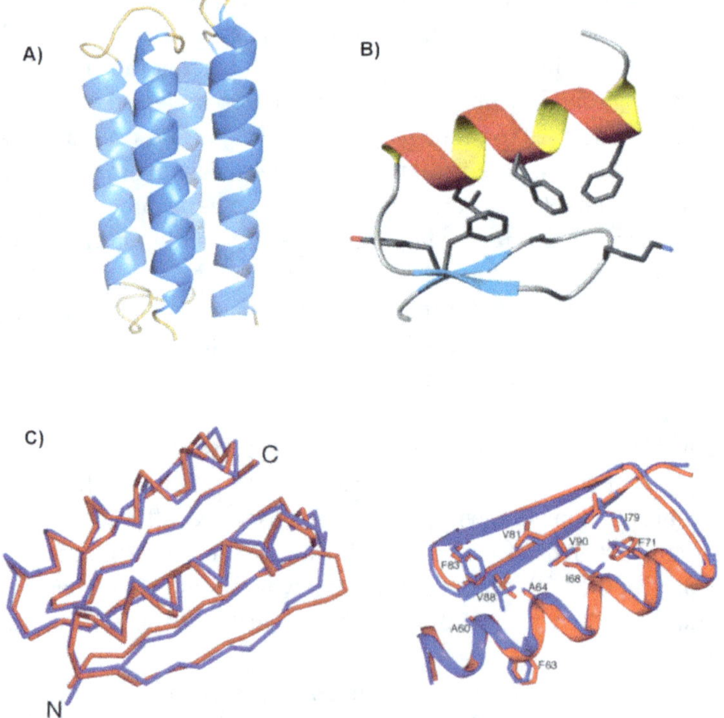


**Figure 6.14:** (A) Backbone of a four-helix bundle structure, bearing a hydrophobic interior and a hydrophilic exterior, by Regan and DeGrado [134]. (B) Designed zinc-finger protein by Dahiyat and Mayo [135]. (C) Left: Predicted backbone structure of Top7 (blue) superimposed with the X-ray structure (red); right, superimposed side chains (similar color code) [136]. Reprinted from Åqvist. J. Computational protein design and protein structure prediction. Scientific Background on the Nobel Prize in Chemistry 2024. The Royal Swedish Academy of Sciences; 9 October 2024. Available from: https://www.nobelprize.org/uploads/2024/10/advanced-chemistryprize2024.pdf (accessed 7 March 2026).

drophobic surfaces inside the protein. This concept was realized by Regan and DeGrado in 1988, who engineered a stable, highly helical four-helix bundle protein by connecting four α-helices with three loops (Figure 6.14A), adhering to these design principle [134].

The first successful computational design of a small protein was accomplished by Dahiyat and Mayo in 1997, marking a milestone in de novo protein design [135]. Their target was the zinc-finger motif, a small protein domain of about 30 amino acids, typically stabilized by binding one or two $Zn^{2+}$ ions. The objective was to design a new amino acid sequence that would fold into the same structure as the zinc finger but without relying on metal ions for stability. To achieve this, Dahiyat and Mayo fixed the protein backbone structure (derived from the zinc finger) and computationally searched for amino acid sequences that would adopt the desired 3D fold in the absence of metal ions. This required evaluating an immense number of possible sequences (on the order of $10^{27}$), as well as optimizing the orientation (rotamers) of amino acid side chains. They used a combination of the dead-end elimination algorithm and Monte Carlo simulations, guided by an empirical scoring function to estimate conformational energies. The resulting designed protein, called FSD-1, shared only 6 out of 28 residues (21% identity) with the original zinc-finger sequence, indicating that it was a novel sequence unrelated to natural proteins. Experimental validation using NMR spectroscopy showed that FSD-1 folded into a compact, well-ordered structure that closely matched the computationally predicted model and the original zinc-finger fold, despite lacking any metal-binding sites (Figure 6.14B).

In 2003, David Baker and his team achieved a landmark in de novo protein design by creating Top7, a 93-amino-acid α/β protein whose experimentally determined crystal structure matched the computational model down to individual side-chain placements [136]. This construct combined two α-helices with a five-strand β-sheet into a fold never before observed in any globular protein and bore no sequence resemblance to known proteins (Figure 6.14C). In other words, Top7 represented a wholly novel fold and sequence, generated entirely through automated computation that simultaneously optimized backbone conformations and side-chain packing. The cornerstone of this success was Rosetta, a program Baker's group introduced in 1999 [44], already mentioned in Section 6.2. Rosetta builds new proteins by stitching together short structural fragments drawn from unrelated proteins in the PDB that share similar local sequences. It then applies Monte Carlo sampling and an energy function – including a 6–12 Lennard-Jones term for van der Waals forces, hydrogen-bonding potentials, and solvation models – to optimize both the amino acid sequence and the 3D structure for the target backbone. Side-chain geometries are selected from an extensive rotamer library, and thousands of candidate designs are ranked by computed energy. Rosetta was conceived as a general platform for both predicting natural protein folds and engineering entirely new ones, and it has since grown through contributions from a large community of developers. Building on earlier fragment-assembly ideas pioneered by Jones and Thirup for automated model building in crystallography

[137], Baker's team demonstrated that Rosetta could produce a vast array of stable protein architectures.

While initial efforts focused on designing static structures, more recent work has sought to imbue proteins with advanced functions – tasks that demand deeper insight into dynamics, conformational changes, allosteric regulation, and catalytic mechanisms [138]. In fact, it is even possible to design and create novel-to-nature enzymes (de novo design), biocatalysts specifically engineered to drive chemical transformations that no enzyme in nature performs. They expand the repertoire of biocatalysis by enabling entirely new reaction pathways, offering tools for sustainable chemistry, pharmaceuticals, and materials science [139]. For this purpose, the Rosetta de novo enzyme design workflow proceeds through a series of coordinated stages, shown in Figure 6.15: first, a minimalist active-site model ("theozyme") is constructed to define the key catalytic groups; next, this transition-state arrangement is grafted into candidate protein scaffolds; then, an iterative Rosetta-driven design phase refines the surrounding amino-acid sequence to optimize both backbone and side-chain interactions; finally, molecular-dynamics-based virtual screening guides selection of top designs for experimental characterization.

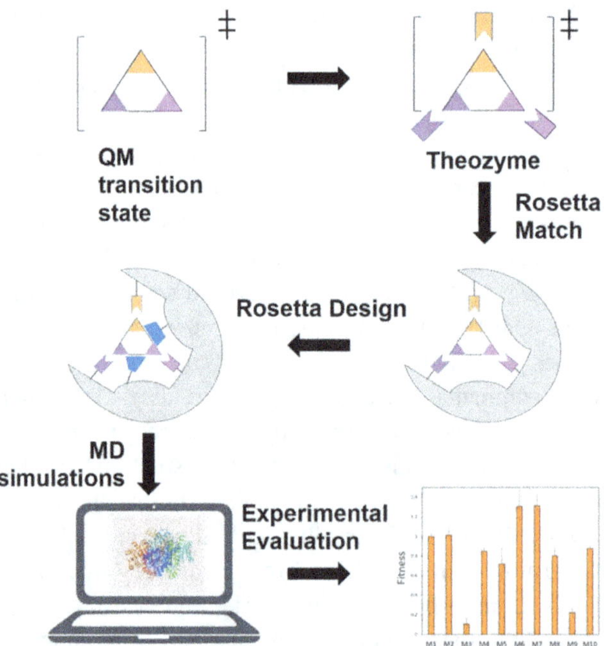


**Figure 6.15:** A schematic of the in silico enzyme-design workflow, covering the creation of a theozyme model, grafting the transition-state geometry into a protein scaffold, iterative Rosetta-driven sequence optimization, and downstream molecular-dynamics-based virtual screening prior to experimental validation. Taken from Song et al. [19]. OPEN ACCESS.

Throughout this protocol, Monte Carlo-based sampling explores residue identities and rotamer conformations around the bound intermediate (with tools like Foldit enabling targeted mutagenesis), while the Rosetta energy function – which integrates van der Waals, electrostatics, hydrogen-bonding, solvation, and rotamer-probability terms – calculates overall protein stability and enzyme-transition-state binding energies to rank and prioritize designs.

In 2008, Baker's group reported the first examples of de novo enzyme design, creating catalysts for reactions converting nonnatural substrates [140] and even reactions for which enzymes in nature did not exist. This is the case of Kemp elimination (Figure 6.16A), a chemical reaction involving the base-catalyzed ring-opening of benzoisoxazoles to produce o-cyanophenolate ions [141]. In fact, Baker and coworkers reported the design of a tailor-made "Kemp-eliminase" [142, 143], based on the methodology shown in Figure 6.15.

The researchers used the Rosetta software suite to design active sites that could catalyze the Kemp elimination. Two distinct catalytic motifs were employed: one based on a carboxylate side chain and the other on a histidine-aspartate pair acting as bases to remove the hydrogen atom from the aromatic ring, and some other residues for donating the final proton atom to the intermediate alkoxide, as well for stabilizing the aromatic structure *via* π-stacking [142]. Quantum mechanical calculations helped define idealized active site geometries (the "theozyme" models), which were then computationally grafted into protein scaffolds. Rosetta optimized the surrounding amino acid sequences to stabilize the transition state and enhance catalysis (Figure 6.16B). Subsequently, the designed enzymes were expressed, purified, and tested for catalytic activity. Several designs showed measurable Kemp elimination activity, with rate enhancements over the uncatalyzed reaction. One variant, KE07, was crystallized, and its structure closely matched the computational model, confirming the accuracy of the design process [142]. These novel enzymes exhibited rate enhancements over uncatalyzed reactions but lagged behind natural enzymes in overall turnover, a gap that directed-evolution strategies (see Section 6.3) helped narrow (Figure 6.16C) [143]. Another archetypal example of de novo design of proteins was also reported by Baker and coworkers in 2010, dealing with the computer design of a "Diels-Alderase" [144], a tailor-made enzyme able to catalyze the [4 + 2] cycloaddition of dienes and dienophiles to furnish cyclohexenes, originally reported in 1928 [145, 146]. Since then, many other examples have been reported; we recommend some recent reviews [147, 148] for a very detailed explanation of the methodology, graphically depicted in Figure 6.17A.

A new step in the design of biocatalysts is the creation of catalytical proteins containing noncanonical amino acids (ncAAs), which have emerged as powerful tools in enzyme engineering, enabling the rational design of biocatalysts with enhanced properties and novel functionalities [149, 150]. By expanding the genetic code beyond the 20 canonical amino acids, researchers can introduce unique chemical groups that refine enzyme active sites, improve stability, and unlock nonnatural catalytic mecha-

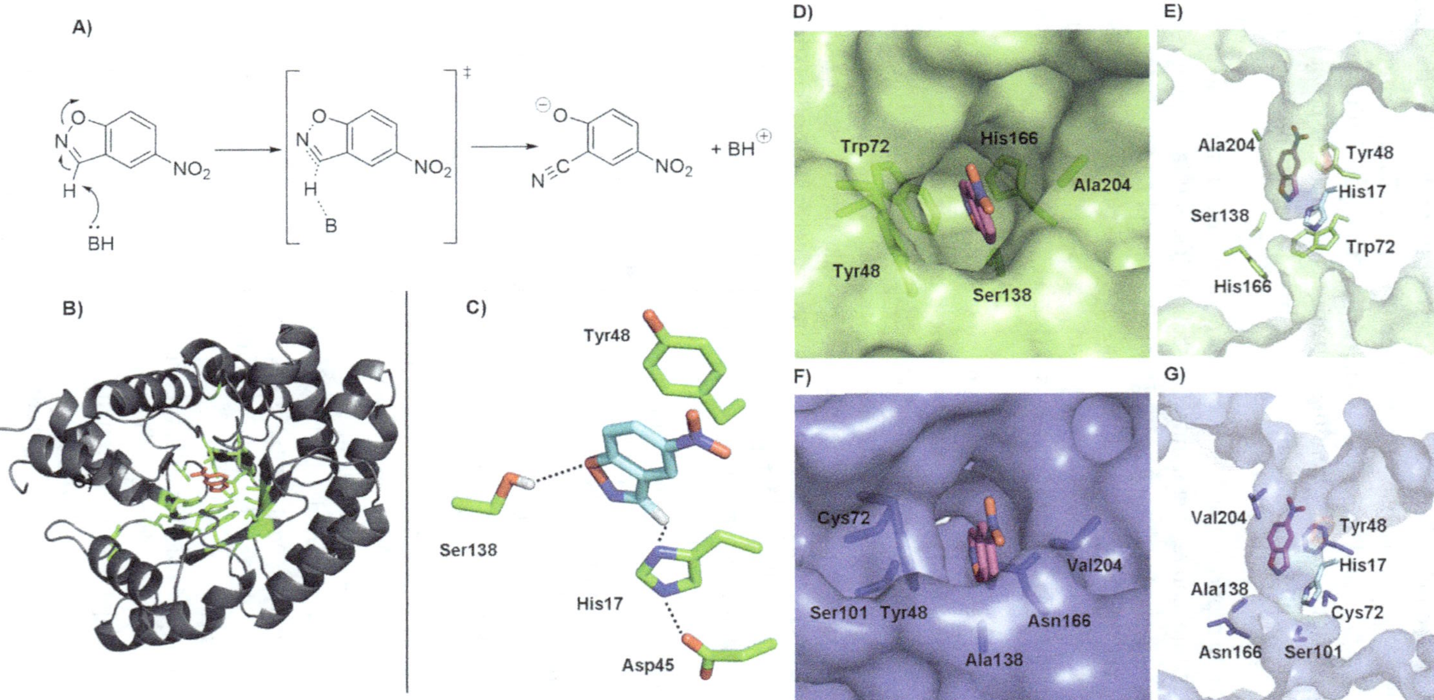


**Figure 6.16:** (A) Kemp elimination, showing the transition state. (B) The KE70 design: gray, natural scaffold (PDB 1JCL); red, 5-nitrobenzisoxazole substrate (red): green, side chains of 16 residues replaced to form the designed Kemp eliminase active site. (C) Key features of KE70's active site, showing substrate, the His17-Asp45 dyad, the H-bond donor (Ser138), and the stacking Tyr48. (D) The active-site cavity of the KE70 designed model, top view; (E) cross-sectional view of the active-site cavity of the KE70. (F) The active-site cavity of the evolved variant R6 6/10A, top view. (G) Cross-sectional view of variant R6 6/10A. Reprinted from Khersonsky et al. [143] with permissions.

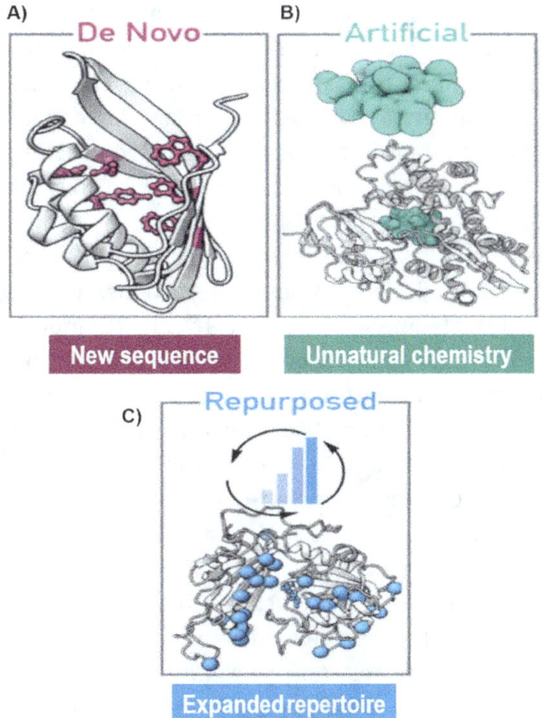


**Figure 6.17:** Strategies for the expansion of enzyme function. (A) Design of de novo sequences from scratch with novel active-site amino acid constellations. (B) Addition of unnatural components (e.g., to produce artificial enzymes). (C) Repurposing of natural or engineered enzymes *via* directed evolution. Reprinted from Leveson-Gower RB. Designing Enzymatic Reactivity with an Expanded Palette. Reprinted from Leveson-Gower, R. B [148].

nisms (Figure 6.17B). Thus, genetic code expansion serves as the primary technique for site-specific ncAA integration. This approach utilizes orthogonal aminoacyl-tRNA synthetase/tRNA pairs to incorporate ncAAs at amber stop codon sites, allowing precise modification of enzyme structures [151]; another option is the use of in vitro methods, such as flexizymes (artificially engineered ribozymes (RNA enzymes) that catalyze the aminoacylation of transfer RNAs (tRNAs) with a wide range of amino acids, including many nonnatural and nonproteinogenic amino acids [152]) or commercial cell-free kits (e.g., PURExpress [153]) for coupled transcription-translation, achieving yields up to 0.5 mg/mL for ncAA-containing enzymes, though industrial scalability remains limited.

At this point, rational design strategies combine computational modeling and structural analysis to identify target residues for substitution with ncAAs that introduce favorable interactions, such as halogen bonds or hydrophobic packing [149], essential for improving thermal stability [154]. In fact, halogenated residues fill internal cavities, reducing conformational flexibility; for instance, the introduction of *p*-benzoyl phenylalanine (*p*BpA) improves thermostability through increased hydrophobic interactions, as shown for an (*R*)-amine transaminase, for which the F86A/F88*p*BpA variant exhibited 30% higher stability at 55 °C while maintaining activity [155]. Interestingly, substituting Phe88 with *p*BpA in the same enzyme led to 15-fold higher activity for 1-phenylpropan-1-amine, demonstrating how ncAAs also optimize substrate binding, in this case by expanding the hydrophobic active site. In another example, the introduction of 4-fluorophenylalanine in *Pseudomonas fluorescens* esterase enhanced substrate orientation through halogen-bond interactions, increasing thermostability (an increase of melting temperature, $T_m$, by 8 °C) [154]. Another example of this hydrophobic packing can be found in azoreductases bearing brominated tyrosines, showing a 13-fold increase in half-life at 78 °C [154]. Additional reasons for increased stability when incorporating ncAAs are the formation of intramolecular covalent crosslinks (*O*-2-benzoyltyrosine (BeTyr) in myoglobin formed intramolecular crosslinks, raising $T_m$ by 10 °C without disrupting function [156]) or β-sheet strengthening; for instance, proline analogs (e.g., 4*S*-fluoroproline) rigidified loops in *Thermoanaerobacter* lipase, improving solvent tolerance [154].

For sure, the presence of ncAAs inside the enzyme structure enables the catalysis of nonnaturally biocatalyzed reactions. For instance, MD simulations helped identify key residues in *Lactococcal multidrug resistance Regulator* (LmrR) that were replaced with fluorinated ncAAs, boosting the biocatalyzed nitroaldolic condensation (Henry reaction) activity by 184% [157]. Even artificial metalloenzymes can be built; in this sense, the transcription factor LmrR has emerged as a highly adaptable scaffold for engineering. Researchers, including Drienovská et al., leveraged its promiscuous hydrophobic binding cavity to incorporate the ncAA (2,2′-bipyridin-5-yl)alanine (BpyAla) at position 89, creating copper-binding sites within the protein dimer [158]. This modification enabled the development of copper-loaded LmrR_M89BpyAla metalloproteins capable of catalyzing enantioselective vinylogous Friedel-Crafts alkylations. In another recent example, also based on LmrR, a novel copper-dependent artificial Michaelase (Cu_Michaelase) was engineered using a genetically encoded BpyA, marking the first successful optimization of an ArM containing a noncanonical metal-binding amino acid *via* directed evolution [159]. This ArM was able to efficiently catalyze the asymmetric Michael addition of 2-acetyl azaarenes to nitroalkenes, enabling efficient access to γ-nitro butyric acid derivatives – key precursors for several pharmaceuticals. For a detailed revision of the preparation and uses of artificial enzymes, we recommend a recent book chapter from Reetz et al. [15].

Another strategy (Figure 6.17C) to generate enhanced or even newer catalytic enzymatic performance is the repurposing of natural or engineered enzymes *via* directed evolution, which will be commented on in the next section.

## 6.3 Directed Evolution of Enzymes

### 6.3.1 Introduction

Directed evolution of enzymes is a laboratory technique that mimics natural selection to engineer enzymes with enhanced or novel functions. It involves iterative cycles of genetic diversification (creating a library of enzyme variants), selection or screening (identifying variants with desired traits), and amplification (replicating successful variants for further rounds) [20, 160, 161]. This method enables the optimization of biocatalysts for specific industrial, medical, or environmental applications without requiring prior structural or mechanistic knowledge of the enzyme. A schematic representation of this technique is depicted in Figure 6.18 [162].

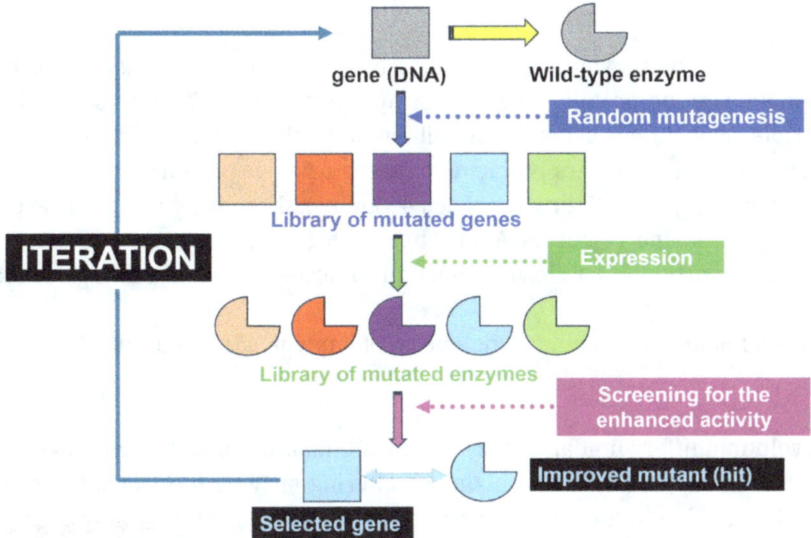


**Figure 6.18:** Schematic representation of the iterative cycle of directed evolution, adapted from Reetz and Krebs [164].

Basically, directed evolution emulates the principles of natural selection but operates under controlled laboratory conditions rather than in ecological environments. The methodology follows a multistep workflow (Figure 6.18), initiating with the introduction of genetic diversity into the target enzyme's coding sequence, using different methodologies [163]. These mutated genes are subsequently cloned into microbial hosts such as *E. coli* or *Pichia pastoris* and plated on solid growth media. Following incubation, individual colonies, each representing a unique genetic variant, are isolated, ensuring a direct correlation between genotype (DNA sequence) and phenotype (enzyme function). A significant challenge arises from the presence of WT sequences and redundant clones in the library, which inflate screening efforts [57]. Precise colony harvesting, whether automated or manual, is critical to avoid cross-contamination that could obscure results. Selected colonies are transferred to multiwell plates for culturing, with parallel plates used to assay enzymatic activity.

In ideal scenarios, initial libraries yield improved variants. More commonly, however, iterative cycles are required: the top-performing variant's gene serves as the template for subsequent mutagenesis rounds, applying evolutionary pressure to accumulate beneficial mutations. Progress halts if libraries plateau at local fitness optima, though several strategies exist to escape these minima. In any case, a central challenge lies in navigating the immense sequence space of proteins. While random mutagenesis generates diversity, the combinatorial explosion of possible variants (*N*) is described by the following equation:

$$N = 19^M X / [(X - M)!M!]$$

where *M* represents the enzyme's total residues, and *X* denotes simultaneous substitutions. For a 300-residue protein, single substitutions produce 5,700 variants, while dual and triple substitutions generate ~16 million and ~30 billion possibilities, respectively. Addressing this "numbers problem" requires balancing library diversity with practical screening capacity. Overly restricted diversity risks missing hits, while excessive diversity overwhelms resources. Alternative strategies include:

- Selection systems: Linking enzyme function to host survival, bypassing manual screening.
- Display technologies: Coupling genotype to phenotype *via* surface-displayed enzymes.

Directed evolution differs fundamentally from traditional rational protein engineering methods in its approach to optimizing or altering enzyme functions. From the core methodology, directed evolution uses iterative cycles of random mutagenesis (e.g., error-prone PCR (epPCR), DNA shuffling) and HTS/selection to mimic natural evolution. This method does not require prior structural or mechanistic knowledge of the target enzyme, as it is mandatory for rational design, which, as already commented in Section 6.2, relies on targeted mutations based on structural insights (e.g., X-ray crystallography, computational modeling) to predict and engineer specific functional changes, requiring a detailed understanding of the protein's active site and catalytic mechanism.

A historical perspective of directed evolution has been extensively reviewed by Reetz et al. [20]. For many years, scientists have aimed to replicate the process of natural evolution within laboratory settings. In the mid-1960s, Spiegelman and colleagues conducted pioneering experiments with self-replicating RNA molecules outside living cells, which were initially thought to model early precellular evolutionary events [164]. Subsequent studies, however, revealed that these RNA molecules were not truly self-replicating, but Spiegelman's work nonetheless laid the foundation for the burgeoning field of RNA evolution, which was further advanced by researchers like Blain and Szostak [165] and Joyce [166]. Unlike protein engineering, directed evolution at the RNA level focuses on the selection of aptamers, catalytic RNAs, or the continuous evolution of ribozymes and RNA polymerases – a field that has been comprehensively reviewed elsewhere [166].

The term "directed evolution" was first applied in the context of protein engineering in the early 1970s, when Francis and Hansche described a system in yeast (*S. cerevisiae*) where spontaneous mutations affecting an acid phosphatase were monitored over a thousand generations [167]. This research demonstrated how sequential mutations could incrementally improve enzyme performance, such as increasing metabolic efficiency or shifting pH optima. Around the same time, Hall's group used genetic complementation to evolve new functions in β-galactosidase (ebgA) in *E. coli*, showing that stepwise mutations could enable the utilization of new carbon sources while sometimes retaining or even improving original enzyme activities [168].

The 1980s and 1990s saw significant advances in laboratory evolution. Kim and colleagues extended these approaches [169], inspiring further work such as Lenski's long-term evolution experiments in bacteria [170] and Liu's innovations in continuous evolution [171], including phage-assisted methods [172]. Meanwhile, developments in mutagenesis, such as the Kunkel method [173, 174] leading to error-prone rolling circle amplification [175] provided new tools for generating mutant libraries. A major conceptual advance came from Eigen and Gardiner, who emphasized the importance of self-replication in molecular evolution and outlined the logic of laboratory evolution cycles involving mutagenesis, amplification, and selection [176]. However, these early approaches predated the widespread use of PCR, which, after its introduction by Mullis in the 1980s [24] (already mentioned in Section 6.2), revolutionized molecular biology and greatly expanded the possibilities for directed evolution.

As researchers experimented with various mutagenesis strategies to generate and screen for improved enzyme variants – often focusing on properties like thermostability – important milestones were achieved. For instance, Matsumura and Aiba used chemical mutagenesis to enhance the stability of kanamycin nucleotidyltransferase [177], and Hageman's group later demonstrated the power of iterative cycles of mutagenesis and selection to further increase thermostability, identifying key mutations that conferred resistance to high temperatures and harsh solvents [178]. This chapter can be considered a real hallmark event, together with the implementation of site-specific mutagenesis (SSM), initially developed by Smith [21, 23] (mentioned in Section 6.2), allowed for targeted amino acid substitutions but was labor-intensive when multiple variants were needed. Innovations such as cassette mutagenesis (targeted protein engineering technique that replaces a specific DNA segment (a "cassette") with a synthetic oligonucleotide duplex containing desired mutations [179]) and saturation mutagenesis (also mentioned in Section 6.2 [25, 26]), which enabled the simultaneous randomization of specific residues, streamlined the process. Further improvements included methods like overlap extension PCR ([180]) and splicing by overlap extension ([181]), which facilitated the creation of chimeric genes and the introduction of random mutations. The introduction of epPCR [182] by Leung et al. [183] enabled the efficient generation of libraries with random point mutations, a technique that became widely used after its application to antibody affinity maturation by Hawkins et al. [184].

In the early 1990s, Frances Arnold and colleagues applied random mutagenesis to subtilisin E, aiming to improve its performance in organic solvents rather than just thermostability [185]. This work, highlighted during Arnold's Nobel Prize recognition [186] and generally considered as the first example of directed evolution, marked a turning point in the field, showing that directed evolution could address challenges beyond simple activity improvements. Anyhow, some researchers [20] identify the seminal paper of Liao et al. [178] as the real starting point of directed evolution. Recombination-based methods also emerged, with DNA shuffling, introduced by Stemmer in 1994 [187], simulating sexual recombination to combine beneficial mutations from different variants. Family shuffling and other recombinant techniques further expanded the toolkit for generating diversity [20].

Throughout the 1990s and beyond, researchers increasingly applied directed evolution not only to study enzyme mechanisms but also to engineer enzymes for practical applications. Techniques like phage display, flow cytometry, and water-in-oil emulsion technology were adapted for protein engineering, although their use in evolving highly selective enzymes remained limited [20]. The development of statistical and algorithmic approaches for designing mutant libraries, such as those by Patrick and Firth [188] or Reetz et al. [20], improved the efficiency and effectiveness of directed evolution campaigns. These methods, often freely available to the scientific community, facilitated the evolution of enzyme selectivity, activity, and stability.

A significant shift occurred when researchers began to focus on evolving enzymes for stereoselectivity, a key goal in organic synthesis. Early and seminal work by Reetz and colleagues [189–192] demonstrated that stepwise cycles of random mutagenesis could incrementally enhance enantioselectivity, even though initial improvements were modest. The introduction of focused saturation mutagenesis [193], CAST, and ISM, already mentioned in Section 6.2, provided systematic strategies for evolving enzyme selectivity and robustness [20].

Recent advancements have included the evolution of artificial metalloenzymes, the use of reduced amino acid alphabets, improved saturation mutagenesis methods, and the integration of machine learning to guide directed evolution, already commented on in Section 6.2. These innovations have enabled the fine-tuning of enzyme regio- and stereoselectivity, the reversal of selectivity, and the engineering of enzymes for entirely new catalytic activities.

Together, these developments have established directed evolution as a foundational methodology in protein engineering, enabling the creation of enzymes with tailored properties for a wide range of scientific and industrial applications. Although impossible to mention all of them, some significant examples will be illustrated below.

### 6.3.2 Directed Evolution for Improving Enzyme Stability

The current understanding in the field of protein engineering emphasizes that a single round of mutagenesis and screening does not qualify as directed evolution. Instead, iterative cycles of mutagenesis, expression, and screening, depicted in Figure 6.18, are required to apply evolutionary pressure and accumulate beneficial mutations, mirroring the stepwise nature of natural selection. Following this criterion, the first reported case of directed evolution was the improvement of the thermal stability of kanamycin nucleotidyltransferase reported in 1985 by Matsumura and Aiba [177]. These researchers employed a mutator strain strategy (using *E. coli* XL1-red) to facilitate random mutagenesis in plasmid libraries due to deficiencies in DNA repair pathways [194]. In this approach, the gene encoding the enzyme was isolated from a mesophilic organism and inserted into the thermophilic bacterium *Bacillus stearothermophilus*. Selection was then conducted at elevated temperatures, leveraging the host's inherent kanamycin resistance at 47 °C, which diminishes above 55 °C. By cycling the plasmid through the *E. coli* mutD5 mutator strain and reintroducing it into *B. stearothermophilus*, a mutation (Asp80Tyr) conferring kanamycin resistance at 63 °C was identified. Subsequent rounds under heightened thermal stress (70 °C) revealed a second mutation (Thr130Lys). The double mutant (Asp80Tyr/Thr130Lys) exhibited markedly enhanced thermostability, illustrating the iterative nature of evolutionary optimization (Figure 6.18). Interestingly, further investigations demonstrated that the optimized variant also displayed increased resilience to denaturing agents, including dimethylformamide (DMF) and urea, outperforming the WT enzyme [195]. These findings highlighted the broader applicability of directed evolution in tailoring enzymes for extreme environments.

Since that moment, a plethora of examples stating improvement of enzymatic thermostability have been reported [196, 197]. Just to mention recent cases, the bacterium *Chromohalobacter salixigens* produces an uronate dehydrogenase, CsUDH, which catalyzes the conversion of uronic acids into C1,C6-dicarboxy aldaric acids, a reaction of significant biotechnological interest for synthesizing sugar acid derivatives. To enhance the enzyme's thermal resilience for industrial processes, Wagschal et al. [198] implemented a directed evolution strategy utilizing gene family shuffling, schematized in Figure 6.19.

This approach generated a library of variants, from which candidates were identified through a two-tiered screening process. The most stabilized variant, CsUDH-inc, incorporated 16 mutations. While individual substitutions showed negligible impacts on stability, their synergistic combination markedly improved thermotolerance, increasing the thermal denaturation midpoint by 18 °C. Remarkably, CsUDH-inc retained full catalytic efficiency even after incubation at 70 °C for 1 h. Structural analysis *via* X-ray crystallography revealed no significant differences in overall conformation or flexibility (as indicated by comparable B-factors [88], see Section 6.2.3) between the WT and CsUDH-inc. However, mutated residues in the engi-

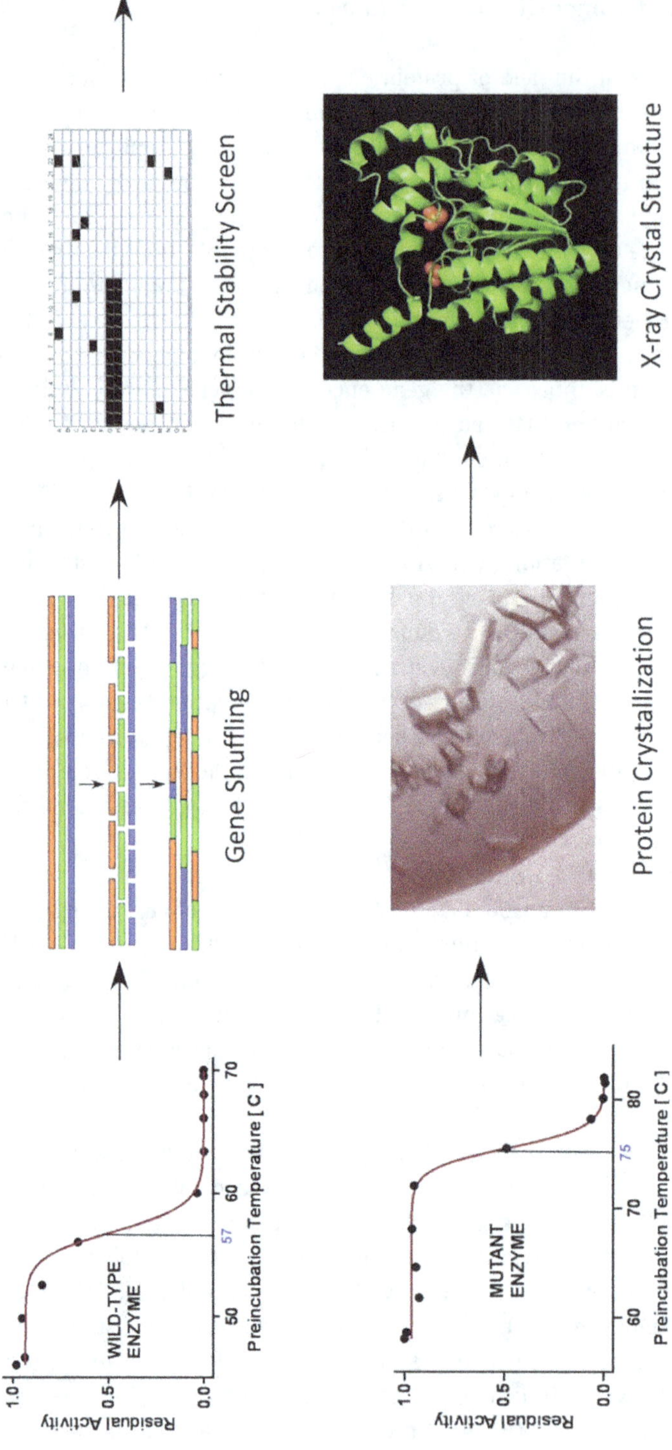


**Figure 6.19:** Directed evolution of CsUDH (uronate dehydrogenase from *Chromohalobacter salixigens*) increasing its thermostability. Reprinted from Wagschal et al. [198] with permissions.

neered variant exhibited tighter side-chain packing, suggesting enhanced structural rigidity. Despite these stability gains, CsUDH-inc displayed drastically reduced activity toward glucuronic and galacturonic acids. Subsequent combinatorial testing revealed that a subset of three mutations could partially restore substrate turnover while maintaining a thermal denaturation midpoint improvement of ~10 °C.

Stability in the presence of organic solvents is also frequently enhanced by directed evolution. The already mentioned paper of Chen and Arnold in 1993 [185] was dealing with this. In fact, these authors employed random mutagenesis using epPCR (although this term was not mentioned in the paper) to engineer the protease subtilisin E for enhanced functionality in extreme nonbiological conditions – specifically, high concentrations of polar organic solvents like DMF. Through iterative cycles of mutagenesis and screening, a variant termed PC3 was developed, demonstrating a 256-fold increase in catalytic efficiency for peptide substrate hydrolysis in 60% DMF compared to the WT enzyme. This engineered variant, along with others containing distinct mutation combinations, proved effective in organic media for applications such as transesterification and peptide synthesis. The PC3 variant incorporated 10 amino acid substitutions, with six additional mutations (G131D, E156G, N181S, S182G, S188P, and T255A) introduced over three rounds of evolution, building upon an initial quadruple-mutant scaffold (D60N, D97G, Q103R, and N218S). These substitutions clustered on the enzyme's surface near the active site and substrate-binding pocket, primarily within flexible loop regions connecting core structural elements. Notably, seven of PC3's substitutions correspond to residues found in naturally occurring subtilisins, highlighting the evolutionary plasticity of these regions.

Fungal unspecific peroxygenases (UPOs) are versatile biocatalysts capable of inserting oxygen into nonactivated C–H bonds with high selectivity, making them attractive for synthetic applications. However, their use in organic synthesis is limited by poor tolerance to organic solvents, which are often required for substrate solubility and product extraction. In a recent example, Martín-Diaz et al. [199] reported directed evolution of the UPO secretion mutant from *Agrocybe aegerita* (PaDa-I variant) aiming to enhance the activity and stability of this UPO in organic solvents. Thus, mutant libraries were generated using random mutagenesis, neutral genetic drift, and site-directed recombination, following the scheme shown in Figure 6.20A.

These libraries were subsequently screened for activity in increasing concentrations of organic solvents (acetone, acetonitrile, methanol, and dimethyl sulfoxide) with varying polarities. The key metric was the ratio of enzyme activity in the presence of organic solvent to that in aqueous buffer, expressed as a percentage ("tolerance"). The best variant (termed WamPa mutant), generated by site-directed recombination, combined mutations from both neutral drift and adaptive evolution. These mutations, depicted in Figure 6.20B, were located in several areas of the new protein:

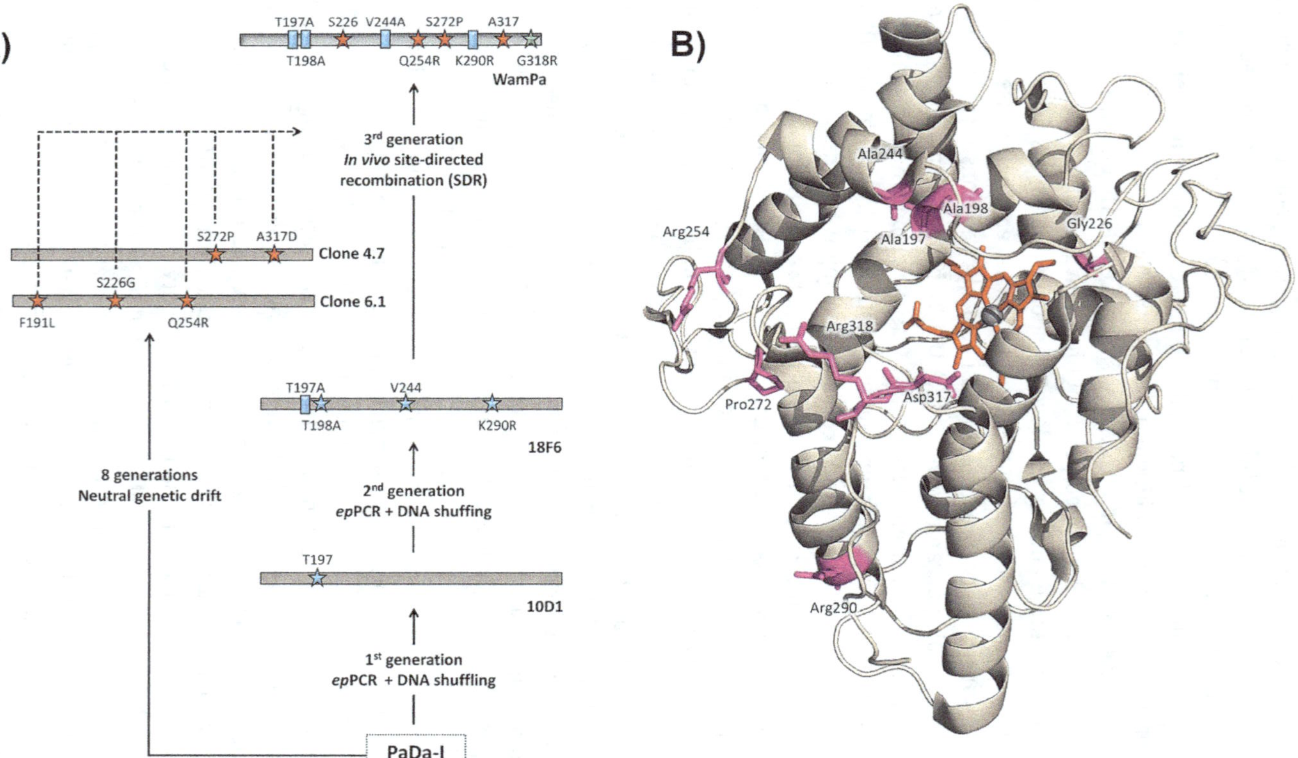


**Figure 6.20:** (A) Evolution pathway from PaDaI leading to WamPa for activity in the presence of organic cosolvents. Codes: stars, new mutations; rectangles, previous accumulated mutations; blue, mutations introduced along the adaptive evolution process; red, mutations by neutral genetic drift; green, mutation introduced by a mis-step during PCR amplification. (B) General view of the mutations (in pink) in the WamPa variant; heme is shown in red, with $Fe^{3+}$ in gray. Reprinted from Martín-Diez et al. [199], OPEN ACCESS.

- Surface mutations: S226G, Q254R, S272P, K290R – mainly increased rigidity and stabilization *via* proline and arginine residues, and new salt bridges.
- Access to heme channel: V244A, A317D, G318R – affected substrate and solvent access.
- Inner heme channel: T197A, T198A – improved activity in DMSO, likely by reducing DMSO's inhibitory effect on the aromatic triad near the heme.

Thus, the WamPa mutant showed up to 23-fold higher activity in 30% acetonitrile compared to PaDa-I, retaining high activity in 30% acetone, methanol, and 15% DMSO, and displaying tolerance (C50 values) of 24.9% (acetone), 27.0% (ACN), 14.0% (methanol), and 15.9% (DMSO), substantially higher than the parental enzyme. Also, using ABTS and veratryl alcohol as substrates in both aqueous and solvent-rich conditions, WamPa retained 29.6% of its aqueous activity in 30% ACN and 25.2% in 15% DMSO, compared to only 6.0% and 4.2% for PaDa-I, respectively.

### 6.3.3 Directed Evolution for Improving Enzyme Activity/Stereoselectivity

Once again, so many cases have been reported that it would be impossible to cover all of them, so that some reviews can be recommended [200, 201]. The initial paper from Reetz et al. in 1997 [189], schematized in Figure 6.21, could be considered the starting point.

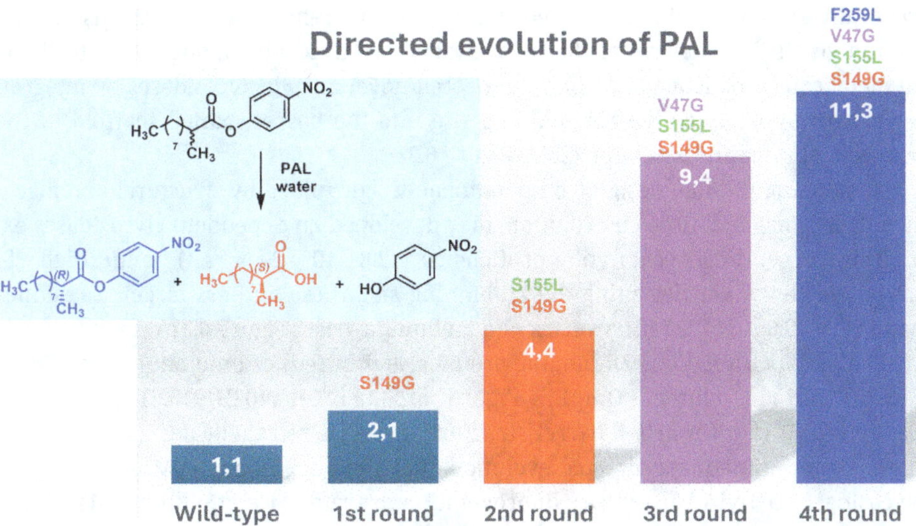


**Figure 6.21:** Graphical representation of the stereoselectivity increase of lipase from *Pseudomonas aeruginosa* by using four cycles of directed evolution. Adapted from Reetz et al. [20].

In this work, the authors explored the use of *P. aeruginosa* lipase (PAL) in the hydrolytic kinetic resolution of a model racemic ester substrate (*p*-nitrophenyl decanoate). Initial experiments revealed that the WT enzyme exhibited minimal enantioselectivity, with a selectivity factor (*E*, enantiomeric ratio, the ratio between the specificity constants ($k_{cat}/K_M$) for the transformation of both enantiomers of substrate [202]) of just 1.1, showing only a slight bias toward the (*R*)-enantiomer. To enhance catalytic performance, the researchers applied four iterative cycles of epPCR under controlled mutagenesis conditions. This process yielded a modified enzyme (Variant A) with significantly improved selectivity ($E = 11$), a breakthrough that sparked considerable interest at the time, though such incremental gains might not meet modern publication standards. This Variant A contained four cumulative amino acid substitutions (S149G, S155L, V47G, and F259L), introduced progressively during successive mutagenesis rounds. A notable challenge in this work was the absence of established high-throughput methods for e.e. analysis. To address this, the team developed an innovative screening platform combining a preliminary plate-based assay with a UV/Vis detection system for tracking *p*-nitrophenolate release. This hybrid approach enabled daily analysis of 300–600 transformants, laying groundwork for later advancements in screening methodologies detailed in subsequent chapters and by other research groups.

Thus, to finish this section, an example showing how the combination of rational design (Section 6.2) and directed evolution could be synergically used for engineering enzymes, in this case, metallohydrolases. These enzymes, found across nearly all hydrolase subclasses, employ metal ions to activate water molecules, enabling the cleavage of diverse chemical bonds. However, this metal ion-mediated catalysis is uncommon in glycosidases, enzymes that hydrolyze sugar glycosidic bonds without relying on metal cofactors. Jeong and Song engineered novel metalloglycosidases by integrating a hydrolytically active zinc-binding site into the barrel-shaped scaffold of the outer membrane protein OmpF [203] (Figure 6.22).

Combining rational design (computational design guided by structural and mechanistic insights) and directed evolution, they developed Zn-dependent glycosidases exhibiting exceptional catalytic proficiency ($2.8 \times 10^9$ $M^{-1}$ $s^{-1}$) and high β-stereoselectivity. For designing the Zn-binding site, these authors targeted residues within 3.8–10.6 Å of R132 (first vertex of a Zn-binding triad, Figure 6.22A) for geometric compatibility with natural Zn-binding proteins, and introduced mutations (e.g., R82H/Y102H/R132H in mutant OmpF1, Figure 6.22C; L83H/Y102H/R132H in OmpF2, Figure 6.22D) into OmpF's β-barrel structure to create tetrahedral Zn-coordination sites. For the subsequent directed evolution, they used epPCR and SSM, screening for glycosidase activity using a fluorogenic substrate (4-methylumbelliferyl-β-D-glucopyranoside (4-β-MUG)). Biochemical characterization revealed that the engineered $Zn^{2+}$ site was acted as a catalytic cornerstone, working synergistically with adjacent acidic residues to position substrates and stabilize transition states. In fact, OmpF1 and OmpF2 showed Zn-dependent hydrolytic activity absent in WT OmpF:

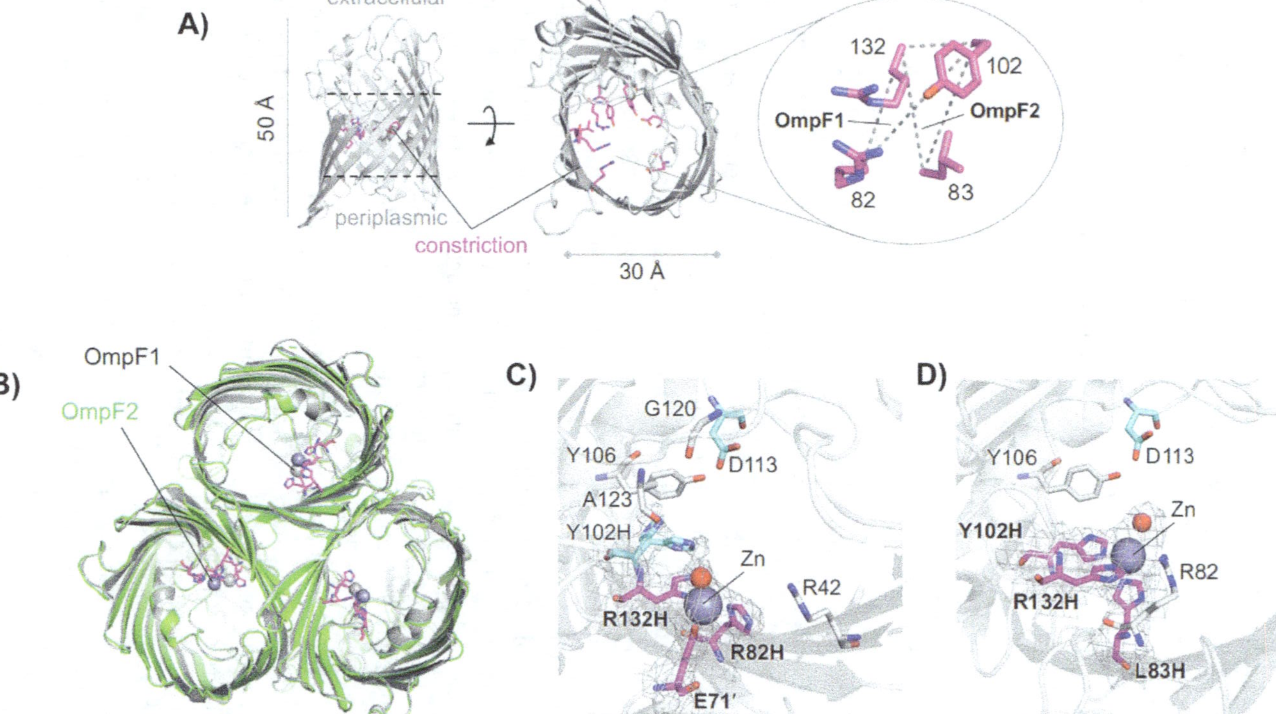


**Figure 6.22:** (A) Structural representation of only one protomer of the trimeric OmpF (PDB 2OMF), showing the sets of three residues selected for installing putative Zn-binding motifs as dotted triangles. (B) Overlaid structures of OmpF1 and OmpF2 shown from the periplasmic side. (C) Active site of b OmpF1; (D) Active site of OmpF2. Codes for (C) and (D): Zn atoms in light navy; metal-bound water molecule as a red sphere; metal-ligating residues in magenta, labeled in bold; rationally redesigned residues are shown as cyan sticks; iteratively optimized residues, gray sticks, respectively. Reprinted from Jeong and Song [203]. OPEN ACCESS.

D113E mutation enhanced glycosidase activity by stabilizing transition states *via* hydrogen bonding with substrate hydroxyl groups, and the OmpF1/E Variant achieved >95% β-selectivity for 4-β-MUG hydrolysis. Structural analysis revealed that Zn-binding sites were coordinated by histidine residues (3His or 2His/1Glu), while adjacent acidic residues (e.g., D113) were critical for substrate positioning. Finally, OmpF variants expressed in *E. coli*'s outer membrane retained activity, enabling Zn-dependent glycoside hydrolysis in vivo, proving scalability for industrial applications.

## 6.4 Conclusions and Future Outlook

From all the information exposed so far, it is clear that enzyme engineering stands at the intersection of molecular biology, chemistry, and computational science, offering transformative solutions for a wide range of industries. The field has evolved from early, labor-intensive mutagenesis experiments to a sophisticated discipline employing rational design, semirational strategies, and directed evolution, as shown in this chapter. Each approach brings unique strengths: rational design leverages deep structural and mechanistic insights, semirational design combines computational prediction with focused mutagenesis, and directed evolution harnesses the power of natural selection in the laboratory to access novel functions and properties [18, 19].

Recent years have witnessed remarkable progress, driven by advances in protein structure determination, HTS, and computational modeling. Rational design, once limited by the availability of structural data and our understanding of enzyme mechanisms, is now empowered by powerful algorithms and AI-driven predictions, making it increasingly feasible to design enzymes with improved activity, selectivity, and stability [19]. Semirational approaches, which integrate bioinformatics and experimental methods, have become mainstream, enabling the efficient exploration of sequence space with smaller, smarter libraries.

Directed evolution remains a cornerstone of enzyme engineering, with its capacity to deliver robust improvements in enzyme properties without the need for detailed mechanistic knowledge. The integration of automated screening, microfluidics, and continuous evolution systems has dramatically increased the speed and scale at which improved enzymes can be discovered and optimized [20, 33, 160].

The impact of enzyme engineering is evident in the growing adoption of biocatalysts across sectors such as pharmaceuticals, food, agriculture, biofuels, and environmental remediation. Enzymes are now central to green chemistry initiatives, enabling more sustainable and efficient processes that reduce waste, energy consumption, and reliance on hazardous chemicals.

A long-term forecast can be envisioned from several points of view:

1. **Ubiquity of enzyme-based solutions**
The next decade will see enzyme engineering become even more central to industrial innovation. As the global enzyme market is projected to double in value by 2030, driven by demand for sustainable and efficient processes, enzymes will increasingly replace traditional chemical catalysts in manufacturing, food processing, and energy production. Customized enzymes tailored for specific applications will become commonplace, supported by robust development pipelines and rapid prototyping capabilities.

2. **Integration of AI, machine learning, and big data**
Artificial intelligence and machine learning will revolutionize enzyme engineering. These technologies will enable the prediction of enzyme function, stability, and substrate scope with unprecedented accuracy, drastically reducing the trial-and-error aspect of enzyme design. Large-scale integration of omics data, structural databases, and experimental results will fuel the development of predictive algorithms that can suggest optimal mutations, design de novo enzymes, and even anticipate evolutionary trajectories.

3. **Acceleration of custom enzyme development**
HTS, automation, and continuous evolution systems will shorten enzyme development timelines from years to months or even weeks. Gene-editing tools like CRISPR will facilitate the rapid integration of engineered enzymes into microbial hosts, streamlining the path from design to application. The rise of custom enzyme development services will empower even small and medium enterprises to access tailored biocatalysts, democratizing the benefits of enzyme engineering.

4. **Expansion into new frontiers**
- **Personalized medicine:** Enzyme engineering will play a pivotal role in personalized healthcare, enabling the development of therapeutic enzymes tailored to individual patient needs, targeted drug activation, and precision diagnostics.
- **Synthetic biology:** Engineered enzymes will be foundational components in synthetic metabolic pathways, biosensors, and cell-free systems, unlocking new possibilities in biomanufacturing, diagnostics, and environmental sensing.
- **Bioremediation and sustainability:** Robust enzymes will be deployed for environmental cleanup, including the degradation of plastics, pollutants, and toxic compounds, contributing to circular economy models and global sustainability goals.

5. **Overcoming challenges**
Despite these advances, several challenges will persist:
- **Scalability and cost:** The complexity of enzyme production and purification at industrial scale will require continued innovation in expression systems, fermentation technology, and downstream processing.

- **Regulatory and safety considerations:** As enzyme applications expand into food, medicine, and the environment, rigorous safety and regulatory frameworks will be essential to ensure public trust and compliance.
- **Raw material and supply chain constraints:** The increasing demand for specialty enzymes may strain supply chains, necessitating the development of alternative feedstocks and more efficient production methods.

6. **Collaboration and interdisciplinary innovation**
The future of enzyme engineering will be shaped by deep collaboration between academia, industry, and government. Interdisciplinary partnerships will foster the exchange of knowledge, resources, and expertise, accelerating breakthroughs and enabling the rapid translation of discoveries into real-world solutions. Open-access databases, shared platforms, and global research networks will further democratize innovation in this space.

7. **Societal and environmental impact**
Enzyme engineering will be a key enabler of the bioeconomy, supporting the transition to renewable resources, waste valorization, and greener production methods. The societal benefits will include reduced environmental footprints, improved public health through cleaner processes and personalized therapies, and new economic opportunities in biotechnology and related fields.

### 6.4.1 Vision for the Next Decades

Looking beyond 2030, enzyme engineering is poised to become a foundational technology for the twenty-first century. The convergence of computational design, synthetic biology, and advanced manufacturing will enable the creation of "designer enzymes" with capabilities far beyond those found in nature. Enzymes will not only catalyze chemical reactions but also sense, regulate, and adapt to their environments in real time.

In the long term, we can anticipate:

- **On-demand enzyme synthesis** using portable, cell-free systems for field diagnostics, environmental monitoring, and emergency response.
- **Self-evolving biocatalysts** that adapt to changing process conditions or new substrates through *in situ* evolution.
- **Integration with digital manufacturing** to enable fully automated, feedback-controlled bioprocesses.
- **Expansion into extraterrestrial environments**, supporting life support, resource extraction, and manufacturing in space missions.

## References

[1] Newton MS, Arcus VL, Gerth ML, Patrick WM Enzyme evolution: Innovation is easy, optimization is complicated. Curr Opin Struct Biol 2018;48:110–116.

[2] Ribeiro AJM, Riziotis IG, Borkakoti N, Thornton JM Enzyme function and evolution through the lens of bioinformatics. Biochem J 2023;148:1845–1863.

[3] Robinson PK Essays Biochem. Essays Biochem 2015;59:1–41.

[4] Gatenby R, Roy Frieden B Investigating information dynamics in living systems through the structure and function of enzymes. PLoS One 2016;11.

[5] Wohlgemuth R Biocatalysis – Key enabling tools from biocatalytic one-step and multi-step reactions to biocatalytic total synthesis. New Biotechnol 2021;60:113–123.

[6] Buller R, Lutz S, Kazlauskas RJ, Snajdrova R, Moore JC, Bornscheuer UT From nature to industry: Harnessing enzymes for biocatalysis. Science 2023;382:eadh8615.

[7] Hanefeld U, Hollmann F, Paul CE Biocatalysis making waves in organic chemistry. Chem Soc Rev 2022.

[8] Alcántara AR, Domínguez de María P, Littlechild J, Schürmann M, Sheldon RA, Wohlgemuth R Biocatalysis as Key to Sustainable Industrial Chemistry. ChemSusChem 2022;e202102709.

[9] Urbieta MS, Donati ER, Chan KG, Shahar S, Sin LL, Goh KM Thermophiles in the genomic era: Biodiversity, science, and applications. Biotechnol Adv 2015;33:633–647.

[10] Kudalkar GP, Tiwari VK, Berkowitz DB Exploiting Archaeal/Thermostable enzymes in synthetic chemistry: Back to the future?. ChemCatChem 2024;16.

[11] Reetz MT What are the limitations of enzymes in synthetic organic chemistry?. Chem Rec 2016;16:2449–2459.

[12] Romero EO, Saucedo AT, Hernández-Meléndez JR, Yang D, Chakrabarty S, Narayan ARH Enabling Broader Adoption of Biocatalysis in Organic Chemistry. JACS Au 2023;3:2073–2085.

[13] Jain S, Ospina F, Hammer SC A New Age of Biocatalysis Enabled by Generic Activation Modes. JACS Au 2024;4:2068–2080.

[14] EU TVV, Santomasi G Tailor-made enzymes poised to propel plastic recycling into a new era. Nature 2022;604:631–633.

[15] Reetz MT, Sun Z, Qu G Artificial enzymes as promiscuous catalysts in organic and pharmaceutical chemistry. In Reetz MT, Sun Z, Qu G, editors. Enzyme Engineering 2023, Weinheim, Germany: WILEY-VCH GmbH; p. 279–316.

[16] Buller R, Damborsky J, Hilvert D, Bornscheuer UT Structure Prediction and Computational Protein Design for Efficient Biocatalysts and Bioactive Proteins. Angew Chem Int Ed n/a:e202421686.

[17] Sheldon RA, Brady D The limits to biocatalysis: Pushing the envelope. Chem Commun 2018;54:6088–6104.

[18] Ndochinwa GO, Wang QY, Okoro NO, Amadi OC, Nwagu TN, Nnamchi CI, et al. New advances in protein engineering for industrial applications: Key takeaways. Open Life Sci 2024;19.

[19] Song Z, Zhang Q, Wu W, Pu Z, Yu H Rational design of enzyme activity and enantioselectivity. Front Bioeng Biotechnol 2023;11.

[20] Reetz MT, Sun Z, Qu G Introduction to directed evolution and rational design as protein engineering techniques. In Reetz MT, Sun Z, Qu G, editors. Enzyme Engineering 2023, Weinheim, Germany: WILEY-VCH GmbH; p. 1–28.

[21] Smith M Site-directed mutagenesis. Trends Biochem Sci 1982;7:440–442.

[22] Hutchison CA III, Phillips S, Edgell MH, Gillam S, Jahnke P, Smith M Mutagenesis at a specific position in a DNA sequence. J Biol Chem 1978;253:6551–6560.

[23] Smith M Synthetic DNA and Biology (Nobel Lecture). Angew Chem Int Ed 1994;33:1214–1221.

[24] Mullis KB The Polymerase Chain Reaction (Nobel Lecture). Angew Chem Int Ed 1994;33:1209–1213.

[25] Chronopoulou EG, Labrou NE Site-saturation mutagenesis: A powerful tool for structure-based design of combinatorial mutation libraries. Curr Protoc Protein Sci 2011.
[26] Siloto RMP, Weselake RJ Site saturation mutagenesis: Methods and applications in protein engineering. Biocatal Agric Biotechnol 2012;1:181–189.
[27] Phintha A, Chaiyen P Rational and mechanistic approaches for improving biocatalyst performance. Chem Catal 2022;2:2614–2643.
[28] Pometun TVI, Stepashkina AA, Fedorchuk VV AV, Zarubina SA, Kargov IS, et al. Rational Design of Practically Important Enzymes. Moscow Univ Chem Bull 2018;73:1–6.
[29] Kinshuk S, Li L, Meckes B, Chan CTY Sequence-based protein design: A review of using statistical models to characterize coevolutionary traits for developing hybrid proteins as genetic sensors. Int J Mol Sci 2024;25.
[30] Alley EC, Khimulya G, Biswas S, AlQuraishi M, Church GM Unified rational protein engineering with sequence-based deep representation learning. Nat Methods 2019;16:1315–1322.
[31] McConnell A, Hackel BJ Protein engineering via sequence-performance mapping. Cell Syst 2023;14:656–666.
[32] Wang T, Liang C, Hou Y, Zheng M, Xu H, An Y, et al. Small design from big alignment: Engineering proteins with multiple sequence alignment as the starting point. Biotechnol Lett 2020;42:1305–1315.
[33] Yu H, Ma S, Li Y, Dalby PA Hot spots-making directed evolution easier. Biotechnol Adv 2022;56.
[34] Sternke M, Tripp KW, Barrick D Consensus sequence design as a general strategy to create hyperstable, biologically active proteins. Proc Natl Acad Sci U S A 2019;166:11275–11284.
[35] Prakinee K, Phaisan S, Kongjaroon S, Chaiyen P Ancestral sequence reconstruction for designing biocatalysts and investigating their functional mechanisms. JACS Au 2024;4:4571–4591.
[36] Xiang C, Ao YF, Höhne M, Bornscheuer UT Shifting the pH Optima of (R)-Selective Transaminases by Protein Engineering. Int J Mol Sci 2022;23.
[37] Xu SY, Zhou L, Xu Y, Hong HY, Dai C, Wang YJ, et al. Recent advances in structure-based enzyme engineering for functional reconstruction. Biotechnol Bioeng 2023;120:3427–3445.
[38] Choi JM, Kim HS Structure-guided rational design of the substrate specificity and catalytic activity of an enzyme. In Tawfik DS, editor. Enzyme Engineering and Evolution: General Methods 2020, London: Academic Press Ltd-Elsevier Science Ltd; p. 181–202.
[39] Crystallography: Protein data bank. Nat New Biol 1971;233:223–23.
[40] Burley SK, Bhatt R, Bhikadiya C, Bi C, Biester A, Biswas P, et al. Updated resources for exploring experimentally-determined PDB structures and computed structure models at the RCSB Protein Data Bank. Nucleic Acids Res 2025;53:D564–D74.
[41] Berman HM, Burley SK Protein Data Bank (PDB): Fifty-three years young and having a transformative impact on science and society. Q Rev Biophys 2025;58.
[42] Jumper J, Evans R, Pritzel A, Green T, Figurnov M, Ronneberger O, et al. Highly accurate protein structure prediction with AlphaFold. Nature 2021;596:583–589.
[43] Baek M, DiMaio F, Anishchenko I, Dauparas J, Ovchinnikov S, Lee GR, et al. Accurate prediction of protein structures and interactions using a three-track neural network. Science 2021;373:871–876.
[44] Simons KT, Bonneau R, Ruczinski I, Baker D Ab initio protein structure prediction of CASP III targets using ROSETTA. Proteins 1999;37:171–176.
[45] Rohl CA, Strauss CEM, Misura KMS, Baker D Protein Structure Prediction Using Rosetta. Methods Enzymol 2004;383:66–93.
[46] Abramson J, Adler J, Dunger J, Evans R, Green T, Pritzel A, et al. Accurate structure prediction of biomolecular interactions with AlphaFold 3. Nature 2024;630:493–500.
[47] Chica RA, Doucet N, Pelletier JN Semi-rational approaches to engineering enzyme activity: Combining the benefits of directed evolution and rational design. Curr Opin Biotechnol 2005;16:378–384.

[48] Lutz S Beyond directed evolution-semi-rational protein engineering and design. Curr Opin Biotechnol 2010;21:734–743.
[49] Korendovych IV Rational and semirational protein design. In Methods in Molecular Biology 2018, Humana Press Inc.; p. 15–23.
[50] Strafford J, Payongsri P, Hibbert EG, Morris P, Batth SS, Steadman D, et al. Directed evolution to re-adapt a co-evolved network within an enzyme. J Biotechnol 2012;157:237–245.
[51] Price MN, Arkin AP Interactive analysis of functional residues in protein families. mSystems 2022;7.
[52] Capra JA, Laskowski RA, Thornton JM, Singh M, Funkhouser TA Predicting protein ligand binding sites by combining evolutionary sequence conservation and 3D structure. PLoS Comput Biol 2009;5.
[53] Kumar S, Clarke D, Gerstein MB Leveraging protein dynamics to identify cancer mutational hotspots using 3D structures. Proc Natl Acad Sci USA 2019;116:18962–18970.
[54] Pavelka A, Chovancova E, Damborsky J HotSpot Wizard: A web server for identification of hot spots in protein engineering. Nucleic Acids Res 2009;37:W376–W83.
[55] Li B, Roden DM, Capra JA The 3D mutational constraint on amino acid sites in the human proteome. Nat Commun 2022;13.
[56] Roda S, Terholsen H, Meyer JRH, Cañellas-Solé A, Guallar V, Bornscheuer U, et al. AsiteDesign: A semirational algorithm for an automated enzyme design. J Phys Chem B 2023;127:2661–2670.
[57] Reetz MT, Sun Z, Qu G Screening and selection techniques. In Reetz MT, Sun Z, Qu G, editors. Enzyme Engineering 2023, Weinheim, Germany: WILEY-VCH GmbH; p. 29–58.
[58] Buller R, Damborsky J, Hilvert D, Bornscheuer UT Structure prediction and computational protein design for efficient biocatalysts and bioactive proteins. Angew Chem Int Ed 2025;64.
[59] Sequeiros-Borja CE, Surpeta B, Brezovsky J Recent advances in user-friendly computational tools to engineer protein function. Brief Bioinform 2021;22.
[60] Mia MM, Sultana H, Amin MA, Hossain MS, Imam H, Mohiuddin AKM, et al. Modern approaches to protein constructions: A comprehensive review of computational tools and databases for de novo protein design and engineering. Eng Rep 2025;7.
[61] Ferreira P, Fernandes PA, Ramos MJ Modern computational methods for rational enzyme engineering. Chem Catal 2022;2:2481–2498.
[62] Putignano G, Marino N, Bischof E, Zhavoronkov A, Vanhaelen Q The use of computational biology in protein engineering and drug discovery. In Innovating Health against Future Pandemics Elsevier; Vol. 2024, p. 15–33.
[63] Salam MD, Tripathi S Strategies for discovery and enhancement of enzyme function: Current developments and opportunities. In Microbial Enzymes: Production, Purification, and Industrial Applications: Volumes 1–2 2024, wiley; p. 491–504.
[64] Li A, Acevedo-Rocha CG, Reetz MT Boosting the efficiency of site-saturation mutagenesis for a difficult-to-randomize gene by a two-step PCR strategy. Appl Microbiol Biotechnol 2018;102:6095–6103.
[65] Pines G, Pines A, Eckert CA Highly efficient libraries design for saturation mutagenesis. Synth Biol 2022;7.
[66] Reetz MT, Bocola M, Carballeira JD, Zha D, Vogel A Expanding the range of substrate acceptance of enzymes: Combinatorial active-site saturation test. Angew Chem Int Ed 2005;44:4192–4196.
[67] Reetz MT, Carballeira JD, Vogel A Iterative saturation mutagenesis on the basis of B factors as a strategy for increasing protein thermostability. Angew Chem 2006;118:7909–7915.
[68] Reetz MT, Carballeira JD Iterative saturation mutagenesis (ISM) for rapid directed evolution of functional enzymes. Nat Protoc 2007;2:891–903.
[69] Dennig A, Shivange AV, Marienhagen J, Schwaneberg U. Omnichange: The sequence independent method for simultaneous site-saturation of five codons. PLoS One 2011;6.
[70] Alejaldre L, Pelletier JN, Quaglia D Methods for enzyme library creation: Which one will you choose?: A guide for novices and experts to introduce genetic diversity. BioEssays 2021;43.

[71] Pongsupasa V, Anuwan P, Maenpuen S, Wongnate T Rational-design engineering to improve enzyme thermostability. In Methods in Molecular Biology 2022, Humana Press Inc.; p. 159–178.

[72] Nicholson H, Becktel WJ, Matthews BW Enhanced protein thermostability from designed mutations that interact with α-helix dipoles. Nature 1988;336:651–656.

[73] Cedrone F, Ménez A, Quéméneur E Tailoring new enzyme functions by rational redesign. Curr Opin Struct Biol 2000;10:405–410.

[74] Brannigan JA, Wilkinson AJ Protein engineering 20 years on. Nat Rev Mol Cell Biol 2002;3:964–970.

[75] Eijsink VGH, Bjørk A, Gåseidnes S, Sirevåg R, Synstad B, Burg BVD, et al. Rational engineering of enzyme stability. In Proceedings of J Biotechnol 2004, p. 15380651.

[76] Yu H, Huang H Engineering proteins for thermostability through rigidifying flexible sites. Biotechnol Adv 2014;32:308–315.

[77] Fukuchi S, Nishikawa K Protein surface amino acid compositions distinctively differ between thermophilic and mesophilic bacteria. J Mol Biol 2001;309:835–843.

[78] Ahmed Z, Zulfiqar H, Tang L, Lin H A Statistical analysis of the sequence and structure of thermophilic and non-thermophilic proteins. Int J Mol Sci 2022;23.

[79] Charoenkwan P, Chotpatiwetchkul W, Lee VS, Nantasenamat C, Shoombuatong W A novel sequence-based predictor for identifying and characterizing thermophilic proteins using estimated propensity scores of dipeptides. Sci Rep 2021;11.

[80] Porebski BT, Buckle AM Consensus protein design. Protein Eng Des Sel 2016;29:245–251.

[81] Chan CH, Yu TH, Wong KB Stabilizing salt-bridge enhances protein thermostability by reducing the heat capacity change of unfolding. PLoS One 2011;6.

[82] Lee CW, Wang HJ, Hwang JK, Tseng CP Protein thermal stability enhancement by designing salt bridges: A combined computational and experimental study. PLoS One 2014;9.

[83] Cui H, Eltoukhy L, Zhang L, Markel U, Jaeger KE, Davari MD, et al. Less unfavorable salt bridges on the enzyme surface result in more organic cosolvent resistance. Angew Chem Int Ed 2021;60:11448–11456.

[84] Ban X, Lahiri P, Dhoble AS, Li D, Gu Z, Li C, et al. Evolutionary Stability of Salt Bridges Hints Its Contribution to Stability of Proteins. Comput Struct Biotechnol J 2019;17:895–903.

[85] Qu Z, Chen K, Zhang L, Sun Y Computation-based design of salt bridges in PETase for enhanced thermostability and performance for PET degradation. ChemBiochem 2023;24.

[86] Grigorakis K, Ferousi C, Topakas E Protein engineering for industrial biocatalysis: Principles, approaches, and lessons from engineered PETases. Catalysts 2025;15.

[87] Xu K, Fu H, Chen Q, Sun R, Li R, Zhao X, et al. Engineering thermostability of industrial enzymes for enhanced application performance. Int J Biol Macromol 2025;291.

[88] Sun Z, Liu Q, Qu G, Feng Y, Reetz MT Utility of B-Factors in protein science: Interpreting rigidity, flexibility, and internal motion and engineering thermostability. Chem Rev 2019;119:1626–1665.

[89] Reetz MT, Carballeira JD, Vogel A Iterative saturation mutagenesis on the basis of b factors as a strategy for increasing protein thermostability. Angew Chem Int Ed 2006;45:7745–7751.

[90] Tang H, Shi K, Shi C, Aihara H, Zhang J, Du G Enhancing subtilisin thermostability through a modified normalized B-factor analysis and loop-grafting strategy. J Biol Chem 2019;294:18398–18407.

[91] Gonzalez-Perez D, Molina-Espeja P, Garcia-Ruiz E, Alcalde M Mutagenic organized recombination process by Homologous In vivo Grouping (MORPHING) for directed enzyme evolution. PLoS One 2014;9.

[92] Craig DB, Dombkowski AA Disulfide by Design 2.0: A web-based tool for disulfide engineering in proteins. BMC Bioinformatics 2013;14.

[93] Dani VS, Ramakrishnan C, Varadarajan R MODIP revisited: Re-evaluation and refinement of an automated procedure for modeling of disulfide bonds in proteins. Protein Eng 2003;16:187–193.

[94] Xie DF, Fang H, Mei JQ, Gong JY, Wang HP, Shen XY, et al. Improving thermostability of *(R)*-selective amine transaminase from *Aspergillus terreus* through introduction of disulfide bonds. Biotechnol Appl Biochem 2018;65:255–262.

[95] Bajaj K, Madhusudhan MS, Adkar BV, Chakrabarti P, Ramakrishnan C, Sali A, et al. Stereochemical criteria for prediction of the effects of proline mutations on protein stability. PLoS Comput Biol 2007;3:2465–2475.

[96] Zhou Z, Wang X Rational design and structure-based engineering of alkaline pectate lyase from Paenibacillus sp. 0602 to improve thermostability. BMC Biotechnol 2021;21.

[97] Zhang Y, Zhang T, Li M, Miao M Rational design to improve the catalytic efficiency and stability of arginine deiminase. Int J Biol Macromol 2024;269.

[98] Shi J, Yuan B, Yang H, Sun Z Recent advances on protein engineering for improved stability. BioDes Res 2025;7.

[99] Bassegoda A, Nguyen GS, Schmidt M, Kourist R, Diaz P, Bornscheuer UT Rational Protein Design of Paenibacillus barcinonensis Esterase EstA for kinetic resolution of tertiary alcohols. ChemCatChem 2010;2:962–967.

[100] Yin X, Wu J, Yang L Efficient reductive amination process for enantioselective synthesis of L-phosphinothricin applying engineered glutamate dehydrogenase. Appl Microbiol Biotechnol 2018;102:4425–4433.

[101] Xia Y, Guo W, Han L, Shen W, Chen X, Yang H Significant Improvement of Both Catalytic Efficiency and Stability of Fructosyltransferase from Aspergillus niger by Structure-Guided Engineering of Key Residues in the Conserved Sequence of the Catalytic Domain. J Agric Food Chem 2022;70:7202–7210.

[102] Yao X, Kang T, Pu Z, Zhang T, Lin J, Yang L, et al. Sequence and Structure-Guided Engineering of Urethanase from Agrobacterium tumefaciens d3 for Improved Catalytic Activity. J Agric Food Chem 2022;70:7267–7278.

[103] Liu Y, Chen Q, Zhu BF, Pei XQ, Wu ZL Sequence-guided stereo-enhancing and -inverting of (R)-styrene monooxygenases for highly enantioselective epoxidation. Mol Catal 2022;531.

[104] Li Y, Ou X, Guo Z, Zong M, Lou W Using multiple site-directed modification of epoxide hydrolase to significantly improve its enantioselectivity in hydrolysis of rac-glycidyl phenyl ether. Chin J Chem Eng 2020;28:2181–2189.

[105] Li C, Hu BC, Wen Z, Hu D, Liu YY, Chu Q, et al. Greatly enhancing the enantioselectivity of PvEH2, a Phaseolus vulgaris epoxide hydrolase, towards racemic 1,2-epoxyhexane via replacing its partial cap-loop. Int J Biol Macromol 2020;156:225–232.

[106] Yi S, Park S Enhancing enantioselectivity of *Candida antarctica* lipase B towards chiral sec-alcohols bearing small substituents through hijacking sequence of A homolog. Tetrahedron Lett 2021;75.

[107] Bustos-Baena AS, Quintana-Castro R, Sánchez-Otero MG, Espinosa-Luna G, Mendoza-López MR, Peña-Montes C, et al. Enantioselectivity Enhancement of a Geobacillus thermoleovorans CCR11 Lipase by Rational Design. Catalysts 2025;15.

[108] Chen-Goodspeed M, Sogorb MA, Wu F, Raushel FM Enhancement, relaxation, and reversal of the stereoselectivity for phosphotriesterase by rational evolution of active site residues. Biochemistry 2001;40:1332–1339.

[109] Jaito N, Phetlum S, Saeoung T, Tiyasakulchai T, Srimongkolpithak N, Uengwetwanit T Improving stereoselectivity of phosphotriesterase (PTE) for kinetic resolution of chiral phosphates. Front Bioeng Biotechnol 2024;12.

[110] Vanhooke JL, Benning MM, Raushel FM, Holden HM Three-dimensional structure of the zinc-containing phosphotriesterase with the bound substrate analog diethyl 4-methylbenzylphosphonate. Biochemistry 1996;35:6020–6025.

[111] Chen H, Wu J, Yang L, Xu G Characterization and structure basis of *Pseudomonas alcaligenes* lipase's enantiopreference towards *D,L*-menthyl propionate. J Mol Catal B Enzym 2014;102:81–87.

[112] Chen H, Wu JP, Yang LR, Xu G Improving *Pseudomonas alcaligenes* lipase's diastereopreference in hydrolysis of diastereomeric mixture of menthyl propionate by site-directed mutagenesis. Biotechnol Bioprocess Eng 2014;19:592–604.

[113] Qiao J, Yang D, Feng Y, Wei W, Liu X, Zhang Y, et al. Engineering a *Bacillus subtilis* esterase for selective hydrolysis of *D,L*-menthyl acetate in an organic solvent-free system. RSC Adv 2023;13:10468–10475.

[114] Wang T, Wei R, Feng Y, Jin L, Jia Y, Yang D, et al. Engineering of yeast old yellow enzyme OYE3 enables its capability discriminating of (*E*)-citral and (*Z*)-citral. Molecules 2021;26.

[115] Maldonado MR, Alnoch RC, De Almeida JM, Santos LAD, Andretta AT, Ropaín RDPC, et al. Key mutation sites for improvement of the enantioselectivity of lipases through protein engineering. Biochem Eng J 2021;172.

[116] Jaeger KE, Ransac S, Dijkstra BW, Colson C, Van Heuvel M, Misset O Bacterial lipases. Fems Microbiol Rev 1994;15:29–63.

[117] Jaeger KE, Reetz MT Microbial lipases form versatile tools for biotechnology. Trends Biotechnol 1998;16:396–403.

[118] Pleiss J, Fischer M, Schmid RD Anatomy of lipase binding sites: The scissile fatty acid binding site. In Proceedings of Chem Phys Lipids 1998, Elsevier Ireland Ltd; p. 9720251.

[119] Del Monte-martínez A, Cutiño-Avila B V, González-Bacerio J Rational design strategy as a novel immobilization methodology applied to Lipases and Phospholipases. Methods Protoc 2018;243–283.

[120] Bauer TL, Buchholz PCF, Pleiss J The modular structure of α/β-hydrolases. FEBS J 2020;287:1035–1053.

[121] Li XJ, Zheng RC, Ma HY, Zheng YG Engineering of *Thermomyces lanuginosus* lipase Lip: Creation of novel biocatalyst for efficient biosynthesis of chiral intermediate of Pregabalin. Appl Microbiol Biotechnol 2014;98:2473–2483.

[122] Li DY, Lou YJ, Xu J, Chen XY, Lin XF, Wu Q Electronic Effect-Guided Rational Design of Candida antarctica Lipase B for kinetic resolution towards Diarylmethanols. Adv Synth Catal 2021;363:1867–1872.

[123] Rotticci D, Rotticci-Mulder JC, Denman S, Norin T, Hult K Improved enantioselectivity of a lipase by rational protein engineering. ChemBiochem 2001;2:766–770.

[124] Manetti F, Mileto D, Corelli F, Soro S, Palocci C, Cernia E, et al. Design and realization of a tailor-made enzyme to modify the molecular recognition of 2-arylpropionic esters by *Candida rugosa* lipase. Biochim Biophys Acta Protein Struct Mol Enzymol 2000;1543:146–158.

[125] Maldonado MR, Alnoch RC, Shiratori GYY, De Oliveira CC, Gonçalves MB, Mitchell DA, et al. Modification of the properties of the metagenomic lipase LipC12 by engineering of the hydrophobic cavity. Biocatal Biotransform 2024;42:334–344.

[126] Qu G, Bi Y, Liu B, Li J, Han X, Liu W, et al. Unlocking the stereoselectivity and substrate acceptance of enzymes: Proline-Induced loop engineering test. Angew Chem Int Ed 2022;61.

[127] Yang B, Wang H, Song W, Chen X, Liu J, Luo Q, et al. Engineering of the conformational dynamics of lipase to increase enantioselectivity. ACS Catal 2017;7:7593–7599.

[128] Kendrew JC, Bodo G, Dintzis HM, Parrish RG, Wyckoff H, Phillips DC A three-dimensional model of the myoglobin molecule obtained by x-ray analysis. Nature 1958;181:662–666.

[129] Perutz MF, Rossmann MG, Cullis AF, Muirhead H, Will G, North ACT Structure of Hæmoglobin: A three-dimensional fourier synthesis at 5.5-Å. resolution, obtained by X-ray analysis. Nature 1960;185:416–422.

[130] Anfinsen CB Principles that govern the folding of protein chains. Science 1973;181:223–230.

[131] Zwanzig R, Szabo A, Bagchi B Levinthal's paradox. Proc Natl Acad Sci U S A 1992;89:20–22.

[132] Chou PY, Fasman GD Prediction of protein conformation. Biochemistry 1974;13:222–245.

[133] Richardson JS The anatomy and taxonomy of protein structure. Adv Protein Chem 1981;34:167–339.

[134] Regan L, Degrado WF Characterization of a helical protein designed from first principles. Science 1988;241:976–978.
[135] Dahiyat BI, Mayo SL De novo protein design: Fully automated sequence selection. Science 1997;278:82–87.
[136] Kuhlman B, Dantas G, Ireton GC, Varani G, Stoddard BL, Baker D Design of a Novel Globular Protein Fold with Atomic-Level Accuracy. Science 2003;302:1364–1368.
[137] Jones TA, Thirup S Using known substructures in protein model building and crystallography. EMBO J 1986;5:819–822.
[138] Kuhlman B, Bradley P Advances in protein structure prediction and design. Nat Rev Mol Cell Biol 2019;20:681–697.
[139] Huang PS, Boyken SE, Baker D The coming of age of *de novo* protein design. Nature 2016;537:320–327.
[140] Jiang L, Althoff EA, Clemente FR, Doyle L, Rothlisberger D, Zanghellini A, et al. De novo computational design of retro-aldol enzymes. Science 2008;319:1387–1391.
[141] Casey ML, Kemp DS, Paul KG, Cox DD Physical organic-chemistry of benzisoxazoles .1. Mechanism of base-catalyzed decomposition of benzisoxazoles. J Org Chem 1973;38:2294–2301.
[142] Rothlisberger D, Khersonsky O, Wollacott AM, Jiang L, DeChancie J, Betker J, et al. Kemp elimination catalysts by computational enzyme design. Nature 2008;453:190–194.
[143] Khersonsky O, Röthlisberger D, Wollacott AM, Murphy P, Dym O, Albeck S, et al. Optimization of the in-silico-designed Kemp eliminase KE70 by computational design and directed evolution. J Mol Biol 2011;407:391–412.
[144] Siegel JB, Zanghellini A, Lovick HM, Kiss G, Lambert AR, Clair JLS, et al. Computational design of an enzyme catalyst for a Stereoselective Bimolecular Diels-Alder Reaction. Science 2010;329:309–313.
[145] Diels O, Alder K Synthesen in der hydroaromatischen Reihe. Justus Liebigs Annalen der Chemie 1928;460:98–122.
[146] Houk KN, Liu F, Yang Z, Seeman JI Evolution of the Diels–Alder reaction mechanism since the 1930s: Woodward, Houk with Woodward, and the influence of computational chemistry on understanding Cycloadditions. Angew Chem Int Ed 2021;60:12660–12681.
[147] Albanese KI, Barbe S, Tagami S, Woolfson DN, Schiex T Computational protein design. Nat Rev Methods Primers 2025;5.
[148] Leveson-Gower RB Designing enzymatic reactivity with an expanded palette. ChemBiochem 2025.
[149] Li Y, Dalby PA Engineering of enzymes using non-natural amino acids. Biosci Rep 2022;42.
[150] Brouwer B, Della-Felice F, Illies JH, Iglesias-Moncayo E, Roelfes G, Drienovská I Noncanonical Amino Acids: Bringing new-to-nature functionalities to biocatalysis. Chem Rev 2024;124:10877–10923.
[151] Costello A, Peterson AA, Chen PH, Bagirzadeh R, Lanster DL, Badran AH Genetic code expansion history and modern innovations. Chem Rev 2024;124:11962–12005.
[152] Katoh T, Suga H Reprogramming the genetic code with flexizymes. Nat Rev Chem 2024;8:879–892.
[153] Tuckey C, Asahara H, Zhou Y, Chong S Protein synthesis using a reconstituted cell-free system. Curr Protoc Mol Biol 2014;2014:16.31.1–16.31.22.
[154] Lugtenburg T, Gran-Scheuch A, Drienovská I Non-canonical amino acids as a tool for the thermal stabilization of enzymes. Protein Eng Des Sel 2023;36.
[155] Pagar AD, Jeon H, Khobragade TP, Sarak S, Giri P, Lim S, et al. Non-Canonical Amino Acid-Based Engineering of (R)-Amine Transaminase. Front Chem 2022;10.
[156] Birch-Price Z, Hardy FJ, Lister TM, Kohn AR, Green AP Noncanonical amino acids in biocatalysis. Chem Rev 2024;124:8740–8786.
[157] Wang L, Zhang M, Teng H, Wang Z, Wang S, Li P, et al. Rationally introducing non-canonical amino acids to enhance catalytic activity of LmrR for Henry reaction. Bioresour Bioprocess 2024;11.
[158] Drienovská I, Rioz-Martínez A, Draksharapu A, Roelfes G Novel artificial metalloenzymes by in vivo incorporation of metal-binding unnatural amino acids. Chem Sci 2015;6:770–776.

[159] Jiang R, Casilli F, Thunnissen AMWH, Roelfes G An artificial copper-michaelase featuring a genetically encoded Bipyridine Ligand for Asymmetric additions to Nitroalkenes. Angew Chem Int Ed 2025;64.

[160] Sellés Vidal L, Isalan M, Heap JT, Ledesma-Amaro R A primer to directed evolution: Current methodologies and future directions. RSC Chem Biol 2023;4:271–291.

[161] Wang Y, Xue P, Cao M, Yu T, Lane ST, Zhao H Directed Evolution: Methodologies and applications. Chem Rev 2021;121:12384–12444.

[162] Reetz MT, Krebs GPL Challenges in the directed evolution of stereoselective enzymes for use in organic chemistry. C R Chim 2011;14:811–818.

[163] Reetz MT, Sun Z, Qu G Gene Mutagenesis methods in directed evolution and rational enzyme design. In Reetz MT, Sun Z, Qu G, editors. Enzyme Engineering 2023, Weinheim, Germany: WILEY-VCH GmbH; p. 59–139.

[164] Mills DR, Peterson RL, Spiegelman S An extracellular Darwinian experiment with a self-duplicating nucleic acid molecule. Proc Natl Acad Sci U S A 1967;58:217–224.

[165] Blain JC, Szostak JW Progress toward synthetic cells. Annu Rev Biochem 2014;83:615–640.

[166] Joyce GF Forty years of in vitro evolution. Angew Chem Int Ed 2007;46:6420–6436.

[167] Francis JC, Hansche PE Directed evolution of metabolic pathways in microbial populations. I. Modification of the acid phosphatase pH optimum in S. cerevisiae. Genetics 1972;70:59–73.

[168] Hall BG Changes in the substrate specificities of an enzyme during directed evolution of new functions. Biochemistry 1981;20:4042–4049.

[169] Hwang BY, Oh JM, Kim J, Kim BG. Pro-antibiotic substrates for the identification of enantioselective hydrolases. Biotechnol Lett 2006;28:1181–1185.

[170] Cooper TF, Rozen DE, Lenski RE Parallel changes in gene expression after 20,000 generations of evolution in Escherichia coli. Proc Natl Acad Sci U S A 2003;100:1072–1077.

[171] Esvelt KM, Carlson JC, Liu DR A system for the continuous directed evolution of biomolecules. Nature 2011;472:499–503.

[172] Leconte AM, Dickinson BC, Yang DD, Chen IA, Allen B, Liu DR A population-based experimental model for protein evolution: Effects of mutation rate and selection stringency on evolutionary outcomes. Biochemistry 2013;52:1490–1499.

[173] Kunkel TA Rapid and efficient site-specific mutagenesis without phenotypic selection. Proc Natl Acad Sci U S A 1985;82:488–492.

[174] Kunkel TA, Roberts JD, Zakour RA Rapid and efficient site-specific Mutagenesis without Phenotypic selection. Methods Enzymol 1987;154:367–382.

[175] Fujii R, Kitaoka M, Hayashi K Error-prone rolling circle amplification: The simplest random mutagenesis protocol. Nat Protoc 2006;1:2493–2497.

[176] Eigen M, Gardiner W Evolutionary molecular engineering based on rna replication. Pure Appl Chem 1984;56:967–978.

[177] Matsumura M, Aiba S Screening for thermostable mutant of kanamycin nucleotidyltransferase by the use of a transformation system for a thermophile, Bacillus stearothermophilus. J Biol Chem 1985;260:15298–15303.

[178] Liao H, McKenzie T, Hageman R Isolation of a thermostable enzyme variant by cloning and selection in a thermophile. Proc Natl Acad Sci U S A 1986;83:576–580.

[179] Richards JH Cassette mutagenesis shows its strength. Nature 1986;323:187.

[180] Ho SN, Hunt HD, Horton RM, Pullen JK, Pease LR Site-directed mutagenesis by overlap extension using the polymerase chain reaction. Gene 1989;77:51–59.

[181] Horton RM, Hunt HD, Ho SN, Pullen JK, Pease LR Engineering hybrid genes without the use of restriction enzymes: Gene splicing by overlap extension. Gene 1989;77:61–68.

[182] Pritchard L, Corne D, Kell D, Rowland J, Winson M A general model of error-prone PCR. J Theor Biol 2005;234:497–509.

[183] Leung DW, Chen E, Goeddel DV A method for random mutagenesis of a defined DNA segment using a modified polymerase chain reaction. Technique 1989;1(1):11–15.

[184] Hawkins RE, Russell SJ, Winter G Selection of phage antibodies by binding affinity. J Mol Biol 1992;226:889–896.

[185] Chen K, Arnold FH Tuning the activity of an enzyme for unusual environments: Sequential random mutagenesis of subtilisin E for catalysis in dimethylformamide. Proc Natl Acad Sci U S A 1993;90:5618–5622.

[186] Arnold FH Innovation by evolution: Bringing new chemistry to life (Nobel Lecture). Angew Chem Int Ed 2019;58:14420–14426.

[187] Stemmer WPC Rapid evolution of a protein in vitro by DNA shuffling. Nature 1994;370:389–391.

[188] Firth AE, Patrick WM Statistics of protein library construction. Bioinformatics 2005;21:3314–3315.

[189] Reetz MT, Zonta A, Schimossek K, Liebeton K, Jaeger KE Creation of Enantioselective Biocatalysts for organic chemistry by *In Vitro* evolution. Angew Chem Int Ed 1997;36:2830–2832.

[190] Reetz MT, Jaeger KE Superior biocatalysts by directed evolution. Biocatalysis 1999;200:31–57.

[191] Jaeger K-E, Reetz MT Directed evolution of enantioselective enzymes for organic chemistry. Curr Opin Chem Biol 2000;4:68–73.

[192] Acevedo-Rocha CG, Hollmann F, Sanchis J, Sun Z A pioneering career in catalysis: Manfred T. Reetz. ACS Catal 2020;10:15123–15139.

[193] Reetz MT, Wilensek S, Zha D, Jaeger KE Directed evolution of an enantioselective enzyme through combinatorial multiple-cassette mutagenesis. Angew Chem Int Ed 2001;40:3589–3591.

[194] Muteeb G, Sen R Random mutagenesis using a mutator strain. Methods Mol Biol 2010;634:411–419.

[195] Liao HH Thermostable mutants of kanamycin nucleotidyltransferase are also more stable to proteinase K, urea, detergents, and water-miscible organic solvents. Enzyme Microb Technol 1993;15:286–292.

[196] Reetz MT, Sun Z, Qu G Protein engineering of enzyme robustness relevant to organic and pharmaceutical chemistry and applications in biotechnology. In Reetz MT, Sun Z, Qu G, editors. Enzyme Engineering 2023, Weinheim, Germany: WILEY-VCH GmbH; p. 233–277.

[197] Eijsink VGH, Gåseidnes S, Borchert TV, Van Den Burg B Directed evolution of enzyme stability. Biomol Eng 2005;22:21–30.

[198] Wagschal K, Chan VJ, Pereira JH, Zwart PH, Sankaran B *Chromohalobacter salixigens* uronate dehydrogenase: Directed evolution for improved thermal stability and mutant CsUDH-inc X-ray crystal structure. Process Biochem 2022;114:185–192.

[199] Martin-Diaz J, Molina-Espeja P, Hofrichter M, Hollmann F, Alcalde M Directed evolution of unspecific peroxygenase in organic solvents. Biotechnol Bioeng 2021;118:3002–3014.

[200] Li G, Reetz MT Learning lessons from directed evolution of stereoselective enzymes. Org Chem Front 2016;3:1350–1358.

[201] Reetz MT Witnessing the birth of directed evolution of Stereoselective Enzymes as catalysts in organic chemistry. Adv Synth Catal 2022;364:3326–3335.

[202] Alcántara AR, De Gonzalo G Chapter 1 – Introduction to asymmetric synthesis employing biocatalysts. In Gonzalo GD, Alcántara AR, editors. Biocatalysis in Asymmetric Synthesis 2024, Academic Press; p. 1–41.

[203] Jeong WJ, Song WJ Design and directed evolution of noncanonical β-stereoselective metalloglycosidases. Nat Commun 2022;13.

# Chapter 7
# Process Prediction and Simulation Models and Their Application in the Field of Biocatalysis and Biotransformations

**Abstract:** This chapter explores the fundamental role of computational prediction and simulation models in the field of biocatalysis and biotransformations. These tools enable the understanding and prediction of enzyme behavior, optimization of reaction conditions, and efficient design of new biocatalysts, reducing reliance on traditional, costly, and time-consuming experimental methods. The techniques are classified into two main domains: ligand-based techniques (such as quantitative structure-activity relationship and similarity models) and structure-based techniques (such as molecular docking and molecular dynamics [MD]).

This chapter delves into core molecular modeling methodologies, including molecular mechanics (MM) and quantum mechanics (QM), as well as hybrid methods like QM/MM, which combine the accuracy of QM with the efficiency of MM for studying complex biological systems. Specific applications, such as MD, molecular docking, and the prediction of enantioselectivity, crucial for the synthesis of chiral pharmaceuticals, are discussed.

Finally, the transformative impact of artificial intelligence and machine learning in this field is highlighted, with applications in protein structure prediction (e.g., AlphaFold), virtual screening, chemical synthesis automation, and ADME-Tox property prediction. Collectively, these computational tools represent a cornerstone for the rational design of biocatalysts and the development of more sustainable and efficient biotechnological processes.

**Keywords:** Force field, molecular modeling, molecular dynamics (MD), Molecular Docking, ADME-Tox

## 7.1 Introduction

Prediction and simulation models are fundamental tools in biocatalysis and biotransformations, enabling the study of enzyme behavior and their interactions with substrates.

One strategy involves using predictive models to anticipate enzyme behavior under different conditions and interactions with substrates, based on previous experimental data and knowledge of enzyme kinetics. Simultaneously, simulation models facilitate the virtual recreation of biocatalytic processes, improving the understanding of underlying mechanisms and aiding in the optimization of reaction conditions.

 | https://doi.org/10.1515/9783111122991-007

Overall, prediction and simulation models significantly contribute to the efficient design of processes, the identification of suitable enzymes for specific tasks, and the optimizationof reaction conditions in biocatalysis and biotransformations. This is crucial for both industrial applications and research in the field [1].

The growing need for environmental protection has driven the development of more sustainable synthetic processes, avoiding hazardous reagents, and promoting synthetic routes based on renewable raw materials. However, not all traditional chemical processes can easily be replaced by biocatalytic processes due to the complex integration of these methods and an often unfavorable cost/benefit analysis. Identifying suitable cases presents a substantial challenge for implementing biocatalytic processes.

In this context, the use of modeling and simulation techniques emerges as a powerful tool to address this challenge. These techniques allow for the exploration of alternative routes and hypothetical changes to existing processes in silico before actual experimentation. This approach streamlines and directs experimentation, accelerating the development of bioprocesses. It is important to emphasize that calculated process models can be instrumental in developing and evaluating control strategies, ensuring desired stability and efficiency.

These techniques can be categorized into two domains: ligand- and structure-based.

## 7.2 Ligand-Based Techniques

- Quantitative structure-activity relationship (QSAR): Uses statistical models to predict the activity of new ligands based on their structural similarities to known active compounds
- Similarity models: Identify potential ligands by comparing their structures to those of known active or inactive compounds
- Requirements: Both approaches require pre-labeled active/inactive compounds, typically obtained through experimental methods

## 7.3 Structure-Based Techniques

- Computational docking: It simulates the binding of ligands to the target protein to predict interaction strength and assigns a score that correlates with binding free energy.
- Molecular dynamics: It models the physical movements of atoms and molecules to understand the interaction between the protein and the ligand.
- Requirements: These methods require a three-dimensional (3D) structure of the target protein, but do not need specific bioactivity data for the target.

Molecular modeling is particularly valuable in the pharmaceutical sector due to the immense time and financial investment of standard drug development. This multi-stage process, from initial discovery to clinical trials, can take an average of 15 years and cost nearly $2.6 billion to commercialize a single drug, as reported in studies [2].

In silico methods are now indispensable in the pharmaceutical industry, as they substantially decrease the time and cost of drug development. The advancement of computational algorithms and large-scale databases has enabled the integration of predictive tools into all phases of the drug development pipeline. These computational approaches have been applied successfully to design and identify therapeutic compounds for a range of diseases, including cancer, diabetes, and viral and bacterial infections [3].

Pharmaceutical compounds often exist in stereoisomeric forms, such as enantiomers and diastereomers, each exhibiting unique biological properties. Achieving a high degree of stereoselectivity in drug synthesis is essential to ensure the production of the correct stereoisomer with the desired therapeutic activity.

The preference for one stereoisomer over another can significantly influence a drug's bioavailability, potency, and toxicity. Moreover, many biological processes are highly sensitive to the stereochemical configuration of molecules, which underscores the importance of stereoselectivity in ensuring accurate interactions with biological receptors and minimizing unwanted side effects [4].

Enzymes possess a 3D structure that allows them to recognize and bind reactive species during catalysis. A detailed understanding of these structural factors at the atomic level is essential for addressing stereocontrol in enzyme-catalyzed reactions. The selective binding of enantiomers at the enzyme's active site supports stereospecificity, while the analysis of energy barriers provides insights into the stereoselective preference in catalytic reactions. Therefore, it is crucial to investigate the various factors that influence how enzymes regulate stereoselectivity and stereospecificity at the atomic level [5].

## 7.4 Molecular Modeling

Modeling aims to simulate molecular system behavior at the atomic level, grounded in the discrete nature of the physical world. This reality is organized across multiple scales, wherein the components at each level can be characterized as discrete particles governed by defined interaction models. Thus, molecular modeling can be defined as a simplified or idealized description of a molecular system or the interactions between molecules, designed to facilitate calculations and predictions (Figure 7.1).

Molecular modeling is primarily a tool for calculating the energy of a given molecular structure. Therefore, the first step in researching the design of a molecular model is to define the problem as a relationship between the structure and the energy of the molecule.

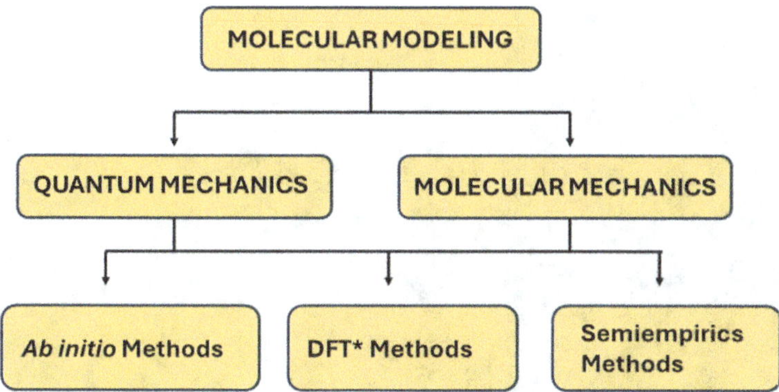


**Figure 7.1:** Molecular modeling. A general term that encompasses theoretical methods and computational techniques for modeling or mimicking the behavior of molecules.

As a preliminary step to studying energy, it is necessary to consider the forces exerted between a target molecule and a substrate:

**van der Waals forces:** These weak attractive forces between atoms or molecules, caused by temporary fluctuations in electron distribution, play a significant role in the interaction between the substrate and the target. These forces promote close proximity and stabilize the substrate within the enzyme's active site, facilitating the formation of transient dipoles.

**Hydrogen bonds:** These are crucial in substrate-target interactions, involving the attraction between a hydrogen atom, bonded to an electronegative atom (such as oxygen or nitrogen), and another electronegative atom. In enzymatic reactions, hydrogen bonds play an essential role in correctly positioning the substrate and facilitating interactions within the active site, thereby contributing to the stabilization of the enzyme-substrate complex (Figure 7.2(A) and 7.2(B)).

**Ionic interactions**: These arise from the attraction between positively and negatively charged ions and also play a role in substrate-target interactions. In the context of enzyme-substrate interactions, ionic forces can emerge when charged amino acid residues in the enzyme's active site interact with oppositely charged groups in the substrate. These ionic interactions contribute to the stabilization of the substrate-enzyme complex and can influence both binding and catalytic processes.

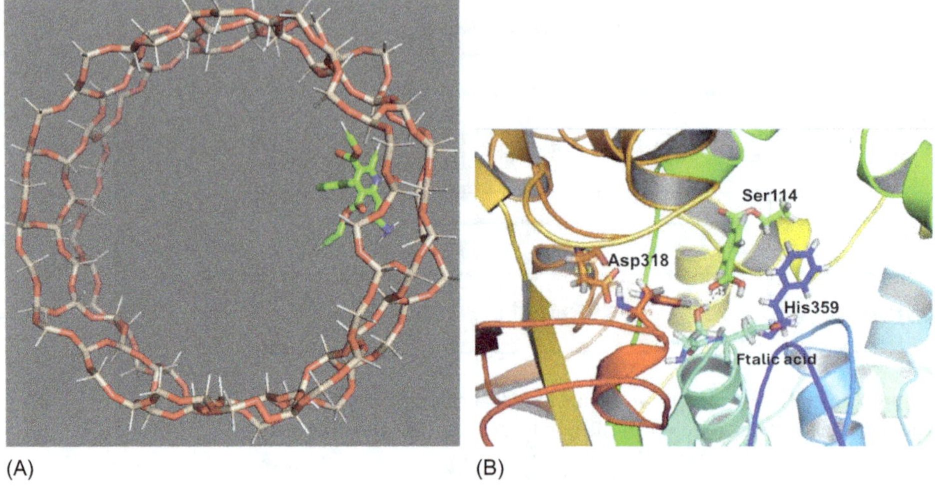


**Figure 7.2:** (A) Hydrogen bonds formed between an amlodipine molecule and a mesoporous silica structure analogous to MCM41. (B) Hydrogen bonds formed between the monoester of terephthalic acid and the thermophilic lipase from *Bacillus thermocatenulatus*.

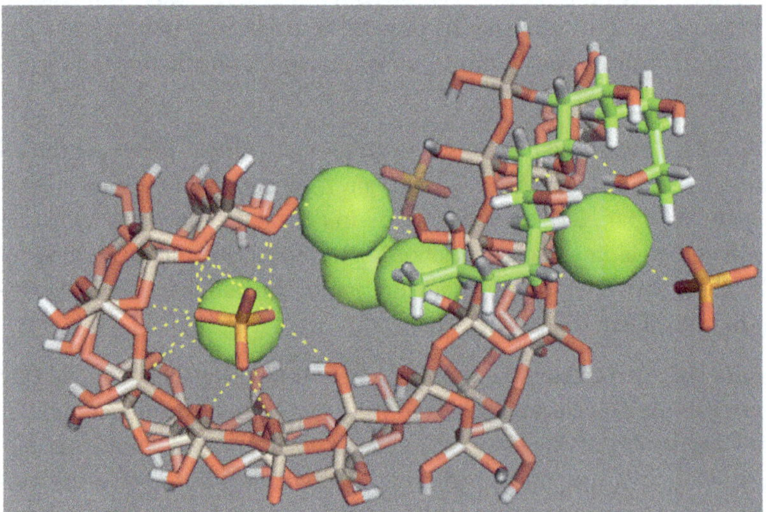

**Figure 7.3:** Simplified model of a Si-Ca-P-PVAL hybrid in which a very strong electrostatic interaction is observed at various levels: P-Ca-O-Si, suggesting the presence of strong ionic and covalent interactions as well as hydrogen bonding.

There are two conceptually different approaches to understanding energy:
- **Molecular mechanics:** This approach describes the energy of a molecule using a simple function that represents the distortion of the molecule relative to ideal bond distances and angles, as well as non-bonded Coulombic and van der Waals interactions.
- **Quantum mechanics:** This approach describes the energy of a molecule in terms of the interactions between nuclei and electrons, as governed by the Schrödinger equation. The solutions to this equation ("wave functions") for the hydrogen atom correspond to atomic orbitals (s, p, d, etc.).

## 7.5 Molecular Modeling Methodologies

### 7.5.1 Molecular Mechanics

The molecular structures generated using the procedures described above must undergo a **geometry optimization process** to reach the **minimum energy state**. This is achieved through methods based on **molecular mechanics** (MM).

MM refers to a broad set of computational techniques used to calculate the energies and geometries of molecular systems. In this approach, atoms are represented as point masses connected by harmonic forces – effectively modeling them as spheres connected by springs. The method assumes that **electrons optimally distribute themselves** around nuclei, allowing the focus to remain on the **arrangement of nuclei**.

The total energy of the system is calculated using a potential function that combines distinct energy contributions. These include bonded interactions (bond stretching and angle bending), torsional rotations, and non-bonded forces, such as van der Waals and electrostatics:

$$E = E_{\text{bonds}} + E_{\text{angles}} + E_{\text{dihedrals}} + E_{\text{vdW}} + E_{\text{electrostatics}} \tag{7.1}$$

**Equation (7.1):** Total energy of the system.

The sum of these components gives the **steric energy** of the system. Although steric energy has no direct physical meaning on its own, it is extremely useful in comparing different conformations and evaluating molecular stability.

One of the primary applications of MM is **energy minimization**, where the force field acts as the optimization criterion. The objective is to find local energy minima that correspond to **stable molecular conformations**. This is carried out using optimization algorithms such as **steepest descent**, **conjugate gradient**, **Monte Carlo (MC)**, and **simulated annealing**. These minima represent the most probable molecular states at a given temperature, and molecular motion is typically modeled as **vibrations and transitions** between these conformers.

At **finite temperatures**, molecules spend most of their time near these low-energy conformations, which significantly influence their **chemical behavior and physical properties**. While energy minimization using force fields considers only the **enthalpic component** of the free energy, the **entropic component** can be estimated using techniques such as **normal mode analysis**, providing a more complete thermodynamic description.

There are several widely used **force fields** tailored for different molecular systems:

- **AMBER (assisted model building with energy refinement)**: Designed for biomolecules like proteins and nucleic acids
- **CHARMM (Chemistry at HARvard Macromolecular Mechanics)**: Extensively applied in simulations of biological macromolecules
- **MM+**: An extended version of classical MM, commonly used in organic chemistry
- **OPLS (optimized potentials for liquid simulations)**: Suitable for both condensed and gas-phase systems
- **GROMOS**: Applied mainly in biomacromolecular simulations
- **UFF (universal force field)**: Covers a broad range of elements, including those used in inorganic and organometallic systems

The **development of a force field** typically involves the following steps:

1. **Defining potential energy functions** that describe the various interactions contributing to steric energy
2. **Parameterizing** these functions using experimental data or results from QM calculations
3. **Refining and reparameterizing** the model until calculated properties agree well with empirical observations

There are no strict rules regarding the number or type of energy functions a force field must contain. Simpler force fields often include only bonded interactions (bonds, angles, and torsions), while more sophisticated models incorporate additional terms to improve accuracy or account for specific molecular behaviors.

### 7.5.2 Energy Optimization Methods

*Steepest descent*: The energy is calculated for the initial geometry using the first derivative (gradient). The calculation is then repeated after moving an atom by a small increment along one of the coordinate axes. This process is performed for all atoms. The procedure stops once a predefined minimum-energy condition is met. This algorithm usually converges slowly and only brings the structure close to a minimum, making it most suitable for refining geometries that are initially far from the minimum.

*Conjugate gradient method*: This approach builds upon the information obtained in one iteration to improve the next. In each minimization step, the gradient is calculated and then combined with information from previous steps to determine a new search direction. Although it requires more computational effort and memory than steepest descent, it is generally more efficient in reaching convergence, especially for moderately optimized structures.

*Powell's method*: Similar to the conjugate gradient approach, Powell's method often reaches convergence more quickly. However, it can cause large changes in torsion angles. Because of this, Powell's method is not recommended for minimization after conformational analysis, as it may significantly alter low-energy conformations.

*Newton-Raphson algorithm*: This method uses the second derivative (Hessian matrix) in addition to the gradient, effectively incorporating curvature information to identify the optimal search direction. The second derivative also helps predict when the function will pass through a minimum. While this approach can be highly accurate, its main drawbacks are high computational cost and memory requirements. It is typically applied only when the geometry has already been pre-optimized using a simpler method such as steepest descent.

MM is also extensively used in protein structure modeling, enzyme catalysis, and drug design. In complex biochemical systems, a quantum mechanics (QM)/MM hybrid approach is often employed: one region – typically the active site – is treated using QM to accurately model bond formation and breaking, while the rest of the system is handled using MM to reduce computational cost. This method provides a balance between precision and efficiency and is particularly valuable in studying enzyme mechanisms and reaction pathways in the biological environment [6, 7].

### 7.5.3 Quantum Mechanics

The determination of total energy is achieved through approximate solutions to the time-dependent Schrödinger equation. The Born-Oppenheimer approximation is applied, allowing the separation of electronic and nuclear motions, which simplifies the Schrödinger equation by excluding relativistic terms. This approximation assumes that the wave function of any polyatomic system can be approximated as a product of one-electron functions, known as orbitals.

This simplification allows the total energy to be evaluated as the sum of the electronic energy at fixed nuclear positions and the nuclear repulsion energy. Unlike MM, ab initio methods can reproduce experimental data without the use of empirical parameters. However, these methods are limited by the significant computational time required for calculations, especially in complex systems:

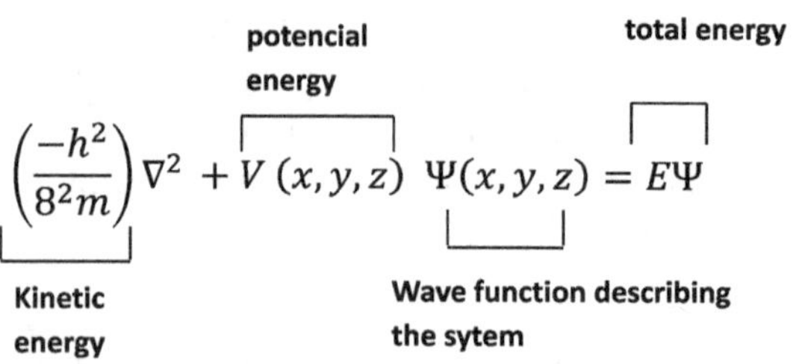


$$\left(\frac{-h^2}{8^2 m}\right)\nabla^2 + V\,(x,y,z)\ \Psi(x,y,z) = E\Psi$$

**Equation (7.2):** Fundamental equation of quantum mechanics: The square of the wave function gives the probability of finding an electron.

This function represents the electron density, analogous to the data obtained from an X-ray diffraction experiment. The significant computational gap between MM and ab initio calculations is addressed by semiempirical methods of molecular orbitals. Semiempirical quantum chemistry methods are based on the formalism of the Hartree-Fock (HF) method but incorporate many approximations and derive some parameters from empirical data. The primary distinction between semiempirical and ab initio methods is the introduction of parameters in semiempirical methods to reduce the high computational time required by ab initio approaches.

Another fundamental concept of semiempirical approximations is the acknowledgment that most interesting molecular properties are largely influenced by the valence electrons of the corresponding atoms used in the calculations.

### 7.5.4 Hybrid QM/MM Methods

The advent of QM/MM approaches has provided computational chemists with tools capable of describing realistic systems with high accuracy, enabling reliable modeling and guiding experimental design. Hybrid QM/MM schemes partition a biomolecular system into quantum and classical regions, which removes the practical limitations associated with treating every atom quantum-mechanically and thus permits simulations of large macromolecular assemblies while retaining an accurate description of the chemically active site.

QM/MM is a hybrid method that combines the precision of QM with the computational efficiency of MM. It is particularly useful for simulating very large molecules, such as enzymes [8].

Early applications of QM/MM focused on evaluating solvation contributions and reaction free-energy changes, typically combining molecular dynamics (MD) or MC sampling with a semiempirical description of the quantum region [9].

It is noteworthy that there are certain approaches, referred to as “direct quantum chemistry,” that address both electrons and nuclei on a common basis. Variants of this approach include density functional methods and semi-empirical methods.

Steps to follow in a QM/MM calculation:
QM/MM investigations of enzyme-catalyzed reactions commonly initiate with the heavy atom coordinates of a protein structure, typically sourced from X-ray crystallography data in the scientific literature. The abundance of available protein crystal structures has established QM/MM as a pivotal methodology in computational chemistry, particularly for structure- and fragment-based drug design [10].

A critical preliminary step for QM/MM or MD simulations is the preparation of the initial Protein Data Bank (PDB) structure. These structures typically lack hydrogen atoms and may contain extraneous molecules, such as crystalline buffer additives. Therefore, essential preprocessing includes adding hydrogen, removing non-essential components, and solvating the system to create a realistic computational model. These preparations are essential to avoid improper folding, particularly when constraints on the protein backbone are relaxed.

The protocol typically ends with an MD run, from which a series of representative structures is sampled to serve as starting geometries for the QM/MM calculations. In this framework, the molecular system is divided into a quantum region and a classical environment, each treated with an appropriate level of theory. The chemically relevant inner portion, such as the active site, is evaluated using a high-level QM method, often density functional theory, whereas the surrounding atoms are modeled with a conventional MM force field.

In density functional theory (DFT), the system's energy is approximated using an electronic density function, rather than directly calculating the system's wave function as in ab initio quantum methods. These semi-empirical DFT methods provide a good approximation of molecular behavior but at a lower computational cost compared to ab initio methods, making them widely used in computational chemistry.

## 7.6 Molecular Dynamics (MD)

Simulation is a computational method that applies the principles of classical statistical physics to predict the behavior of multi-particle systems. In this technique, atoms and molecules are allowed to interact over a period, providing a detailed visualization of particle movement.

In biocatalysis and the pharmaceutical industry, MD simulations are critical for understanding the conformational changes and dynamic mechanisms of proteins. This approach is frequently used in drug design and in studying enzyme-substrate interactions. Additionally, MD simulations can replicate various experimental conditions, such as changes in pH, temperature, and solvent, to study their effects on molecular behavior.

Description:
MD simulates the movement of atoms and molecules over time, allowing researchers to explore MD and behavior in detail (Figure 7.4).

Applications:
- Drug design
- Study of protein-ligand interactions
- Understanding molecular structures and properties

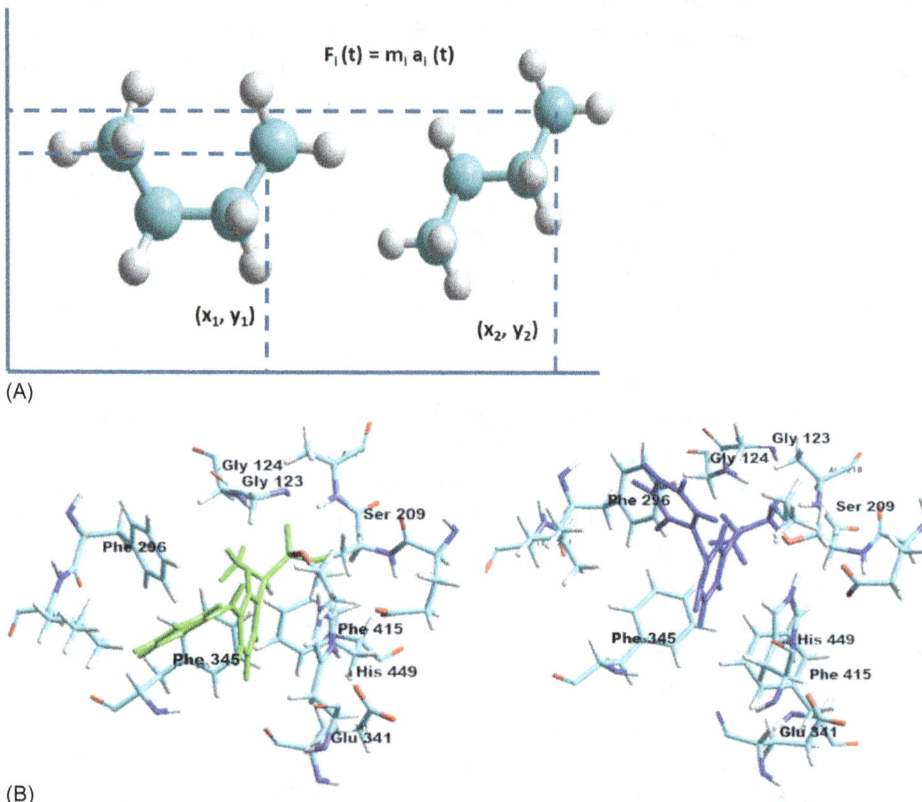


**Figure 7.4: (A)** Illustration of atomic motion in a system according to Newton's second law (Fi(t) = mi ai (t)), showing the displacement of atoms between two positions (x1, y1) and (x2, y2). **(B)** Illustration of the movement in a molecular dynamic of the *S*-ketoprofen molecule in the active center of the *Candida rugosa* lipase.

### 7.6.1 QM/MM Molecular Dynamics and Monte Carlo Simulations

Historically, comprehensive QM/MM simulations were utilized to investigate solvation effects and reaction free energies in explicit solvent environments. These studies em-

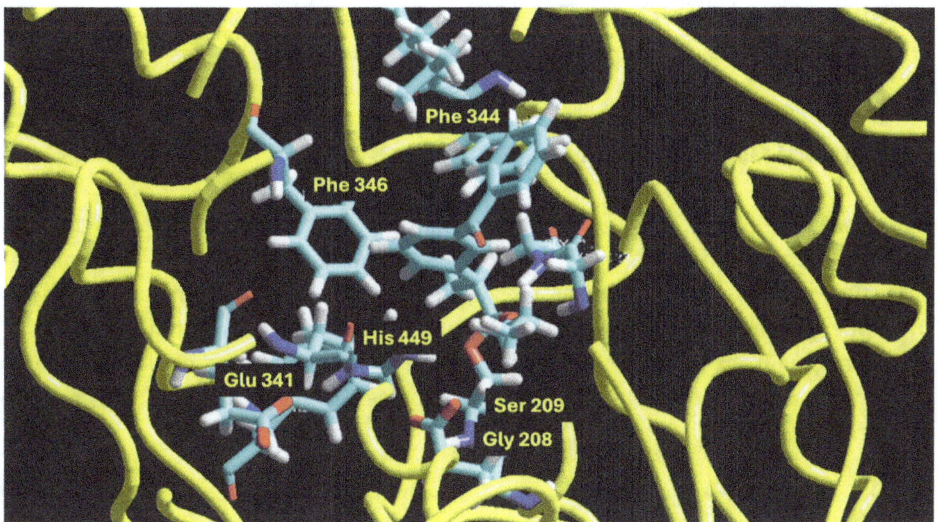


**Figure 7.5:** Molecular dynamics representing the different conformations that a molecule can adopt at the active site of an enzyme.

ployed sampling techniques, such as MD and MC, coupled with free-energy calculation methods such as free-energy perturbation and thermodynamic integration (Figure 7.5).

Although semi-empirical QM methods were the most common choice for the QM region, implementations using first-principles methods like DFT or HF also exist. However, while semi-empirical approaches are computationally tractable, first-principles QM/MM MD simulations remain prohibitively demanding in terms of computational resources.

In these studies, QM energy and forces are derived from a self-consistent field calculation at each step. An alternative method is Car-Parrinello molecular dynamics (CP-MD) [11], where the wave functions are treated as fictitious dynamic variables. QM/MM approaches based on CP-MD have been developed by several research groups, with notable contributions from Röthlisberger and Carloni, who are active in biomolecular simulations [12].

### 7.6.2 QM/MM Free-Energy Perturbation

QM/MM free-energy perturbation is a computational method designed to efficiently calculate free-energy differences along reaction pathways. In this approach, the MM degrees of freedom are simulated while the QM atoms remain fixed, thereby reducing the computational burden. An electrostatic potential charge model is used for the QM density, which helps to avoid the high computational cost of QM-MM electrostatic interactions. This method is particularly valuable for studying enzymatic reactions and other chemical transformations in biomolecular systems [13, 14].

### 7.6.3 QM/MM Modeling of Drug Target Enzymes

The application of QM/MM methods to enzymes that serve as drug targets is expanding, as these simulations offer critical insights for rational drug design. A prominent case study involves cholinesterases, enzymes responsible for catalyzing the hydrolysis of the neurotransmitter acetylcholine (ACh). This reaction is essential for resetting cholinergic neurons to their resting state. Consequently, inhibitors of cholinesterases constitute an important class of therapeutics for neurodegenerative diseases, including Alzheimer's and Parkinson's [15].

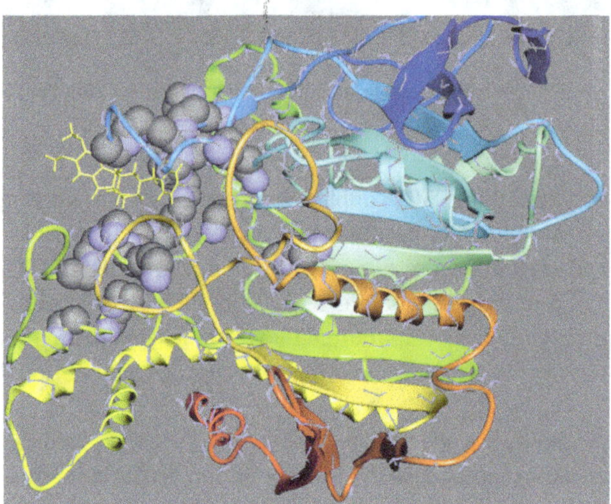

**Figure 7.6:** Human AChE (PDB 1B41) and amino acid catalytic center, the throat and the anionic site peripheral (yellow). In black, the inhibitor donepezil.

Cholinesterases are categorized into two main types: acetylcholinesterase (AChE) and butyrylcholinesterase (BChE). AChE, which demonstrates high catalytic efficiency for ACh hydrolysis, is located predominantly in cholinergic synapses. In contrast, BChE has a broader distribution across various tissues and in plasma. Both enzymes feature a characteristic Ser-His-Glu catalytic triad. The hydrolysis mechanism initiates with a nucleophilic attack by the serine residue on the substrate's carbonyl carbon, facilitated by a proton transfer from the histidine. This step results in the release of choline and the formation of an acyl-enzyme intermediate. The details of this catalytic mechanism have been elucidated through ab initio QM/MM studies

MD simulations have been employed to compute the free-energy profiles for both steps of the reaction. These studies consistently identify acylation as the rate-determining step in the catalytic mechanism of BChE. QM/MM simulations have also been used to investigate enzyme-inhibitor interactions in AChE. Additionally, the dephosphorylation mechanism of paraoxon-methyl with AChE has been extensively modeled, providing insights that could aid in the development of treatments for poisoning [16]. Lastly, QM/MM modeling has been used to design variants with higher catalytic activity in BChE (Figure 7.6).

## 7.7 Molecular Docking

Molecular docking is a computational technique used to predict the optimal binding orientation and conformation (pose) of a small molecule (ligand) within a specific site on a biological target, such as a protein. This method has become a cornerstone of modern drug discovery, with critical applications in virtual screening (VS) of compound libraries, lead optimization, and elucidating structure-activity relationships. The process requires defining the ligand's state variables, such as its position, orientation, and conformation relative to the target.

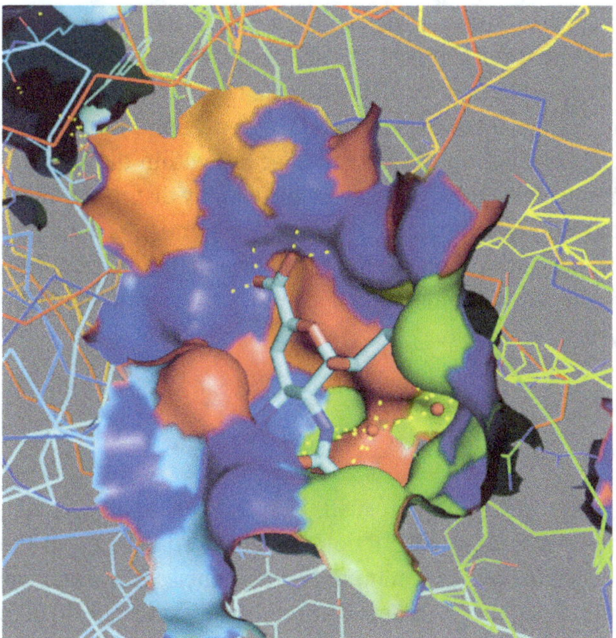

**Figure 7.7:** Partial structure of the HIV protease bound to tipranavir (Aptivus®) (PDB code: 1D4S).

Docking can be performed under rigid or flexible conditions, with flexibility affecting both the speed of the process and the likelihood of finding a complementary fit. Systematic or stochastic search methods are used to explore potential binding modes, often in conjunction with empirical scoring functions based on force fields or knowledge-based approaches. Several programs, such as AutoDock, DOCK, GOLD, and FlexX, offer different strategies for conducting docking simulations [17] (Figure 7.7).

## 7.8 Biomolecular Reaction Modeling

Executing a QM/MM study of a biomolecular reaction, particularly within an enzyme, necessitates extensive preparatory work and system setup prior to the core computational phase. The following section outlines common challenges that can be encountered during this preparatory workflow. It is important to note that many of these issues are not exclusive to QM/MM methodologies but are also prevalent in classical MD simulations.

### 7.8.1 Preparation of a QM/MM Study

Setting up a QM/MM study of a biocatalytic reaction demands careful attention to detail, particularly during the preparation stage. One critical aspect is the development of MM parameters, which is essential for classical MD simulations and includes considerations for the prospective QM part, particularly atomic charges. These partial atomic charges play a critical role in dictating the spatial orientation of polar and charged residues, as well as the distribution of solvent molecules within the system's environment. The various biomolecular force fields available employ distinct methodologies for deriving these charges, often utilizing QM calculations on molecular fragments or analogous model compounds. Consequently, the substantial effort involved in obtaining or generating suitable MM parameters is a significant factor that influences the selection of an appropriate force field.

### 7.8.2 Structure Validation

The initial structure for a QM/MM investigation is usually derived from an experimental source, most often a single-crystal X-ray diffraction model. A critical step in the setup process is to validate this structure for potential inconsistencies. For instance, structures of the same enzyme crystallized under different conditions can display conformational differences that may significantly influence the computational results. Furthermore, a structure solved by crystallography does not necessarily represent the enzyme's catalytically active or productive conformation.

### 7.8.3 Addition of Hydrogen Atoms

Hydrogen atoms are rarely resolved in X-ray structures. Most biomolecular simulation packages can place these atoms with reliable accuracy according to standard bond distances and angles. However, the placement of polar hydrogen atoms is less straightforward. Several programs and algorithms are available to optimize the positions of these hydrogen atoms [18]. It is also important to consider protonation states, which depend on the pH chosen for the simulation. Notably, the optimal pH for enzymatic activity often differs from the pH at which the structure was crystallized, as crystallization generally occurs near the enzyme's isoelectric point.

### 7.8.4 Choice of Simulation System

In QM/MM studies, it is essential to carefully define the QM region and decide which parts of the environment will be included in the simulation. Ideally, the entire protein and the surrounding hydration shell should be considered. The active region, where atoms can move freely, is then determined, while the rest of the system remains fixed.

In MD simulations, the number of degrees of freedom is less critical. However, in optimizations, having more than 2,000 atoms can lead to convergence and consistency issues. A general recommendation is to include residues within 10 Å of the QM region in the active zone to maintain accuracy without overwhelming computational resources.

### 7.8.5 Solvation and Hydration

Water molecules with well-defined positions are typically resolved through crystallography, while mobile water molecules remain undetected. Therefore, for MD simulations or optimizations, it is essential to include additional water molecules in the mobile parts of the system, along with a buffer region, to ensure proper hydration.

There are two common approaches for handling solvation:

*Periodic water box*: The entire system is placed inside a periodic box of water, allowing for more comprehensive solvation.

*Spherical water overlay*: A sphere of water is applied around the active region, with a spherical boundary potential used to prevent water diffusion out of this region.

In both cases, water molecules that are too close to existing atoms are removed, and energy minimizations and MD runs are performed to ensure proper hydration of the protein. Small internal cavities that are not connected to the surface may not be adequately hydrated, but these cavities can be identified and filled using specialized software (Figure 7.8).

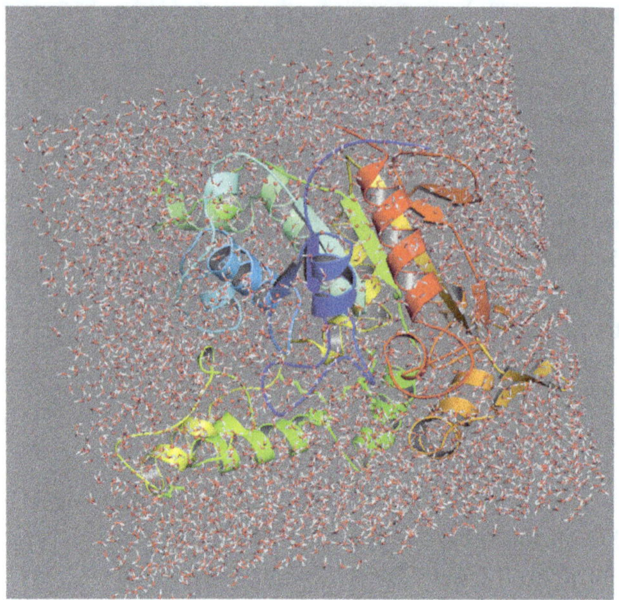

**Figure 7.8:** Box of solvent molecules in *Bacillus thermocatenulatus* lipase in water.

### 7.8.6 Classical MD Simulations

After setting up the system, running classical MD simulations of the entire system or an active region is recommended. This serves two purposes: (i) to validate the configuration by gradually releasing positional restraints and running free MD simulations to identify potential protonation or hydration issues and (ii) to use MD snapshots as initial structures for subsequent optimizations, allowing for the introduction or modification of components not present in the experimental structure, such as replacing an inhibitor with a substrate or generating mutants. Depending on the modifications, it is necessary to verify protonation states and rehydrate the system with another MD run.

Counterions and charge neutralization: A consensus is lacking within the computational community regarding the necessity or benefit of adding counterions to neutralize a system's total charge. For charged surface residues, corresponding counterions (such as sodium or chloride) should be included in the simulation to balance the charges. Neutralization through (de)protonation has also been suggested, especially outside the hydration sphere. Even so, a net charge typically remains due to buried charged groups. For a neutral system, counterions can be added or removed in the hydrated region. A common approach is to leave the total charge as is after assigning protonation states according to the chosen pH. Simulations of the stability of differently charged QM regions require a neutral environment to avoid artificial electrostatic fields that could stabilize one charge state over another.

## 7.9 Prediction of Enzymatic Enantioselectivity

The prediction of enzyme selectivity represents a central objective in many computational biocatalysis studies. This endeavor typically involves calculating the selectivity constant ($k_{cat}/K_M$), a parameter directly governed by the relative free energies of the diastereomeric transition states for each enantiomer's reaction pathway.

A fundamental application of this enantioselectivity is in the kinetic resolution (KR) or dynamic kinetic resolution (DKR) of racemic mixtures. In a typical KR, an enzyme like *Candida rugosa* lipase selectively catalyzes a reaction for one enantiomer from a racemic mixture, while exhibiting significantly lower activity toward the other. This differential conversion effectively purifies the unreactive enantiomer, allowing for the isolation of a homochiral product.

Additionally, several parameters are used in asymmetric synthesis to quantify the stereochemical outcome. One of these parameters is enantiomeric excess (e.e.), which quantifies the absolute difference in the amount of one enantiomer compared to the other in a sample and is typically expressed as a percentage (% e.e.). The enantiomeric ratio ($E$) represents the ratio between the specificity constants for both enantiomers. Unlike e.e., which relates to the relative amounts of reactants or products, $E$ is associated with the enzyme's intrinsic properties.

Undoubtedly, the main challenge in achieving accurate quantitative prediction of enantioselectivity arises from the fact that, while the free energy of the transition state for the preferentially recognized enantiomer within the active site can be precisely determined, the other enantiomer may exhibit different orientations. These orientations involve interactions with distinct regions of the active site, as visualized in the attached figure from the study conducted on the two ketoprofen enantiomers. The distinct orientations of *R*- and *S*-ketoprofen in the enzyme's active site lead to varying

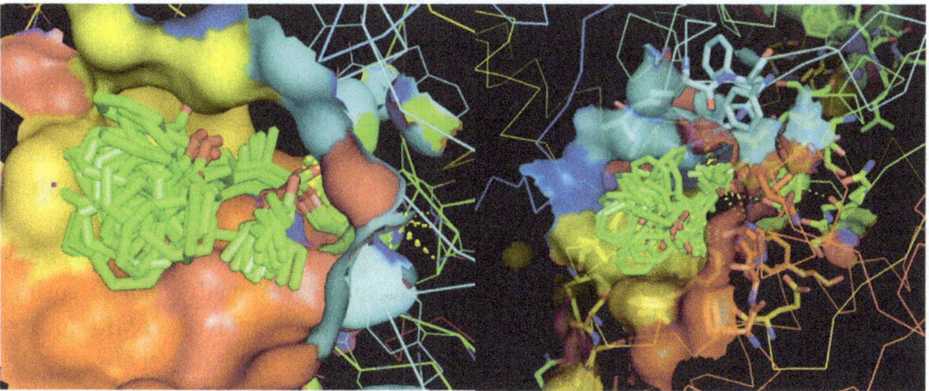

**Figure 7.9:** Superposition of *R*- and *S*-ketoprofen conformers at the active lipase site where the different substrate orientations are revealed.

stereoselectivity, which can be qualitatively predicted in silico, providing insight into which enantiomer will be preferentially recognized (Figure 7.9).

## 7.10 3D-QSAR Methods for the Prediction of Enantioselectivity

3D-QSAR methods (quantitative structure-activity relationships in three dimensions) represent a powerful tool for analyzing the relationship between molecular structure and biological or catalytic activity. These models are based on the generation of 3D descriptors derived from molecular simulations (MD, molecular docking, approximate QM, etc.), which are correlated through multivariate statistical techniques with experimentally measurable parameters such as catalytic rates, kinetic constants, or degrees of enantioselectivity. In this way, 3D-QSAR integrates two disciplines: **Molecular modeling**, which provides the structural information, and **chemometrics**, which supplies the statistical tools for its analysis.

Although the use of 3D-QSAR is well established in the field of rational drug design, its application in **biocatalysis** is relatively more recent. The pioneering work of **Tomić et al.** [19–21] provides a clear example of how it is possible to establish quantitative predictive models of kinetic parameters (such as $k_{cat}/K_M$) through a 3D-QSAR approach, in which the chemical descriptors of the system are correlated with experimental data.

The prediction of enzymatic enantioselectivity is not trivial, as it depends on a complex combination of factors such as the **geometry of the active site**, the **steric and electrostatic interactions** between enzyme and substrate, the **conformational dynamics** of both molecules, and the **reaction conditions**. Given this complexity, computational methodologies such as 3D-QSAR models emerge as an effective alternative to rationalize and predict the enantioselective behavior of biocatalysts.

In particular, it has been demonstrated that the enantioselectivity of **Burkholderia cepacia** lipase can be predicted through a model in which the **binding free energy** is estimated as a linear combination of the enzyme-substrate interaction energy and parameters related to the polar and non-polar solvent-accessible surface area. The weighting coefficients of each term were calculated by **partial least squares analysis**, and the high predictive correlation coefficient obtained ($Q^2 = 0.84$) confirms the validity of the approach [21].

An interesting aspect of 3D-QSAR models is that they do not always require knowledge of the 3D structure of the enzyme. In such cases, methodologies such as **CoMFA (comparative molecular field analysis)** can be employed. CoMFA focuses on the comparative analysis of steric and electrostatic fields around a set of substrates or ligands. The fundamental hypothesis of CoMFA is that variation in biological activity can be explained by the spatial distribution of these fields around the molecule.

A representative example is the study of **ketone bioreduction** mediated by whole cells of *Geotrichum candidum* and *Saccharomyces octosporus*. Using CoMFA, it was possible to identify regions of molecular space relevant to the enzyme-substrate interaction:

- **Low-steric hindrance regions (green):** Areas where the presence of substituents favors the interaction
- **High-steric hindrance regions (yellow):** Areas where bulky groups reduce the affinity of the substrate for the enzyme
- **Electrostatic regions:** Red regions (where high electronic density favors the interaction) and blue regions (where a negative charge density decreases the interaction)

The great advantage of this methodology is that it allows the **prediction of selectivity and affinity of the biocatalyst without the need to know its crystallographic structure or to isolate the enzyme**. Thus, 3D-QSAR constitutes a highly versatile and predictive strategy for understanding and anticipating enantioselectivity phenomena in Biocatalysis, with direct applications in biocatalyst design, optimization of asymmetric reactions, and the development of more efficient processes in green chemistry.

## 7.11 Prediction of Three-Dimensional Protein Structures Using Computational Methods

Computational prediction of 3D protein structures is a crucial tool in modern structural biology. This approach enables the deduction of a protein's spatial conformation based on its amino acid sequence, which is vital for understanding its function, molecular interaction mechanisms, and its involvement in various diseases. Determining the 3D structure of proteins in specific conformational states is essential for uncovering their biological roles and supporting the rational design of targeted therapies. While traditional experimental techniques – such as X-ray crystallography, nuclear magnetic resonance (NMR) spectroscopy, and cryo-electron microscopy – provide high accuracy, they are often time-consuming and expensive [22].

### 7.11.1 Data Entry: The Amino Acid Sequence

In the domain of structural bioinformatics and molecular biology, the unequivocal prerequisite for in silico prediction of a protein's 3D conformation is its primary amino acid sequence. This linear polymer of amino acid residues encodes the intrinsic physicochemical information dictating the protein's folding landscape and thermodynamically stable native conformation under physiological milieus.

The canonical representation format for such sequences is the FASTA format, an industry-standard plaintext format designed for efficient storage and exchange of biomolecular sequences. A typical FASTA entry comprises:

A descriptor line commencing with the character " >," which includes a unique accession code or systematic identifier, often coupled with metadata such as species or protein name.

The amino acid sequence rendered in single-letter IUPAC codes corresponding to the twenty standard amino acids (e.g., M for methionine, R for arginine, and W for tryptophan).

For instance:

```
shell
>sp|P12345|HEM_HUMAN Hemoglobin subunit alpha – Homo sapiens
VLSPADKTNVKAAWGKVGAHAGEYGAEALERMFLSFPTTKTYFPHFDLSHGSAQVKGHG
```

Contemporary computational frameworks – such as AlphaFold2, Rosetta, and I-TASSER – exploit this primary sequence as the sole input to reconstruct high-resolution tertiary structures. These platforms leverage integrative methodologies combining sequence homology inference, ab initio folding simulations, knowledge-based potentials, and cutting-edge deep learning (DL) architectures to model conformational states that approximate experimentally determined structures.

Fundamental to these predictive efforts is the *thermodynamic hypothesis formulated by Christian Anfinsen*, which posits that the native conformation of a globular protein is encoded entirely by its amino acid sequence, assuming a physiologically relevant environment [23, 24]. This postulate rationalizes the use of the primary structure as the exclusive datum required for tertiary structure elucidation, thereby grounding the theoretical and practical basis for algorithmic protein folding prediction.

In homology modeling, high-quality templates are found and aligned using simple sequence-sequence alignment algorithms like Needleman-Wunsch (global) and Smith-Waterman (local). BLAST is also commonly used to detect short matches between the query and template, which are then extended to form full alignments [25–27].

Homology-based methods: There are also tools like I-TASSER, Phyre2, and SWISS-MODEL, which apply traditional homology modeling approaches. These methods compare the protein of interest with known structures to generate a model based on structural similarity. Although less accurate than AlphaFold, they can be useful when homologous experimental structures exist in databases like PDB.

Once the models are generated, it is possible to visualize and analyze them using specialized programs such as PyMOL, UCSF Chimera, or web platforms like Mol. These tools allow for exploring the geometry of the protein, identifying functional regions, analyzing potential active sites, and simulating molecular interactions. The analysis of the structure also enables the rational design of drugs, vaccines, and modified proteins for specific applications.

## 7.12 Artificial Intelligence (AI) in Molecular Modeling

Artificial intelligence (AI) is highly useful for extracting information from datasets and providing practical or valuable insights. Algorithms, which are sequences of finite and well-defined instructions, process and transform images, words, or phrases from natural language into a format readable by computer programs. During this process, algorithms identify rules, extract patterns, and relate data to provide meaningful information (Figure 7.10).

AI is a broad term that refers to computer programs capable of simulating human thought and behavior. Machine learning (ML) goes further by introducing data into the system along with algorithms such as naïve Bayes, decision trees, hidden Markov models, among others, allowing the machine to learn without being explicitly programmed. With the development of neural networks, machines gained the ability to classify and organize input data in a way that mimics the human brain, marking a significant advancement in AI.

The origins of AI in drug discovery and development can be traced back to QSAR studies, which began in the 1960s [28]. In these studies, which continue to be used to this day, the goal is to find a mathematical model that can estimate the intrinsic connections between chemical structures – encoded by a series of descriptors or representations, which are an essential part of any AI model – and biological activity.

The drug discovery and development pipeline tends to be long and expensive, creating significant challenges for pharmaceutical companies. Advances in computer-aided drug design, especially when combined with AI, provide new opportunities to address these issues. AI covers a broad spectrum of methodologies, among them ML and DL, which can operate under supervised, unsupervised, or reinforcement learning schemes. These approaches are increasingly incorporated at multiple decision points in drug design.

A wide variety of algorithms contribute to this process, including artificial and deep neural networks, support vector machines, models for classification and regression, generative adversarial networks, symbolic learning strategies, and meta-learning techniques.

AI tools are employed in many tasks such as synthesizing peptides, designing novel molecules, conducting VS, performing molecular docking, building QSAR models, repositioning existing drugs, investigating protein misfolding, analyzing protein-protein interactions, mapping molecular pathways, and studying polypharmacology.

Furthermore, AI-based principles serve to distinguish active from inactive compounds, track drug release, aid preclinical and clinical development, and enhance both early and late-stage drug screening as well as biomarker identification. Terms like "machine learning," which refers to automatic learning, and "deep learning," which refers to advanced neural network models, are commonly used in the context of AI-driven drug design.

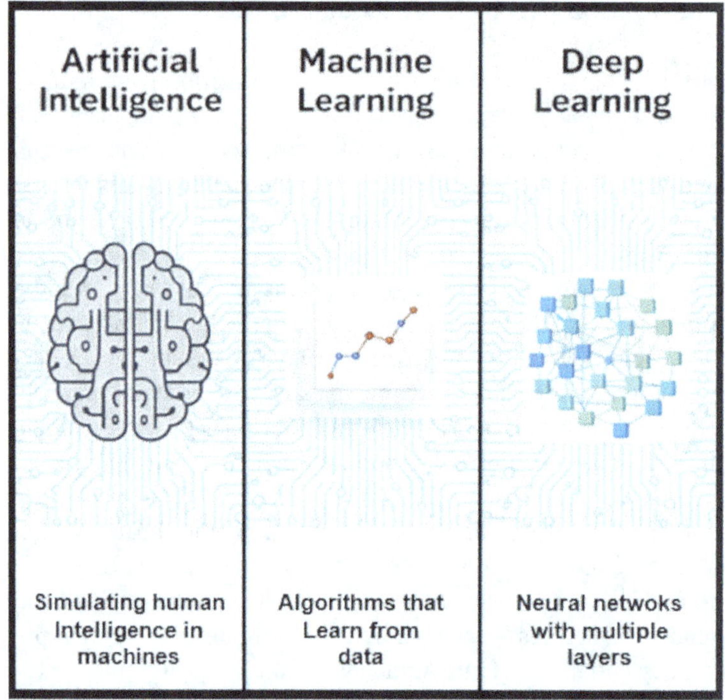


**Figure 7.10:** Diagram showing the relationship between artificial intelligence, machine learning, and deep learning.

ML and DL methodologies in the field of AI have proven effective in optimizing performance and reducing costs in drug design [29]. It is important to note that both ML and DL require large amounts of data, and the quality and accuracy of the models, as well as the insights they provide, are closely tied to the quality and reliability of the data used during the training process.

### 7.12.1 AI Applied to Predict Three-Dimensional Protein Structure

Developing accurate methods for predicting the 3D structures of proteins is of great value, as it aids in the discovery of new drugs and enhances our understanding of diseases related to protein misfolding, such as Alzheimer's, Parkinson's, Huntington's diseases, and amyotrophic lateral sclerosis. AlphaFold, an AI network created by DeepMind (Google), is used to determine the 3D structure of a protein based on its amino acid sequence. AlphaFold utilizes a DL framework to predict a protein's 3D structure directly from its amino acid sequence. At its core, the system employs a convolutional neural network that was trained on the known structures in the PDB. This network learns to predict the spatial distances between pairs of amino acid residues in the sequence. The model's output includes a probabilistic estimation for each 64 × 64 segment of the resulting pairwise distance map, which is then used to determine the final structural model [29].

## 7.13 Artificial Neural Networks (ANNs)

Neural networks are fundamental to the fields of AI and ML. Inspired by the functioning of the human brain, artificial neural networks (ANNs) are composed of layers of interconnected nodes, also known as artificial neurons.

ANNs are typically organized into three types of layers. The first, called the input layer, receives the initial data. This data is then transmitted through the hidden layers, where complex mathematical operations are performed to transform the information. Finally, the output layer provides the result, which could be a prediction or classification.

ANNs are powerful tools for modeling non-linear statistical data, making them useful for simulating complex interactions between inputs and outputs and for identifying patterns. These networks are widely used in applications such as signal processing and pattern recognition.

In the field of biocatalysis, neural networks are revolutionizing the ability to predict enzymatic activity and optimize reaction conditions, improving the efficiency of biotechnological processes. For example, they can determine the optimal conditions for a specific enzymatic reaction, leading to more effective and cost-efficient processes.

In drug discovery, neural networks and DL are employed to analyze large volumes of chemical and biological data, identifying patterns and predicting the biological activity of new compounds. These technologies enable the correlation of physicochemical and molecular properties with specific biological activities, facilitating the prediction of crucial values such as the half-maximal inhibitory concentration (IC50), the inhibition constant ($K_i$), and the half-maximal effective dose (ED50) of new compounds.

The use of neural networks in these fields not only accelerates the discovery and optimization process but also enhances accuracy and efficiency, opening new possibilities for research and development in biotechnology and pharmacology [30].

## 7.14 Virtual Screening

VS is a key computational methodology in drug discovery that involves the automated evaluation of large chemical libraries to prioritize compounds with the highest predicted affinity for a given drug target, such as a protein receptor or enzyme. This process incorporates various computational methods, such as molecular modeling, quantum chemistry, and MD. Although VS itself is not an AI technique, it can integrate AI to enhance efficiency and precision.

ML and DL are particularly useful in VS for analyzing large volumes of chemical and biological data. These techniques help identify patterns and relationships that traditional methods may overlook, enabling more accurate predictions of a compound's biological activity. For example, deep neural networks can be trained to predict the affinity of compounds to target proteins, improving the selection of candidates for experimental assays. This rational lead discovery based on receptor structure makes VS a compelling alternative to traditional methods.

The key steps involved in performing VS are:

1. Defining objectives: Identify and select the biological targets related to the disease under study
2. Constructing the virtual library: Create a collection of potential molecules that can interact with the selected target
3. Molecular docking simulation: Use specialized software to predict how the molecules in the library interact with the biological target
4. Affinity estimation: Apply scoring functions to evaluate the binding strength between the molecules and the target
5. Refinement through bioinformatics: Use bioinformatics tools to improve and refine the predictions obtained
6. Data analysis with artificial intelligence: Employ AI techniques, such as neural networks, to analyze the data and select the most promising compounds
7. Experimental validation: Conduct laboratory experiments to confirm the biological activity of the selected compounds [31].

A highly effective process in ligand VS, which often involves evaluating large compound libraries, is the use of flexible docking techniques. This approach allows both ligands and receptors to adapt and change their conformation during interaction, more accurately reflecting the dynamic conditions present in molecular binding within real biological environments. By enabling both the ligand and the receptor to adjust and flex during the docking process, the accuracy of molecular interaction predictions is improved. This is particularly valuable when exploring more complex or flexible binding sites on the receptor, which can lead to greater efficacy in identifying promising pharmacological candidates. The combination of flexible docking with VS techniques not only enhances the ability to capture relevant interactions but also optimizes process efficiency by reducing false positives and improving the selection of compounds with therapeutic potential. Additionally, ligand VS frequently involves the evaluation of large compound libraries. In this context, AI, particularly ML and DL, can be employed to analyze these vast datasets, identifying patterns or relevant features for molecular interaction.

## 7.15 Chemical Synthesis Automation

The automation of chemical synthesis through AI is a rapidly developing field that holds the promise of transforming the chemical and pharmaceutical industries. AI enables the optimization of processes, reduces development times, and enhances precision in compound synthesis [32]. Here are some recent advances in this field:

RoboChem: An autonomous chemical synthesis robot developed by chemists at the University of Amsterdam, integrating AI-based ML. This device outperforms human chemists in terms of speed and precision, demonstrating a high level of ingenuity. RoboChem uses a robotic needle to mix materials in small volumes, which are then directed into a reactor where LED light activates molecular conversion. Data from the transformed molecules are identified via an automated NMR spectrometer and processed in real time by AI, which autonomously determines the reactions to perform next.

MIT computational model: Chemical engineers and chemists at the Massachusetts Institute of Technology have developed a computational model using generative AI to predict the structures formed during chemical reactions. This model focuses on capturing the transition states of chemical reactions, which are crucial for understanding and manipulating chemical processes. Transition states represent the key moments where reactants transform into products, and understanding them can significantly impact areas such as drug synthesis, catalysis, and materials research.

Augmented intelligence in drug design: AI is combined with human expertise to create augmented intelligence, enhancing decision-making in pharmaceutical chemistry.

Applications in chemistry: AI algorithms are employed to predict chemical properties, design new materials, and accelerate drug discovery. One of the primary uses of AI in chemistry is predicting chemical properties. By training ML models with experimental data, it becomes possible to predict properties such as solubility, chemical reactivity, and toxicity of chemical compounds.

These advancements highlight how AI is facilitating the automation of complex and repetitive tasks, allowing scientists to focus on more innovative aspects of research and the development of new products.

## 7.16 ADME-Tox Property Prediction

Utilizing advanced computational techniques, AI predicts various pharmacokinetic and toxicological characteristics of small molecules, including ADME-Tox properties (absorption, distribution, metabolism, excretion, and toxicity). ML and DL algorithms analyze extensive datasets containing chemical structures and their corresponding

ADME-Tox profiles. Trained on diverse molecular descriptors, biological data, and experimental outcomes, these algorithms develop predictive models capable of estimating a molecule's likelihood of exhibiting specific ADME-Tox properties. By evaluating parameters such as bioavailability, blood-brain barrier permeability, metabolic stability, tissue distribution, pharmacological interactions, and potential toxicity, AI identifies correlations between chemical structures and ADME-Tox properties, enabling predictions for novel or untested compounds [33].

## References

[1] Wijma HJ, Janssen DB. Computational design gains momentum in enzyme catalysis engineering. FEBS J 2013;280:2948–2960.

[2] DiMasi JA, Grabowski HG, Hansen RW. Innovation in the pharmaceutical industry: New estimates of R&D costs. J Health Econ 2016;47:20–33.

[3] Shaker B, Ahmad S, Lee J, Jung C, Na D. In silico methods and tools for drug discovery. Comput Biol Med 2021;137:104851.

[4] Lees P, Hunter RP, Reeves PT, Toutain PL. Pharmacokinetics and pharmacodynamics of stereoisomeric drugs with particular reference to bioequivalence determination. J Vet Pharmacol Ther 2012;35(Suppl 1);17–29.

[5] Song Z, Zhang Q, Wu W, Pu Z, Yu H. Rational design of enzyme activity and enantioselectivity. Front Bioeng Biotechnol 2023;11:1129149.

[6] Molecular mechanics: Basic theory. Comput Chem Using PC 2003;93–130.

[7] Molecular mechanics II: Applications. Comput Chem Using PC 2003;131–168.

[8] Kar RK. Benefits of hybrid QM/MM over traditional classical mechanics in pharmaceutical systems. Drug Discov Today 2023;28:103374.

[9] Senn HM, Thiel W. QM/MM methods for biomolecular systems. Angew Chem Int Ed Engl 2009;48:1198–1229.

[10] Kulkarni PU, Shah H, Vyas VK. Hybrid Quantum Mechanics/Molecular Mechanics (QM/MM) simulation: A tool for structure-based drug design and discovery. Mini Rev Med Chem 2022;22:1096–1107.

[11] Grotendorst J. Modern methods and algorithms of quantum chemistry. 2000.

[12] Rothlisberger U, Carloni P. Drug-target binding investigated by Quantum Mechanical/Molecular Mechanical (QM/MM) methods. In Ferrario M, Ciccotti G, Binder K, editors. Computer Simulations in Condensed Matter Systems: From Materials to Chemical Biology 2006, Berlin, Heidelberg: Springer Berlin Heidelberg; Vol. 2, p. 449–479.

[13] Yang W, Bitetti-Putzer R, Karplus M. Free energy simulations: Use of reverse cumulative averaging to determine the equilibrated region and the time required for convergence. J Chem Phys 2004;120:2618–2628.

[14] Zheng L, Yang W. On the simulated scaling based free energy simulations: Adaptive optimization of the scaling parameter intervals. J Chem Phys 2008;129.

[15] Zhang Y, Kua J, McCammon JA. Role of the catalytic triad and oxyanion hole in acetylcholinesterase catalysis: An ab initio QM/MM study. J Am Chem Soc 2002;124:10572–10577.

[16] Liu J, Zhang Y, Zhan CG. Reaction pathway and free-energy barrier for reactivation of dimethylphosphoryl-inhibited human acetylcholinesterase. J Phys Chem B 2009;113:16226–16236.

[17] Morris GM, Lim-Wilby M. Molecular docking. Methods Mol Biol 2008;443:365–382.

[18] Forrest LR, Honig B. An assessment of the accuracy of methods for predicting hydrogen positions in protein structures. Proteins 2005;61:296–309.
[19] Tomić S, Kojić-Prodić B. A quantitative model for predicting enzyme enantioselectivity: Application to Burkholderia cepacia lipase and 3-(aryloxy)-1,`propanediol derivatives. J Mol Graph Model 2003;21:241–252.
[20] Tomić S, Kojić-Prodić B. A quantitative model for predicting enzyme enantioselectivity: Application to Burkholderia cepacia lipase and 3-(aryloxy)-1,2-propanediol derivatives. J Mol Graph Model 2002;21:241–252.
[21] Tomić S, Bertoša B, Kojić-Prodić B, Kolosvary I. Stereoselectivity of Burkholderia cepacia lipase towards secondary alcohols: Molecular modelling and 3D QSAR approach 2004, Tetrahedron: Asymmetry; Vol. 15, p. 1163–1172.
[22] Wuyun Q, Chen Y, Shen Y, Cao Y, Hu G, Cui W, et al. Recent progress of protein tertiary structure prediction. Molecules 2024;29:832.
[23] Kresge N, Simoni RD, Hill RL. The thermodynamic hypothesis of protein folding: The work of christian anfinsen. J Biol Chem 2006;281:e11–e13.
[24] Anfinsen CB. Principles that govern the folding of protein chains. Science 1973;181:223–230.
[25] Needleman SB, Wunsch CD. A general method applicable to the search for similarities in the amino acid sequence of two proteins. J Mol Biol 1970;48:443–453.
[26] Smith TF, Waterman MS. Identification of common molecular subsequences. J Mol Biol 1981;147:195–197.
[27] Altschul SF, Gish W, Miller W, Myers EW, Lipman DJ. Basic local alignment search tool. J Mol Biol 1990;215:403–410.
[28] Hansch C, Fujita T. p-σ-π Analysis. A method for the correlation of biological activity and chemical structure. J Am Chem Soc 1964;86:1616–1626.
[29] Kolluri S, Lin J, Liu R, Zhang Y, Zhang W. Machine learning and artificial intelligence in pharmaceutical research and development: A review. Aaps j 2022;24:19.
[30] Qamar R, Zardari B. Artificial neural networks: An overview. Mesopotamian J Comput Sci 2023;2023:130–139.
[31] Graff DE, Shakhnovich EI, Coley CW. Accelerating high-throughput virtual screening through molecular pool-based active learning. Chem Sci 2021;12:7866–7881.
[32] Shen Y, Borowski JE, Hardy MA, Sarpong R, Doyle AG, Cernak T. Automation and computer-assisted planning for chemical synthesis. Nat Rev Methods Primers 2021;1:23.
[33] Butina D, Segall MD, Frankcombe K. Predicting ADME properties in silico: Methods and models. Drug Discov Today 2002;7:S83–8.

# Index



 | https://doi.org/10.1515/9783111122991-008

www.ingramcontent.com/pod-product-compliance
Lightning Source LLC
Chambersburg PA
CBHW080849250726
48663CB00003B/394